Studies in Classification, Data Analysis, and Knowledge Organization

For further volumes:
http://www.springer.com/series/1564

Donatella Vicari • Akinori Okada •
Giancarlo Ragozini • Claus Weihs
Editors

Analysis and Modeling of Complex Data in Behavioral and Social Sciences

Springer

Editors
Donatella Vicari
Department of Statistical Science
University of Rome "La Sapienza"
Rome
Italy

Akinori Okada
Graduate School of Management
and Information Sciences
Tama University
Tokyo
Japan

Giancarlo Ragozini
Department of Political Science
University of Naples "Federico II"
Naples
Italy

Claus Weihs
Fakultät Statistik
Technische Universität Dortmund
Dortmund
Germany

ISSN 1431-8814
ISBN 978-3-319-06691-2 ISBN 978-3-319-06692-9 (eBook)
DOI 10.1007/978-3-319-06692-9
Springer Cham Heidelberg New York Dordrecht London

Library of Congress Control Number: 2014943911

Printed on acid-free paper

Springer is part of Springer Science+Business Media (www.springer.com)

Preface

The papers contained in this volume stem from contributions firstly presented at the joint international meeting JCS-CLADAG held in Anacapri (Capri Island, Italy) on September 3–4, 2012, where the Japanese Classification Society and the Classification and Data Analysis Group of the Italian Statistical Society had a stimulating scientific discussion and exchange.

The conference focus was primarily on the Analysis and Modeling of Complex Data in Behavioral and Social Sciences, including theoretical developments, applications, and computational methods. The joint meeting aimed at enhancing the scientific cooperation between Italian and Japanese data analysts and at establishing new cooperation between members of the two societies.

The scientific program encompassed 93 presentations in 24 sessions, including keynote and specialized lectures as well as solicited and contributed session talks, where the contributions mainly from, but not limited to, the two societies established interesting exchanges and debates between Italian and Japanese data analysts. With over 130 attendees mainly from Italy and Japan, and also from other ten countries of the European Community and the USA and Australia, the conference represented a moment of discussion and exchange of knowledge.

After the conference, from the original presentations, a selection of 31 contributions has been chosen after a double peer-review.

The volume presents theoretical developments, applications, and computational methods for the analysis and modeling in behavioral and social sciences, where data are usually complex to explore and investigate. The challenging proposals provide a connection between statistical methodology and social and psychological domain with particular attention to computational issues in order to address effectively complex data analysis problems.

We are also indebted to the Department of Political Science of Federico II, University of Naples, Italy which partially financially supported the publication of this volume. Moreover, we want to thank also the REPOS Project that partially supported the organization of the conference.

We are grateful to the referees for their active, careful, and constructive work which certainly stimulated and improved the scientific content of the contributions.

Finally, a special thanks goes to the authors of the papers who are the real core of the volume for their direct contribution to new and interesting challenges.

Roma, Italy — Donatella Vicari
Tokyo, Japan — Akinori Okada
Napoli, Italy — Giancarlo Ragozini
Dortmund, Germany — Claus Weihs
28 February 2014

Contents

Contributors

Giada Adelfio Dipartimento di Scienze Economiche, Aziendali e Statistiche, Universitá degli Studi di Palermo, Palermo, Italy

Yasumasa Baba The Institute of Statistical Mathematics, Tokyo, Japan

Antonio Balzanella Dipartimento di Studi Europei e Mediterranei, Seconda Universitá di Napoli, Naples, Italy

Alessandro Barbiero Department of Economics, Management and Quantitative Methods, Università di Milano, Milan, Italy

Francesco Bartolucci Department of Economics, Finance and Statistics, University of Perugia, Perugia, Italy

Eric J. Beh School of Mathematical and Physical Sciences, University of Newcastle, Callaghan, NSW, Australia

Lucio Bertoli-Barsotti University of Bergamo, Bergamo, Italy

Giuseppe Boari Dipartimento di Scienze Statistiche, Università Cattolica del Sacro Cuore, Milano, Italy

Mario Bolzan Department of Statistical Sciences, University of Padua, Padua, Italy

Andrea Bonanomi Dipartimento di Scienze Statistiche, Università Cattolica del Sacro Cuore di Milano, Milano, Italy

Giuseppe Bove Dipartimento di Scienze dell'Educazione, Università degli Studi Roma Tre, Roma, Italy

Paula Brito Fac. Economia & LIAAD INESC TEC, Univ. Porto, Porto, Portugal

Raffaella Calabrese Essex Business School, University of Essex, Colchester, UK

Gabriele Cantaluppi Dipartimento di Scienze Statistiche, Università Cattolica del Sacro Cuore, Milano, Italy

Marcello Chiodi Dipartimento di Scienze Economiche, Aziendali e Statistiche, Universitá degli Studi di Palermo, Palermo, Italy

Livio Corain Department of Management and Engineering, University of Padua, Padua, Italy

Antonino D'Alessandro Centro Nazionale Terremoti, Istituto Nazionale di Geofisica e Vulcanologia, Rome, Italy

Valeria De Giuli Department of Industrial Engineering, University of Padua, Padua, Italy

Renato De Leone School of Science and Technology, University of Camerino, Camerino, Italy

Giovanni De Luca University of Naples Parthenope, Naples, Italy

Maria Rosaria D'Esposito Department of Economics and Statistics, University of Salerno, Fisciano (SA), Italy

Domenico De Stefano Department of Economics, Business, Mathematics and Statistics, University of Trieste, Trieste, Italy

Marco Enea Dipartimento di Scienze Economiche, Aziendali e Statistiche, University of Palermo, Palermo, Italy

Pier Alda Ferrari Department of Economics, Management and Quantitative Methods, Università degli Studi di Milano, Milano, Italy

Brunella Fiore Department of Sociology, University of Milano-Bicocca, Milano, Italy

Giuseppe Giordano Department of Economics and Statistics, Università di Salerno, Salerno, Italy

Kuniyoshi Hayashi Graduate School of Environmental and Life Science, Okayama University, Okayama City, Japan

Irene L. Hudson School of Mathematical and Physical Sciences, University of Newcastle, Callaghan, NSW, Australia

Domenica Fioredistella Iezzi Tor Vergata University, Roma, Italy

Alfonso Iodice D'Enza Università di Cassino, Cassino (FR), Italy

Antonio Irpino Dip. di Scienze Politiche "J. Monnet", Seconda Università di Napoli, Caserta, Italy

Fumio Ishioka School of Law, Okayama University, Okayama City, Japan

Kei Kurakawa National Institute of Informatics, Tokyo, Japan

Koji Kurihara Graduate School of Environmental and Life Science, Okayama University, Okayama City, Japan

Tommaso Lando VŠB-Techn., University of Ostrava, Ostrava, Czech Republic

Caterina Liberati DEMS, Università Milano-Bicocca, Milano, Italy

Rosaria Lumino University of Naples Federico II, Naples, Italy

Dario Luzio Dipartimento di Scienza della Terra e del Mare, Universitá degli Studi di Palermo, Palermo, Italy

Giancarlo Manzi Department of Economics, Management and Quantitative Methods, Università di Milano, Milan, Italy

Paolo Mariani DEMS, Università Milano-Bicocca, Milano, Italy

Marina Marino Università degli Studi di Napoli Federico II, Napoli, Italy

Mario Mastrangelo Sapienza University, Roma, Italy

Valentina Minnetti Department of Statistics, Sapienza University of Rome, Rome, Italy

Linda Moore Statistics New Zealand, Wellington, New Zealand

Marta Nai Ruscone Dipartimento di Scienze Statistiche, Università Cattolica del Sacro Cuore, Milano, Italy

Michiko Nakano Waseda University, Tokyo, Japan

Hiromasa Ohba Joetsu University of Education, Joetsu, Japan

Akinori Okada Graduate School of Management and Information Sciences, Tama University, Tokyo, Japan

Silvia Angela Osmetti Università Cattolica del Sacro Cuore di Milano, Milano, Italy

Francesco Palumbo Università degli Studi di Napoli Federico II, Napoli, Italy

Marco Passarotti Centro Interdisciplinare di Ricerche per la Computerizzazione dei Segni dell'Espressione, Università Cattolica del Sacro Cuore, Milano, Italy

Alessandra Petrucci Department of Statistics, Informatics, Applications, University of Firenze, Firenze, Italy

Antonella Plaia Dipartimento di Scienze Economiche, Aziendali e Statistiche, University of Palermo, Palermo, Italy

Laura Prota University of Salerno, Fisciano (SA), Italy

Antonio Punzo University of Catania, Catania, Italy

Emanuela Raffinetti Department of Economics, Management and Quantitative Methods, Università degli Studi di Milano, Milano, Italy

Giancarlo Ragozini Department of Political Sciences, Federico II University of Naples, Naples, Italy

Bhargav Raman Department of Radiology, Stanford University School of Medicine, Stanford, CA, USA

Lidia Rivoli Università di Napoli Federico II, Napoli, Italy

Emilia Rocco Department of Statistics, Informatics, Applications, University of Firenze, Firenze, Italy

Isabella Romeo Department of Statistics and Quantitative Methods, University of Milano-Bicocca, Milano, Italy

Luigi Salmaso Department of Management and Engineering, University of Padua, Padua, Italy

Concetta Scolorato University of Naples Federico II, Naples, Italy

Yuko Shimizu Ritsumeikan University, Kyoto, Japan

Kojiro Shojima National Center for University Entrance Examinations, Tokyo, Japan

Nadia Solaro Department of Statistics and Quantitative Methods, Università di Milano-Bicocca, Milan, Italy

Naoki Sugino Ritsumeikan University, Kyoto, Japan

Hiroshi Suito Graduate School of Environmental and Life Science, Okayama University, Okayama City, Japan

Yuan Sun National Institute of Informatics, Tokyo, Japan

Daniel Y. Sze Department of Radiology, Stanford University School of Medicine, Stanford, CA, USA

Cristina Tortora University of Guelph, Guelph, ON, Canada

Duy Tran School of Mathematical and Physical Sciences, University of Newcastle, Callaghan, NSW, Australia

Hiroyuki Tsurumi College of Business Administration, Yokohama National University, Yokohama-shi, Japan

Takuya Ueda Department of Radiology, St. Luke's International Hospital, Tokyo, Japan

Michel Van de Velden Erasmus University of Rotterdam, PA Rotterdam, The Netherlands

Rosanna Verde Dip. di Scienze Politiche "J. Monnet", Seconda Università di Napoli, Caserta, Italy

Maria Prosperina Vitale University of Salerno, Fisciano (SA), Italy

Kenichi Yamakawa Yasuda Women's University, Hiroshima, Japan

Nagayoshi Yamashita GMO Research, Tokyo, Japan

Paola Zappa Dipartimento di Economia, Metodi Quantitativi e Strategie d'Impresa, Università Milano-Bicocca, Milano, Italy

Paola Zuccolotto University of Brescia, Brescia, Italy

Time-Frequency Filtering for Seismic Waves Clustering

Antonio Balzanella, Giada Adelfio, Marcello Chiodi, Antonino D'Alessandro, and Dario Luzio

Abstract This paper introduces a new technique for clustering seismic events based on processing, in time-frequency domain, the waveforms recorded by seismographs. The detection of clusters of waveforms is performed by a k-means like algorithm which analyzes, at each iteration, the time-frequency content of the signals in order to optimally remove the non discriminant components which should compromise the grouping of waveforms. This step is followed by the allocation and by the computation of the cluster centroids on the basis of the filtered signals. The effectiveness of the method is shown on a real dataset of seismic waveforms.

Keywords Earthquakes clustering • Time-frequency filtering • Waveforms clustering

A. Balzanella (✉)
Dipartimento di Studi Europei e Mediterranei, Seconda Universitá di Napoli, Naples, Italy
e-mail: antonio.balzanella@unina2.it

G. Adelfio • M. Chiodi
Dipartimento di Scienze Economiche, Aziendali e Statistiche, Universitá degli Studi di Palermo, Palermo, Italy
e-mail: giada.adelfio@unipa.it; marcello.chiodi@unipa.it

A. D'Alessandro
Centro Nazionale Terremoti, Istituto Nazionale di Geofisica e Vulcanologia, Rome, Italy
e-mail: antonino.dalessandro@ingv.it

D. Luzio
Dipartimento di Scienza della Terra e del Mare, Universitá degli Studi di Palermo, Palermo, Italy
e-mail: dario.luzio@unipa.it

D. Vicari et al. (eds.), *Analysis and Modeling of Complex Data in Behavioral and Social Sciences*, Studies in Classification, Data Analysis, and Knowledge Organization, DOI 10.1007/978-3-319-06692-9_1,

1 Introduction

The catalogs of the instrumental tectonic, volcanic or induced seismicity in mines or in geothermal or oil fields, often contain groups of similar events, i.e. events that have produced similar vectorial seismograms at the observation points. This similarity involves the near coincidence of the seismogenic volumes location and of the geometrical and physical parameters that controlled the energy release. The search for families of similar events is more and more applied in seismology with two fundamental aims. The first relates directly to the seismicity characterization, the second to the creation of databases for the inverse problems of relative hypocenter location and determination of the focal mechanism parameters. In the study of the tectonic seismicity the detection of seismic families allows to distinguish events produced by different tectonic structures. In particular, structures with similar geometric characteristics but with different locations, or structures with nearly coincident locations but with different focal mechanisms (Ferretti et al. 2005; D'Alessandro et al. 2013). Other applications of searching similar events were aimed at the identification of foreshocks and aftershocks and to obtain instrumental catalogues that are cleaned of dependent events (Barani et al. 2007).

The methods of hypocenter parameters and focal mechanisms absolute estimation are not accurate enough for reliable modeling of the seismogenic processes in all geological environments above mentioned, due to the low energy that characterizes, in general, the events that occur with greater frequency. Recent studies have shown substantial improvements in the precision of the earthquake location and focal mechanism determination when clustering techniques are used to select sets of events a priori characterized by a high level of similarity (Zhang 2003). Accurate estimates of differential arrival times, relative to similar events, in fact, allow a direct estimate of the differences between the parameters of their hypocenters and focal mechanisms by solving an inverse problem approximately linear.

The topic of seismic waveform clustering has been dealt in several papers. Adelfio et al. (2011) combined the aim of finding clusters from a set of seismograms with the functional nature of data, applying a variant of a k-means algorithm based on the principal component rotation of data. Adelfio et al. (2012) optimize an internal homogeneity criterion, finding simultaneously a partition of the seismograms and an optimal alignment of the curves observed over time. Furthermore, in Shumway (2003) a technique which uses the Kullback-Leibler discrimination measures for clustering non-stationary seismic time series has been proposed.

This paper aims at analyzing seismic events recorded, over time, by a single seismographic station. The clustering problem becomes, in this case, a very challenging task. As for the datasets of our application, traditional methods fail to find clusters because the similarity between waveforms is strongly dominated by a wide common behavior. This motivates the need to develop appropriate methods which process the seismograms in order to highlight the features which allow to get a better discrimination among the signals.

As in Shumway (2003), we consider the use of time-varying frequency spectra for clustering seismograms to analyze data which, unlike to Adelfio et al. (2012), do not need any alignment.

As first step, we propose to perform a Discrete Short-time Fourier transform on the time varying seismogram records (separately on the three spatial directions x, y, z) in order to obtain a representation in time and frequency of the data. This allows to study the frequency content of the signals over time.

The clustering algorithm used in the second step is a variant of the k-means algorithm where a time-frequency filtering stage is introduced in the iterative part of the algorithm, before the allocation and the centroid computation. The filtering is based on detecting and removing, from the computation of distances, the time frequency components which provide a strong increasing of the ratio between the within variability and the total variability. This involves that only the most discriminant time-frequency components will be kept in the clustering process.

2 Seismic Waves Clustering

Let $\{\mathbf{x_1}, \ldots, \mathbf{x_i}, \ldots, \mathbf{x_n}\}$ be a set of multivariate time series $\mathbf{x_i} = \left\{x_i^p(t)\right\}$ with $t = 1, \ldots, T$ indicating the time stamp and $p = 1, \ldots, P$ identifying the variables. Usually P is set to 3 since p identifies the time series associated to one of the spatial directions x, y, z.

Unlike to Adelfio et al. (2011, 2012) where the analysis is performed on $\{\mathbf{x_1}, \ldots, \mathbf{x_i}, \ldots, \mathbf{x_n}\}$ addressing the problem of the temporal misalignments among the series, our approach is based on performing the analysis in the time-frequency domain in order to consider the frequency content of the seismograms. With this aim, we propose to represent the set of time series in the time-frequency domain through the Discrete Short Time Fourier Transform (STFT). This transformation splits the temporal data into frames (which usually overlap each other) and then computes the Fourier transform of the data of each frame. This allows to record the value for each point as a function of the time τ and of the frequency ω. This is performed on each single time series $x_i^p(\cdot)$, so that:

$$y_i^p(\tau, \omega) = STFT(x_i^p(t)) \tag{1}$$

STFT is performed using the Fast Fourier Transform, so $\tau = 1, \ldots, \Xi$ and $\omega = 1, \ldots, \Omega$ are discrete and quantized. From this transformation step we get a new set $\mathbf{Y} = \{\mathbf{y_1}, \ldots, \mathbf{y_i}, \ldots, \mathbf{y_n}\}$ where $\mathbf{y_i} = \left\{y_i^p(\tau, \omega)\right\}$ (with $p = 1, \ldots, P$).

The aim of our proposal is to find a partition $\mathbf{C} = \{C_1, \ldots, C_k, \ldots, C_K\}$ of the n elements of the set $\mathbf{Y}$ in K exhaustive clusters with low internal heterogeneity. To reach this aim, in the following, we define the measure Δ for the internal heterogeneity:

$$\Delta(\mathbf{C},\mathbf{G},\mathbf{I}) = \sum_{k=1}^{K} \sum_{\mathbf{y_i} \in C_k} \sum_{p=1}^{P} \delta\left(y_i^p; g_k^p\right) \tag{2}$$

where $\mathbf{G} = \{\mathbf{g_1}, \ldots, \mathbf{g_k}, \ldots, \mathbf{g_K}\}$ (with $\mathbf{g_k} = \{g_k^p\}$, $p = 1, \ldots, P$), is the set of centroids and $\delta(.)$ is defined as follows:

$$\delta(y_i^p; g_k^p) = \frac{\sum_{\tau=1}^{\Xi} \sum_{\omega=1}^{\Omega} I_{\tau,\omega}(y_i^p(\tau,\omega) - g_k^p(\tau,\omega))^2}{\sum_{\tau=1}^{\Xi} \sum_{\omega=1}^{\Omega} I_{\tau,\omega}} \tag{3}$$

The indicator function $\mathbf{I} = \{I_{\tau,\omega}\}$ (with $\tau = 1, \ldots, \Xi$ and $\omega = 1, \ldots, \Omega$) assumes the value 0 or 1 so that $\delta(.)$ is an averaged squared Euclidean distance computed on a subset of the time-frequency content of the waveforms.

In the following we define the measure Θ for the total heterogeneity:

$$\Theta(\mathbf{Y},\mathbf{g},\mathbf{I}) = \sum_{i=1}^{n} \sum_{p=1}^{P} \delta\left(y_i^p; g^p\right) \tag{4}$$

where $\mathbf{g} = \{g^p\}$ is the general centroid.

Consistently with the definitions above, we propose to optimize the following criterion function in order to obtain the optimal partition $\mathbf{C}^*$ and the optimal indicator functions $\mathbf{I}^*$:

$$W(\mathbf{C}^*,\mathbf{G},\mathbf{I}^*) = \min_{\mathbf{C},\mathbf{I}} \frac{\Delta(\mathbf{C},\mathbf{G},\mathbf{I})}{\Theta(\mathbf{Y},\mathbf{g},\mathbf{I})} \tag{5}$$

To reach this aim, we use a variant of the k-means algorithm (Gan et al. 2007) which includes a step where the value of the indicator function $I_{\tau,\omega}$ is set to 1 or 0 according to the contribution of each time-frequency component to the separation among the clusters.

In particular, given a time-frequency (τ', ω'), the following expressions introduce $\Delta_{\tau',\omega'}$ and $\Theta_{\tau',\omega'}$ to measure the internal and total heterogeneity induced by (τ', ω'):

$$\Delta_{\tau',\omega'} = \sum_{k=1}^{K} \sum_{\mathbf{y_i} \in C_k} \sum_{p=1}^{P} (y_i^p(\tau',\omega') - g_k^p(\tau',\omega'))^2 \tag{6}$$

$$\Theta_{\tau',\omega'} = \sum_{i=1}^{n} \sum_{p=1}^{P} (y_i^p(\tau',\omega') - g^p(\tau',\omega'))^2 \tag{7}$$

On the basis of the previous formulas, the following criterion allows to set the values of each $I_{\tau,\omega}$:

$$\mathit{if}\ \frac{\Delta(\mathbf{C},\mathbf{G},\mathbf{I}) - \Delta_{\tau',\omega'}}{\Theta(\mathbf{Y},\mathbf{g},\mathbf{I}) - \Theta_{\tau',\omega'}} < \frac{\Delta(\mathbf{C},\mathbf{G},\mathbf{I})}{\Theta(\mathbf{Y},\mathbf{g},\mathbf{I})} \Rightarrow I_{\tau',\omega'} = 1 \quad \mathit{Else}\ \ I_{\tau',\omega'} = 0 \tag{8}$$

Algorithm 1 Clustering algorithm on time frequency spectra

INITIALIZATION:
Set $I_{\tau,\omega} = 1$ for all pairs (τ, ω)
Get an initial partition $\mathbf{C} = \{C_1, \ldots, C_k, \ldots, C_K\}$ of $\{\mathbf{y_1}, \ldots, \mathbf{y_i}, \ldots, \mathbf{y_n}\}$
CurrentIter=1
repeat
  $\mathbf{C}' = \mathbf{C}$
  CENTROIDS COMPUTATION:
  Compute the cluster centroids $\mathbf{g_k}$ for each $k = 1, \ldots, K$
  TIME-FREQUENCY FILTERING:
  Compute $\Delta(\mathbf{C}, \mathbf{G}, \mathbf{I})$ and $\Theta(\mathbf{Y}, \mathbf{g}, \mathbf{I})$
  for all τ', ω' **do**
    Compute $\Delta_{\tau',\omega'}$ and $\Theta_{\tau',\omega'}$
    if $\frac{\Delta(\mathbf{C},\mathbf{G},\mathbf{I}) - \Delta_{\tau',\omega'}}{\Theta(\mathbf{Y},\mathbf{g},\mathbf{I}) - \Theta_{\tau',\omega'}} < \frac{\Delta(\mathbf{C},\mathbf{G},\mathbf{I})}{\Theta(\mathbf{Y},\mathbf{g},\mathbf{I})}$ **then**
      $I_{\tau',\omega'} = 1$
    else
      $I_{\tau',\omega'} = 0$
    end if
  end for
  ALLOCATION:
  Allocate $\{\mathbf{y_1}, \ldots, \mathbf{y_i}, \ldots, \mathbf{y_n}\}$ to the nearest group C_k using the distance function in (3) and update the partition $\mathbf{C}$ accordingly
  CurrentIter=CurrentIter+1
until $\mathbf{C}' = \mathbf{C}$ OR CurrentIter=MaxIter

The pseudocode of the algorithm is shown in Algorithm 1. After the initialization, performed randomly or by using some other generic clustering algorithm such as the K-means, the algorithm iterates three steps until the convergence to a stable partition or until the maximum number of iterations has been reached. The first step is the computation of the clusters centers $\mathbf{g_k}$, component by component, as average of the time-frequency spectra allocated to C_k. The second step is the filtering of the non discriminant time-frequency components by means of the criterion in (8). By means of this step, at each iteration only a subset of time-frequencies will be selected and considered in the allocation step. The latter realizes the assignment of the spectra $\{\mathbf{y_1}, \ldots, \mathbf{y_i}, \ldots, \mathbf{y_n}\}$ to the clusters $\{C_1, \ldots, C_k, \ldots, C_K\}$ according to the distance, computed component by component, to the centers $\mathbf{g_k}$ (with $k = 1, \ldots, K$).

3 Results on Real Data

The proposed method has been tested on three real datasets which collect the recording of a sequence of seismic events made by three stations in Capo D'Orlando area (Italy) between 2011 and 2012. We have $n = 306$ events recorded by each station along three directions x, y, z so that each event will correspond to a multivariate time series $\mathbf{x_i} = \{x_i^p(t)\}$, with $P = 3$. Each event has a duration of 30 s; however, for two stations, the sampling frequency has been 100 Hz while for

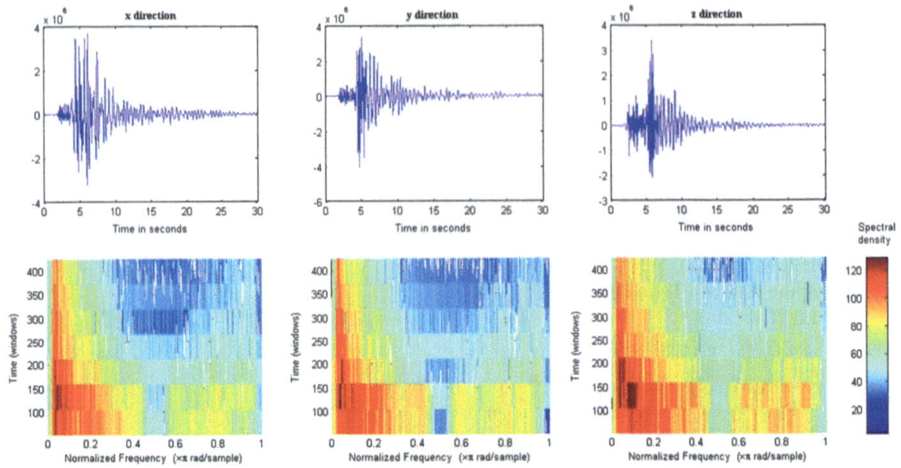

Fig. 1 Single seismic event plotted, for the three spatial directions, in time domain and time-frequency domain

the third station it has been 50 Hz. All the signals have been cut, aligned and filtered in order to make the analysis more effective.

The aims of the test on the three datasets are: (1) To evaluate if the proposed method has been able to provide a better discrimination among the clusters of spectra when compared to the k-means; (2) To measure the agreement between the partitions obtained on the three datasets.

The first step has been to compute the time frequency spectra of the curves in the three datasets. For this aim we have used a Short Time Fourier Transform with Hamming windows which has been performed on each direction of each single seismic event of the datasets. In Fig. 1 we have an example of the transformation for a seismic event of the first dataset.

The next step is to run the proposed algorithm on the three datasets. The only parameter to set is the number of clusters K.

Due to the very high similarity among the data, we have been able to run the algorithm only for $K = 2, \ldots, 5$ clusters. By evaluating the criterion function $W(\mathbf{C}^*, \mathbf{G}, \mathbf{I}^*)$ for each value of K, we have chosen $K = 4$ since for this value we have the highest decreasing.

Although there is a high homogeneity among the curves in the datasets, our method, based on filtering time-frequency components, is able to provide a good partitioning of the spectra. In Table 1, we show this issue by comparing the number of spectra allocated to each cluster using our algorithm and a standard k-means on the spectra:

It is evident how the proposed method is able to provide a better splitting of the data into clusters while the k-means tends to allocate the most of time-frequency spectra into a single group. Still, we show in Figs. 2 and 3 the centroids of the

Table 1 Number of spectra allocated to each cluster using the proposed algorithm and the k-means on the spectra

	Filtered clustering			k-means		
Cluster ID	Dataset 1	Dataset 2	Dataset 3	Dataset 1	Dataset 2	Dataset 3
1	8	11	13	12	4	8
2	62	67	66	35	37	45
3	165	185	157	203	255	196
4	71	43	70	56	10	57

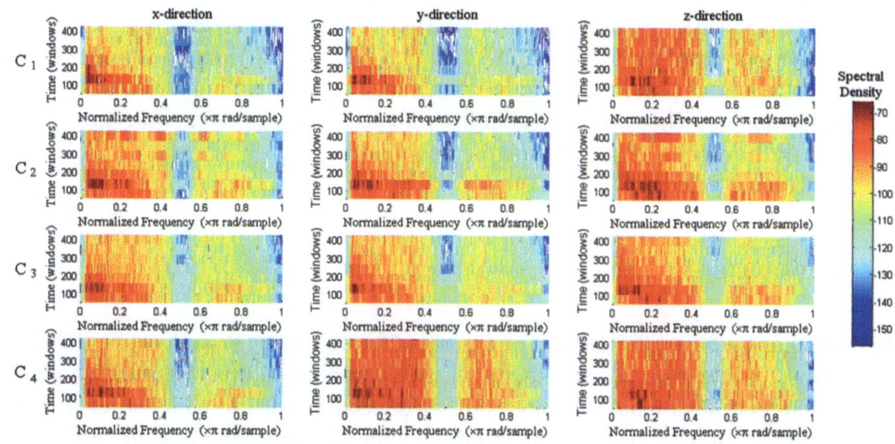

Fig. 2 Centroids of the clusters, on the three spatial directions, for the dataset 1 in time-frequency domain

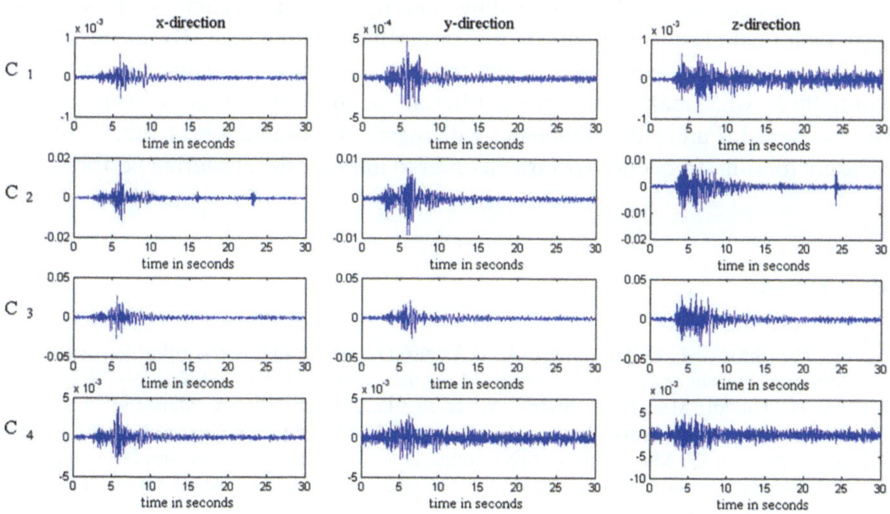

Fig. 3 Centroids of the clusters, on the three spatial directions, for the dataset 1 in time domain

Table 2 Rand index

Rand index(Dataset 1;Dataset 2)	0.59
Rand index(Dataset 1;Dataset 3)	0.69
Rand index(Dataset 2;Dataset 3)	0.63

obtained clusters, related to the first dataset, respectively in time-frequency and time domain. These are useful to summarize the main behaviors in the data.

Since the three datasets concern the monitoring of the same sequence of seismic events at three seismographic stations we expect to observe some agreement between the obtained partitions. In our tests, we have evaluated this issue using the well known Rand Index (Rand 1971) which measures the consensus between a couple of partitions providing a value in the range 0–1, where higher values denote a high agreement. The results, available in Table 2, show the pairwise agreement between the partitions obtained from the three datasets.

4 Conclusions

In this paper we have introduced a new method for clustering seismic waves. Unlike to methods in literature, we have dealt with this problem in time-frequency domain rather than in time domain. We have still introduced a heuristic which selects, optimally, a subset of the time-frequency content with the aim to improve the separation among signals. We have evaluated the performance of the proposed method on three real world datasets in which a seismograph records the events of a geographic area. These datasets are characterized by very high similarities among the waves so that traditional clustering methods fail in partitioning the data. By means of our approach, we have been able remove the non discriminant frequencies allowing to highlight what makes the waves different so that a partitioning structure can be discovered. Future developments will regard the evaluation of other distance measure for time-frequency spectra, in order to improve the clustering performance.

References

Adelfio, G., Chiodi, M., D'Alessandro, A., & Luzio, D. (2011). FPCA algorithm For waveform clustering. *Journal of Communication and Computer, 8*(6), 494–502. ISSN:1548–7709.

Adelfio, G., Chiodi, M., D'Alessandro, A., Luzio, D., D'Anna, G., & Mangano G. (2012). Simultaneous seismic wave clustering and registration. *Computers & Geosciences, 44*, 60–69.

Barani, S., Ferretti, G., Massa, M., & Spallarossa, D. (2007). The waveform similarity approach to identify dependent events in instrumental seismic catalogues. *Geophysical Journal International, 168*(1), 100–108.

D'Alessandro, A., Mangano, G., D'Anna, G., & Luzio, D. (2013). Waveforms clustering and single-station location of microearthquake multiplets recorded in the northern Sicilian offshore region. *Geophysical Journal International*. doi:10.1093/gji/ggt192.

Ferretti, G., Massa, M., & Solarino, S. (2005) An improved method for the recognition of seismic families: application to the Garfagnana-Lunigiana area, Italy. *Bulletin of the Seismological Society of America, 95*(5), 1903–1015.

Gan, G., Ma, C., & Wu, J. (2007). *Data clustering: theory, algorithms, and applications*. Philadelphia: SIAM.

Rand, W. M. (1971). Objective criteria for the evaluation of clustering methods. *Journal of the American Statistical Association, 66*(336), 846–850.

Shumway, R. H. (2003). Time-frequency clustering and discriminant analysis. *Statistics & Probability Letters, 63*, 307–314

Zhang, H. (2003). *Double-difference seismic tomography method and its applications* (Ph.D. thesis, Department of Geology and Geophysics, University of Wisconsin-Madison).

Modeling Longitudinal Data by Latent Markov Models with Application to Educational and Psychological Measurement

Francesco Bartolucci

Abstract I review a class of models for longitudinal data, showing how it may be applied in a meaningful way for the analysis of data collected by the administration of a series of items finalized to educational or psychological measurement. In this class of models, the unobserved individual characteristics of interest are represented by a sequence of discrete latent variables, which follows a Markov chain. Inferential problems involved in the application of these models are discussed considering, in particular, maximum likelihood estimation based on the Expectation-Maximization algorithm, model selection, and hypothesis testing. Most of these problems are common to hidden Markov models for time-series data. The approach is illustrated by different applications in education and psychology.

Keywords Forward and Backward recursions • Expectation-Maximization algorithm • Hidden Markov models • Rasch model

1 Introduction

Among the statistical models for the analysis of longitudinal data which are available in the literature (Diggle et al. 2002; Frees 2004; Fitzmaurice et al. 2009), those based on a latent Markov chain have a special role when the main interest is on individual characteristics which are not directly observable. Such models are based on assumptions similar to those of hidden Markov models; for a recent review see Zucchini and MacDonald (2009). Applications focusing on individual characteristics which are not directly observable usually arise in education and psychology, when these characteristics are measured through the responses to a

F. Bartolucci (✉)
Department of Economics, University of Perugia, Perugia, Italy
e-mail: bart@stat.unipg.it

D. Vicari et al. (eds.), *Analysis and Modeling of Complex Data in Behavioral and Social Sciences*, Studies in Classification, Data Analysis, and Knowledge Organization,
DOI 10.1007/978-3-319-06692-9_2,

series of items, also corresponding to specific tasks, administered at consecutive occasions.

One of the first authors who dealt with models for longitudinal data which are based on a latent Markov chain, LM models for short, was Wiggins (1973); see Bartolucci et al. (2013) for a review. The basic assumption of these models is that the response variables, corresponding to test items in the present context, are conditionally independent given the Markov chain. Due to the adoption of suitable parametrizations, such as that characterizing the Rasch model (Rasch 1961) or that characterizing the graded response model (Samejima 1969, 1996), the states of the chain may be interpreted as different levels of ability or tendency towards a certain behavior; these states identify different latent classes in the population from that the observed sample comes. Of particular interest is the possibility of estimating probabilities of transition between the classes; these probabilities may be allowed to depend on individual covariates or experimental factors.

Aim of the present paper is to review LM models in the context of educational and psychological measurement, also discussing likelihood based inference. In particular, maximum likelihood estimation may be performed by an Expectation-Maximization (EM) algorithm implemented by the Baum-Welch forward and backward recursions (Baum et al. 1970), which have been developed in the literature on hidden Markov models for time-series data. Moreover, model selection, regarding in particular the number of states (or latent classes), is based on information criteria, such as the Akaike Information Criterion (AIC) (Akaike 1973) or the Bayesian Information Criterion (BIC) (Schwarz 1978). Finally, hypothesis testing may be based on the likelihood ratio statistic which, under certain regularity conditions (Bartolucci 2006), has a null asymptotic distribution of chi-bar-squared type, that is, a finite mixture of chi-squared distributions.

The remainder of this paper is organized as follows. Basic assumptions of LM models for longitudinal data are illustrated in Sect. 2 with focus on educational and psychological contexts. Likelihood inference, regarding in particular parameter estimation, model selection, and testing, is dealt with in Sect. 3. Finally, some applications of the discussed statistical models in educational and psychological fields are briefly described in Sect. 4.

2 Model Assumptions and Likelihood Inference

Suppose that a set of J items is administered at T consecutive occasions to a sample of n subjects and let $Y_{ij}^{(t)}$, $i = 1,\ldots,n$, $j = 1,\ldots,J$, $t = 1,\ldots,T$, denote the random variable for the response to item j at occasion t by subject i. This random variable is binary in the case of dichotomously-scored items and categorical, with more than two categories, in the case of polytomously-scored items. In the second case the response categories are typically ordered. In any case, the number of response categories is denoted by c and they are labelled from 0 to $c - 1$.

In the above context, the basic assumption of LM models is that, for every sample unit i, the random variables $Y_{ij}^{(t)}$, $j = 1,\ldots,J$, $t = 1,\ldots,T$, are conditionally independent given a sequence of latent variables $U_i^{(1)},\ldots,U_i^{(T)}$. These latent variables identify latent classes of subjects sharing common characteristics. Moreover, the distribution of these variables is assumed to follow a first-order Markov chain with k states and homogeneous or non-homogenous transition probabilities. In particular, the initial probabilities are denoted by $\pi_u = p(U_i^{(1)} = u)$, $u = 1,\ldots,k$, whereas the transition probabilities are denoted by $\pi_{v|u}^{(t)} = p(U_i^{(t)} = v|U_i^{(t-1)} = u)$, $u,v = 1,\ldots,k$, $t = 2,\ldots,T$, in the time non-homogenous case. These probabilities are collected in the $k \times k$ *transition matrix* $\boldsymbol{\Pi}^{(t)}$, with the index u running by row and the index v running by column. In the time-homogenous case we have $\pi_{v|u}^{(t)} = \pi_{v|u}$ for $t = 2,\ldots,T$.

In educational and psychological measurement, the interpretation of the model is enforced by letting every manifest variable $Y_{ij}^{(t)}$ to depend only on the corresponding latent variable $U_i^{(t)}$ according to a suitable parametrization for the conditional distribution of the former given the latter. For instance, with dichotomously-scored test items ($c = 2$), we may adopt a Rasch parametrization (Rasch 1961; Bartolucci et al. 2008) by requiring that

$$\log \frac{p(Y_{ij}^{(t)} = 1|U_i^{(t)} = u)}{p(Y_{ij}^{(t)} = 0|U_i^{(t)} = u)} = \psi_u - \beta_j, \quad j = 1,\ldots,J, \ \ u = 1,\ldots,k,$$

for all t, where ψ_u is a parameter interpreted as the ability level of the examinees in latent class u. Moreover, β_j is the difficulty level of item j. The constraint that this difficulty level does not vary with t makes sense only if the same battery of items is administered at all occasions and then, in certain contexts, this constraint must be relaxed in a suitable way.

The above parametrization may be extended in a natural way to the case of $c > 2$ response categories. If these categories are ordered, we may assume a parametrization which is also adopted in the graded response model (Samejima 1969, 1996), that is,

$$\log \frac{p(Y_{ij}^{(t)} \geq y|U_i^{(t)} = u)}{p(Y_{ij}^{(t)} < y|U_i^{(t)} = u)} = \psi_u - \beta_{jy},$$

$$j = 1,\ldots,J, \ \ u = 1,\ldots,k, \ \ y = 1,\ldots,c-1,$$

where the parameters have an interpretation similar to the previous one. In particular, ψ_u is still interpreted as the ability level of subjects in latent class u. Note that, in order to avoid a wrong model specification, the β_{jy} parameters are increasing ordered in y for all j. Under such a parametrization, the initial probabilities of the latent Markov chain allow us to study the distribution of the ability (or another latent trait of interest) among the examinees at the beginning of the period of observation.

Moreover, the probabilities of transition between the latent classes allow us to study the evolution of the ability, even depending on particular individual covariates or factors (e.g., teaching method).

It has to be clear that we can also use the probabilities

$$\phi_{jy|u} = p(Y_{ij}^{(t)} = y|U_i^{(t)} = u), \quad t = 1, \dots, T, \ u = 1, \dots, k, \ y = 0, \dots, c-1,$$

as free parameters, without assuming any specific parametrization. In this case, if covariates are ruled out and the transition probabilities are not constrained to be homogeneous, then a multivariate version of the so-called *basic LM model* (Bartolucci et al. 2013) results. However, the interpretation of the latent classes may be more difficult in terms, for instance, of different levels of ability, since it is not ensured that these classes are monotonically ordered according to the conditional distribution of the response variables.

More complex formulations of the LM models, with respect to those described above, are available in the literature; see Bartolucci et al. (2013) for a complete review about these models. A typical extension of interest is for the inclusion of individual covariates, possibly time varying, that affect the initial and the transition probabilities of the latent Markov chain. A more complex extension is to deal with multilevel longitudinal data in which subjects are collected in clusters, such as students in school. In this case, further latent variables are used to account for the effect of each cluster on the response variables; see Bartolucci and Lupparelli (2012) and the references therein.

Finally, it is worth noting that the modeling framework illustrated in this section may be also applied to the case of unbalanced panel settings in which the number of time occasions is not the same for all subjects, due to some forms of ignorable drop-out. In this case, the number of time occasions for subject i is indicated by T_i and must be substituted to T in the expressions above.

3 Likelihood Inference

Estimation of an LM model is usually performed by maximizing its likelihood. In the case of independent sample units and when individual covariates are ruled out, this likelihood has logarithm

$$\ell(\boldsymbol{\theta}) = \sum_i \log p(\mathbf{y}_i), \tag{1}$$

where $\boldsymbol{\theta}$ is the vector of all model parameters and $p(\mathbf{y}_i)$ is the manifest probability of the sequence of item responses provided by subject i, which are collected in the vector $\mathbf{y}_i$. This probability is in practice computed by a suitable recursion which is well known in the hidden Markov literature and has been set up by Baum and Welch (Baum et al. 1970; Welch 2003; Zucchini and MacDonald 2009).

In order to maximize the log-likelihood in (1), the main tool is the Expectation-Maximization (EM) algorithm (Baum et al. 1970; Dempster et al. 1977), which is based on the so-called *complete data log-likelihood*, that is, the log-likelihood that we could compute if we knew the value of $U_i^{(t)}$ for every subject i and time occasion t. When the conditional response probabilities do not depend on t, as in the parametrizations illustrated in Sect. 2, this function may be expressed as:

$$\ell^*(\boldsymbol{\theta}) = \sum_j \sum_t \sum_u \sum_y a_{juy}^{(t)} \log \phi_{jy|u} + \sum_u b_u^{(1)} \log \pi_u + \sum_{t>1} \sum_u \sum_v b_{uv}^{(t)} \log \pi_{v|u}^{(t)}, \quad (2)$$

where $a_{juy}^{(t)}$ is the frequency of subjects responding by y to item j at occasion t and belonging to latent state u at the same occasion, $b_u^{(t)}$ is the frequency of subjects in latent state u at occasion t, and $b_{uv}^{(t)}$ is the number of transitions from latent state u at occasion $t-1$ to state v at occasion t.

The EM algorithm alternates two steps until convergence in the model log-likelihood $\ell(\boldsymbol{\theta})$. The E-step consists of computing the conditional expected value of every frequency in (2) given the observed data and the current value of the parameters. The M-step consists of maximizing the function $\ell^*(\boldsymbol{\theta})$ in which these frequencies have been substituted by the corresponding expected values. An implementation of this algorithm, and of a bootstrap algorithm to compute standard errors for the parameter estimates, is available in the package `LMest` for `R` (Bartolucci 2012).

With more complex formulations of the LM model, the EM algorithm is still used for parameter estimation. In particular, in the presence of individual covariates affecting the initial and transition probabilities of the Markov chain, the log-likelihood to be maximized is expressed as in (1) with $p(\mathbf{y}_i)$ substituted by $p(\mathbf{y}_i|\mathbf{x}_i)$. The latter may be computed by the same recursion mentioned above (Baum et al. 1970), while an extended version of the complete log-likelihood in (2) is used within the algorithm; then the EM algorithm is not much more complex than the one used for the model without covariates. On the other hand, estimating a multilevel LM model requires a more complex implementation of the EM algorithm and then we refer the reader to specific articles, see in particular (Bartolucci et al. 2011; Bartolucci and Lupparelli 2012), for a detailed description.

A crucial point in applying an LM model is the choice of the number of latent states, k. This choice is usually accomplished by an information criterion based on penalizing the maximum value of the log-likelihood. For instance, BIC leads to selecting the value of k which minimizes $BIC = -2\ell(\hat{\boldsymbol{\theta}}) + g\log(n)$, where g is the number of free parameters (Schwarz 1978). Alternatively, we can use AIC (Akaike 1973), which is based on an index similar to the previous one, where the penalization term is $2g$ instead of $g\log(n)$. BIC usually leads to more parsimonious models and it is typically preferred to AIC.

Another important point concerns how to test hypotheses of interest on the parameters. In many cases, the standard asymptotic theory may be employed to test these hypotheses on the basis of a likelihood ratio statistic. In particular, the null asymptotic distribution turns out to be of chi-squared type. However, in certain cases, and in particular when the hypothesis is expressed through linear constraints on the transition probabilities, a more complex asymptotic distribution results, that is, the chi-bar-squared distribution. This distribution may be expressed as a mixture of standard chi-squared distributions with suitable weights, which may be computed analytically in certain simple cases or by a suitable Monte Carlo method in general (Shapiro 1988; Bartolucci 2006). In particular, this distribution arises when testing that transitions between latent states are not allowed and then an LM model specializes into a latent class model (Bartolucci 2006). This hypothesis is simply expressed by constraining the transition matrices to be equal to an identity matrix. A less restrictive constraint is that these matrices are triangular, so that a certain type of evolution of the latent trait is considered. For instance, with $k = 3$ states, these two constraints are expressed as follows:

$$\boldsymbol{\Pi}^{(t)} = \begin{pmatrix} 1 & 0 & 0 \\ 0 & 1 & 0 \\ 0 & 0 & 1 \end{pmatrix}, \quad \boldsymbol{\Pi}^{(t)} = \begin{pmatrix} \pi_{1|1}^{(t)} & \pi_{2|1}^{(t)} & \pi_{3|1}^{(t)} \\ 0 & \pi_{2|2}^{(t)} & \pi_{3|2}^{(t)} \\ 0 & 0 & 1 \end{pmatrix}.$$

When the latent states correspond to increasing levels of ability in an educational application, the second matrix corresponds to the hypothesis that the ability never decreases during the period of observations. Many other constraints on the transition matrices may be of interest, such as that of symmetry.

4 Applications in Educational and Psychological Measurement

In order to illustrate how LM models may be used in the educational and psychological measurement, in the following we describe some benchmark applications.

4.1 Evolution of Ability Level in Mathematics

The first application concerns testing the hypothesis of absence of tiring or learning-through-training phenomena during the administration of a series of 12 items on Mathematics (Bartolucci 2006; Bartolucci et al. 2008). In this case, we are not properly dealing with longitudinal data since all items were administered at the same occasion; however, an LM model makes sense since these items were administered in the same order to all examinees. In particular, the adopted LM model is based on a Rasch parametrization.

The main conclusion of the study is that there is not an evolution of the ability during the administration of the test items and then there is no evidence of the existence of tiring or learning-through-training phenomena. This conclusion is reached by comparing, by a likelihood ratio test statistic, the LM Rasch model with homogenous transition probabilities with a constrained version of this model in which the transition matrices are equal to an identity matrix (null hypothesis). The corresponding p-value, which is larger than 0.05, is computed on the basis of the chi-bar-squared distribution and leads to the conclusion that there is not enough evidence against the null hypothesis.

4.2 Evolution of Psychological Traits in Children

The second application (Bartolucci and Solis-Trapala 2010) concerns data collected through a psychological experiment based on tests which were administered at different occasions to pre-school children in order to measure two types of ability: inhibitory control and attentional flexibility. In this case, the model is more complex than the one adopted for the first application since it is multidimensional and then subjects are classified in latent classes according to different abilities. Moreover, more complex parametrizations than the Rasch parametrization are adopted and transition probabilities are suitably formulated so as to account for certain experimental features. In particular, the maximum number of items administered to the same child is 132; these items were administered in three separated periods of time.

This application led to the conclusion that the two abilities must be conceptualized as distinct constructs, and so a unidimensional latent variable model cannot be validly used in this context. Moreover, it was demonstrated that these abilities develop at an early age and that there are different dynamics within different sequences of task administration, with mild tiring effects within certain sequences and learning-through-training phenomena concerning other sequences.

In dealing with this applications, the authors also fitted an extended version of the LM model in which the Markov chain is of second order. This was reformulated as a first-order model with an extended state space having k^2 elements. However, the data did not provide evidence in favor of this second-order extension, meaning that the ability level at a given occasion only depends on that at the previous occasion.

4.3 Evaluation of School Effectiveness on Proficiency of Students

The third application (Bartolucci et al. 2011) involves a multilevel LM model based on a Rasch parameterization, which is used to analyze data collected by a series of test items administered to middle-school students. The overall number of items is

97; these items were administered at the end of each of the 3 years of schooling. The adopted model takes into account that students (level 1 units) are collected in schools (level 2 units) by the inclusion of further latent variables. This extension, already mentioned in Sect. 2, allows us to evaluate the effect of every school and then allows us to perform an analysis of the school effectiveness, also considering characteristics such as the type of school (e.g., public or not).

As a main result, the study found evidence of four different typologies of school which have a different effect on the ability level of their students and on the way in which this ability evolves across time. These typologies cannot be easily ordered because the effect is not in general constant across time. However, it emerges that most public schools have an intermediate positive effect on the proficiency of their students. On the other hand, a polarization is observed for non-public schools with some of them which have poor performance, while the others have very good performance.

This application may be considered as one attempt to implement an LM model having a causal perspective, which may be then used for evaluation purposes; for a similar application see Bartolucci et al. (2009). This is a promising field of application of LM models since, in a single framework, these models allow us to evaluate the effect of certain factors, not only on the distribution of a characteristic of interest, but also on its evolution, even when in a longitudinal setting this characteristic is not directly observable but it is observed through a series of time-specific response variables.

References

Akaike, H. (1973). Information theory and extension of the maximum likelihood principle. In B. N. Petrov & Csaki, F. (Eds.), *Second international symposium on information theory* (pp. 267–281). Budapest: Akademiai Kiado.

Bartolucci, F. (2006). Likelihood inference for a class of latent Markov models under linear hypotheses on the transition probabilities. *Journal of the Royal Statistical Society: Series B, 68*, 155–178.

Bartolucci, F. (2012). Package LMest for R, available via CRAN at http://cran.r-project.org/web/packages/LMest/index.html.

Bartolucci, F., & Lupparelli, M. (2012). Nested hidden Markov chains for modeling dynamic unobserved heterogeneity in multilevel longitudinal data. arXiv:1208.1864.

Bartolucci, F., & Solis-Trapala, I. (2010). Multidimensional latent Markov models in a developmental study of inhibitory control and attentional flexibility in early childhood. *Psychometrika, 75*, 725–743.

Bartolucci, F., Farcomeni, A., & Pennoni, F. (2013). *Latent markov models for longitudinal data*. Boca Raton: Chapman and Hall/CRC.

Bartolucci, F., Lupparelli, M., & Montanari, G. E. (2009). Latent Markov model for longitudinal binary data: an application to the performance evaluation of nursing homes. *Annals of Applied Statistics, 3*, 611–636.

Bartolucci, F., Pennoni, F., & Lupparelli, M. (2008). Likelihood inference for the latent Markov Rasch model. In C. Huber, N. Limnios, M. Mesbah, & M. Nikulin (Eds.), *Mathematical methods for survival analysis, reliability and quality of life* (pp. 239–254). London: Wiley.

Bartolucci, F., Pennoni, F., & Vittadini, G. (2011). Assessment of school performance through a multilevel latent Markov Rasch model. *Journal of Educational and Behavioral Statistics, 36*, 491–522.

Baum, L. E., Petrie, T., Soules, G., & Weiss, N. (1970). A maximization technique occurring in the statistical analysis of probabilistic functions of Markov chains. *Annals of Mathematical Statistics, 41*, 164–171.

Dempster, A. P., Laird, N. M., & Rubin, D. B. (1977). Maximum likelihood from incomplete data via the EM algorithm (with discussion). *Journal of the Royal Statistical Society: Series B, 39*, 1–38.

Diggle, P. J., Heagerty, P. J., Liang, K.-Y., & Zeger, S. L. (2002). *Analysis of longitudinal data* (2nd ed.). Oxford: Oxford University Press.

Fitzmaurice, G., Davidian, M., Verbeke, G., & Molenberghs, G. (2009). *Longitudinal data analysis*. London: Chapman and Hall/CRC.

Frees, E. W. (2004). *Longitudinal and panel data: analysis and applications in the social sciences*. Cambridge: Cambridge University Press.

Rasch, G. (1961). On general laws and the meaning of measurement in psychology. *Proceedings of the IV Berkeley Symposium on Mathematical Statistics and Probability, 4*, 321–333.

Samejima, F. (1969). *Estimation of latent ability using a response pattern of graded scores* (Psychometrika monograph, 17). Richmond, VA: Psychometric Society.

Samejima, F. (1996). Evaluation of mathematical models for ordered polychotomous responses. *Behaviormetrika, 23*, 17–35.

Schwarz, G. (1978). Estimating the dimension of a model. *The Annals of Statistics, 6*, 461–464.

Shapiro, A. (1988). Towards a unified theory of inequality constrained testing in multivariate analysis. *International Statistical Review, 56*, 49–62.

Welch, L. R. (2003). Hidden Markov models and the Baum-Welch algorithm. *IEEE Information Theory Society Newsletter, 53*, 1–13.

Wiggins, L. M. (1973). *Panel analysis: latent probability models for attitude and behaviour processes*. Amsterdam: Elsevier.

Zucchini, W., & MacDonald, I. L. (2009). *Hidden Markov models for time series: an introduction using R*. New York: Springer.

Clustering of Stratified Aggregated Data Using the Aggregate Association Index: Analysis of New Zealand Voter Turnout (1893–1919)

Eric J. Beh, Duy Tran, Irene L. Hudson, and Linda Moore

Abstract Recently, the Aggregate Association Index (AAI) was proposed to identify the likely association structure between two dichotomous variables of a 2×2 contingency table when only aggregate, or equivalently the marginal, data are available. In this paper we shall explore the utility of the AAI and its features for analysing gendered New Zealand voting data in 11 national elections held between 1893 and 1919. We shall demonstrate that, by using these features, one can identify clusters of homogeneous electorates that share similar voting behaviour between the male and female voters. We shall also use these features to compare the association between gender and voting behaviour across each of the 11 elections.

Keywords 2×2 contingency tables • Aggregate association index • Aggregate data • New Zealand voting data

1 Introduction

The study of the association between two dichotomous variables that form a 2×2 contingency table when only marginal information is known has a long and rich history. Fisher (1935) asked us to "blot out the contents of the table, leaving only the marginal frequencies" and concluded that the marginal information "by themselves, supply no information ... as to the proportionality of the frequencies in the body of the table". Others to have considered this issue include (Aitkin and

E.J. Beh (✉) • D. Tran • I.L. Hudson
School of Mathematical and Physical Sciences, University of Newcastle, Callaghan 2308, NSW, Australia
e-mail: eric.beh@newcastle.edu.au; duy.tran@newcastle.edu.au; irene.hudson@newcastle.edu.au

L. Moore
Statistics New Zealand, Wellington, New Zealand

D. Vicari et al. (eds.), *Analysis and Modeling of Complex Data in Behavioral and Social Sciences*, Studies in Classification, Data Analysis, and Knowledge Organization,
DOI 10.1007/978-3-319-06692-9_3,

Table 1 Cross-classification of registered voters by gender and voting status for electorate 1 in 1893

First electorate 1893	Vote	No vote	Total
Women	1,443	289	1,732
Men	1,747	842	2,589
Total	3,190	1,131	4,321

Table 2 Summary of information for the first 11 New Zealand federal elections, 1893–1919

Year	Electorates	Number of registered men	Number of registered women	Men votes	Women votes
1893	57	175,915	147,567	126,183	88,484
1894	62	191,881	157,942	74,366	47,862
1896	62	197,002	142,305	149,471	108,783
1899	59	202,044	157,974	159,780	119,550
1902	68	229,845	185,944	180,294	138,565
1905	76	263,597	212,876	221,611	175,046
1908	76	294,073	242,930	238,534	190,114
1911	76	321,033	269,009	271,054	221,878
1914	76	335,697	280,346	286,799	234,726
April 1919	76	321,773	304,859	241,524	241,510
December 1919	76	355,300	328,320	289,244	261,083

Hinde 1984; Barnard 1984; Plackett 1977). Much of the focus of these issues has been on determining the cell values, or some function of them (such as a conditional proportion, or odds ratio). In particular, one may consider the variety of ecological inference (EI) techniques that exist to perform such estimation (Brown and Payne 1986; Chambers and Steel 2001; Forcina et al. 2012; Goodman 1953; King 1997; Steel et al. 2004; Wakefield 2004). An alternative strategy, described in Beh (2008, 2010) is to consider the aggregate association index, or AAI. Given the margins of a 2×2 table, the AAI reflects how likely a statistically significant association exists between the dichotomous variables, and (unlike all of the EI techniques) does so for a single table. In this paper we shall demonstrate the characteristics of the AAI, in particular the "AAI curve", for the study of multiple, or stratified, 2×2 tables.

In 1893, New Zealand became the first self-governing nation in the world to grant women the right to vote in federal elections; even though they were not eligible to stand as candidates until 1919; this trend quickly spread across the globe. One may consult the following URL www.elections.org.nz/study/education-centre/history/votes-for-women.html for an extensive history of the voting status for women in NZ.

The data from the New Zealand federal elections held between 1893 and 1919 provides a wealth of information for the analysis of early voting behaviour. A detailed investigation of the data based only on the marginal information was discussed in Hudson et al. (2010). Fortunately for analysts studying issues concerned with aggregate (or marginal) information, data at the electorate level were also kept that records the gender of those that voted and those that did not; for example Table 1 provides a summary of the number of registered men and women who voted or did not vote in the 1893 election. A more comprehensive overview of the data is given in Table 2. It provides a summary of the number of men and women voters as well as the number of registered voters for each gender. This table is derived from Table 1 of Hudson et al. (2010).

2 The Aggregate Association Index

For each election, the electorate data can be summarised as a 2×2 contingency table. Therefore, for a particular election, denote the total number of registered voters in the gth electorate by n_g. Suppose that the number of voters in the ith row and jth column (for i = 1, 2 and j = 1, 2) of the 2×2 table is n_{ijg} with an electorate proportion of $p_{ijg} = n_{ijg}/n_g$, the ith and jth marginal proportions can be denoted as $p_{i\bullet g}$ and $p_{\bullet jg}$. For the study of the NZ voting data, the row variable consists of the gender categories "Women" (for i = 1) and "Men" (i = 2). Similarly, the column variable reflects whether a registered individual voted or not with categories "Vote" (j = 1) for a registered individual who voted and "No Vote" (j = 2) for a registered individual who did not vote.

Let $P_{1g} = n_{11g}/n_{1\bullet g}$ be the (conditional) probability of a woman, who resides in the gth electorate, who votes. Beh (2008, 2010) showed that the Pearson chi-squared statistic can be expressed as a quadratic function of P_{1g}

$$X_g^2\left(P_{1g}\middle|p_{1\bullet g},\ p_{\bullet 1g}\right) = n_g\left(\frac{P_{1g} - p_{\bullet 1g}}{p_{2\bullet g}}\right)^2\left(\frac{p_{1\bullet g}p_{2\bullet g}}{p_{\bullet 1g}p_{\bullet 2g}}\right).$$

Here P_{1g} bounded by

$$L_{1g} = \max\left(0,\ \frac{n_{\bullet 1g} - n_{2\bullet g}}{n_{1\bullet g}}\right) \le P_{1g} \le \min\left(\frac{n_{\bullet 1g}}{n_{1\bullet g}},\ 1\right) = U_{1g}.$$

Refer to Duncan and Davis (1953) for further details. More narrow bounds of P_{1g} can be obtained given a chi-squared test of the association is made at the α level of significance. These bounds, derived by Beh (2008), are

$$L_{\alpha g} = \max\left(0,\ p_{\bullet 1g} - p_{2\bullet g}\sqrt{\frac{\chi_\alpha^2}{n_g}\left(\frac{p_{\bullet 1g}p_{\bullet 2g}}{p_{1\bullet g}p_{2\bullet g}}\right)}\right) < P_{1g}$$
$$< \min\left(1,\ p_{\bullet 1g} + p_{2\bullet g}\sqrt{\frac{\chi_\alpha^2}{n_g}\left(\frac{p_{\bullet 1g}p_{\bullet 2g}}{p_{1\bullet g}p_{2\bullet g}}\right)}\right) = U_{\alpha g}$$

Since L_{1g} and U_{1g} only depend only on the marginal information, the chi-squared statistic $\chi^2(P_{1g}|\ p_{1\bullet g},\ p_{\bullet 1g})$ can also be investigated by using only the margins. By taking into account the above properties of P_{1g}, Beh (2008) proposed the *Aggregate Association Index*, or AAI. For the gth electorate, this index is defined as:

$$A_{\alpha g} = 100\left(1 - \frac{\chi_\alpha^2\left[\left(L_{\alpha g} - L_{1g}\right) + \left(U_{1g} - U_{\alpha g}\right)\right] + \mathrm{Int}\left(L_{\alpha g}, U_{\alpha g}\right)}{\mathrm{Int}\left(L_{1g}, U_{1g}\right)}\right)$$

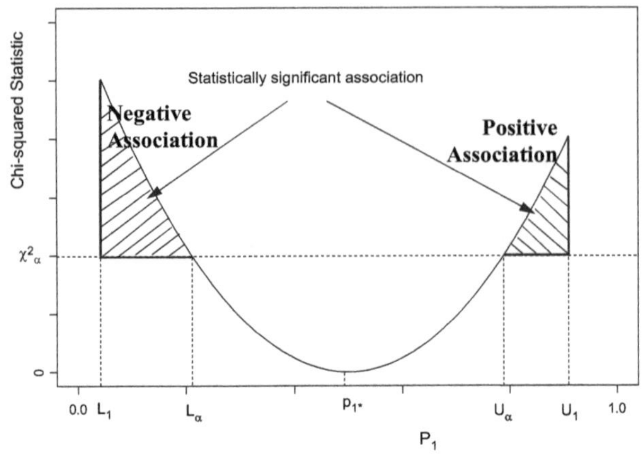

Fig. 1 A graphical interpretation of the aggregate association index. The *shaded regions* reflect the magnitude of the AAI

where $\text{Int}(a, b) = \int_a^b X^2(P_1 | p_{1\bullet}, p_{\bullet 1})dP_1$. This index is bounded by [0, 100] where a value of zero indicates that, at the α level of significance, there is no evidence of an association between the variables. A value close to 100 indicates that, at the α level of significance, there is strong evidence to conclude (based on the available aggregate data) that there exists an association between the variables. When analysing the New Zealand voting data in Sect. 3 we shall be considering a level of significance of $\alpha = 0.05$.

An alternative, but equivalent, expression for the AAI is (Beh 2010)

$$A_{\alpha g} = 100\left(1 - \frac{\chi^2_\alpha\left[(L_{\alpha g} - L_{1g}) + (U_{1g} - U_{\alpha g})\right]}{k_g n_g\left[(U_{1g} - p_{\bullet 1g})^3 - (L_{1g} - p_{\bullet 1g})^3\right]} - \frac{(U_{\alpha g} - p_{\bullet 1g})^3 - (L_{\alpha g} - p_{\bullet 1g})^3}{(U_{1g} - p_{\bullet 1g})^3 - (L_{1g} - p_{\bullet 1g})^3}\right)$$

where

$$k_g = \frac{1}{3p^2_{2\bullet g}}\left(\frac{p_{1\bullet g}p_{2\bullet g}}{p_{\bullet 1g}p_{\bullet 2g}}\right).$$

We shall now consider the graphical depiction of the AAI for the New Zealand voting data. The curve that describes the quadratic relationship between the chi-squared statistic and P_{1g} is referred to as the "AAI curve". Figure 1 provides a visual description of the AAI; the magnitude of the AAI is reflected by the shaded area as a proportion of the total area under the curve.

It can be shown (Beh 2010) that $A_{\alpha g}$ can be partitioned into an aggregate negative association index $A^-_{\alpha g}$ and an aggregate positive association $A^+_{\alpha g}$; these terms reflect how likely a negative, or positive, association will result given the known marginal information. They may be quantified by considering

$$A^{+}_{\alpha g} = 100\left(\frac{(U_{1g} - p_{\bullet 1g})^3 - (U_{\alpha g} - p_{\bullet 1g})^3}{(U_{1g} - p_{\bullet 1g})^3 - (L_{1g} - p_{\bullet 1g})^3} - \frac{\chi^2_{\alpha}(U_{1g} - U_{\alpha g})}{k_g n_g\left[(U_{1g} - p_{\bullet 1g})^3 - (L_{1g} - p_{\bullet 1g})^3\right]}\right)$$

and

$$A^{-}_{\alpha g} = 100\left(\frac{(L_{\alpha g} - p_{\bullet 1g})^3 - (L_{1g} - p_{\bullet 1g})^3}{(U_{1g} - p_{\bullet 1g})^3 - (L_{1g} - p_{\bullet 1g})^3} - \frac{\chi^2_{\alpha}(L_{\alpha g} - L_{1g})}{k_g n_g\left[(U_{1g} - p_{\bullet 1g})^3 - (L_{1g} - p_{\bullet 1g})^3\right]}\right)$$

so that $A_{\alpha g} = A^{-}_{\alpha g} + A^{+}_{\alpha g}$. Here $A^{+}_{\alpha g}$ and $A^{-}_{\alpha g}$ are termed the aggregate positive association index and aggregate negative association index, respectively.

3 The Early New Zealand Elections

Consider, as did Tran et al. (2012), the analysis of the electorates in the first (1893) New Zealand national election where women were granted the right to vote. For each electorate the AAI is at least 98 using a 0.05 level of significance. This indicates that it is very likely that, given only the marginal information, a statistically significant association exists between the two dichotomous variables at each of the electorates. Figure 2a provides a graphical view of the AAI curve for each of the 1893 electorates.

One can see that, by observing Fig. 2a, since the sample size in each electorate is very large, the AAI for each electorate is very close to 100. This indicates that, for each electorate, and given only the marginal information, there is very strong evidence to suggest that gender and voter turnout is associated. Due to the impact of the sample size on the AAI, the magnitude of these AAI values may seem obvious. However it is very apparent that, for the 1893 election, there are definite clusters of electorates that share common association structures (given only their marginal information). This can be seen by observing the clustering of the AAI curves. Figure 2a shows five distinct clusters that have been identified via cluster analysis using Ward's method (Lattin et al. 2003). Alternative clustering algorithms could also be considered and have the potential to reveal any sensitivity issues that may be encountered; we shall leave issues for future consideration. These five clusters were formed on the basis of the bounds of P_{1g} (L_{1g} and U_{1g}), the vertex of the parabola—existing at the point ($p_{1\bullet g}$, 0)—and the curvature of the parabola. It is clear there are four electorates that appear to have curvatures very different from the rest suggesting that the association for these four electorates is somehow different when compared with the remaining electorates across New Zealand. Further studies to incorporate electorate level covariate information needs to be carried out to identify what, if any, features of the electorates that lead to these clusters being formed.

By considering the AAI, a "clustered" AAI curve that summarises the association structure for clusters of homogeneous electorates (as determined based on the

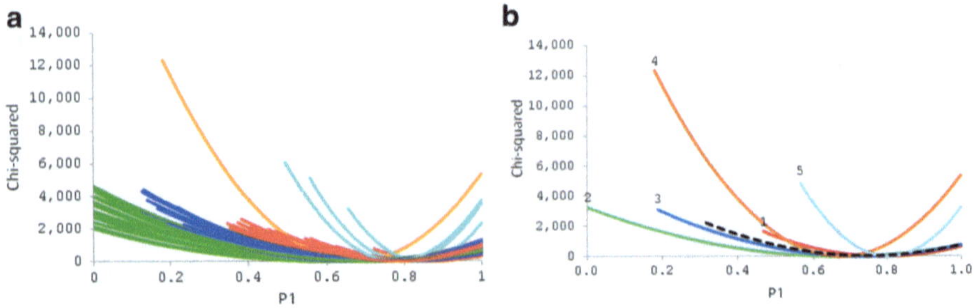

Fig. 2 (**a**) The AAI curve for each electorate of the 1893 election. The five colours indicate the five clusters of homogeneous association for electorates, (**b**) "Clustered" AAI curves for each of the five clusters of homogeneous electorates in the 1893 election. The *dashed line* depicts the "Yearly" AAI curve for 1893

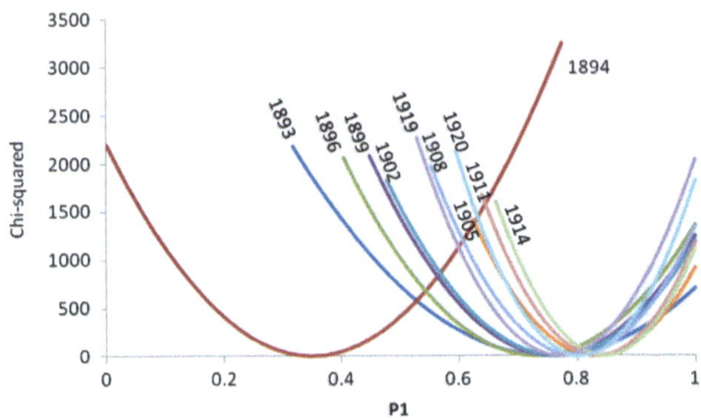

Fig. 3 Overall AAI curves for each of the 11 early New Zealand elections

marginal information) can be obtained. For the 1893 election these five curves are depicted in Fig. 2b and are obtained by considering the weighted mean of each AAI curve contained in the cluster. One may also determine the overall AAI curve for the 1893 election. The dashed line in Fig. 2b represents this overall curve. Clustered AAI curves, for each of the 11 elections are depicted in Fig. 3 and provide a comparison of the overall association structure between gender and voter turnout (note that the last election is labelled "1920" for convenience). It can be seen that, based only on the marginal information, the voting behaviour of each of the elections (with the exception of 1894) are fairly similar. However, it is as yet unclear why the difference in the gender turnout for the 1894 election has arisen. Further studies incorporating covariate information are expected to shed additional light on this issue.

Given only the marginal information for each electorate, a more detailed investigation of the association structure between gender and voting data can be revealed by calculating $A^{+}_{0.05g}$ and $A^{-}_{0.05g}$ based on the yearly AAI curves given in Fig. 3. Table 3 summarises these aggregate positive and aggregate negative

Table 3 Summary of the AAI, aggregate positive association index and aggregate negative association index for each of the 11 NZ elections, 1893–1919

Year	$A_{0.05g}$	$A^{+}_{0.05g}$	$A^{-}_{0.05g}$
1893	99	15	84
1894	100	64	36
1896	99[a]	35	65
1899	99	31	68
1902	99	34	65
1905	99	34	65
1908	99[a]	36	64
1911	99[a]	37	63
1914	99	36	63
April 1919	100	46	54
December 1919	99	43	56

[a]Indicates error due to rounding

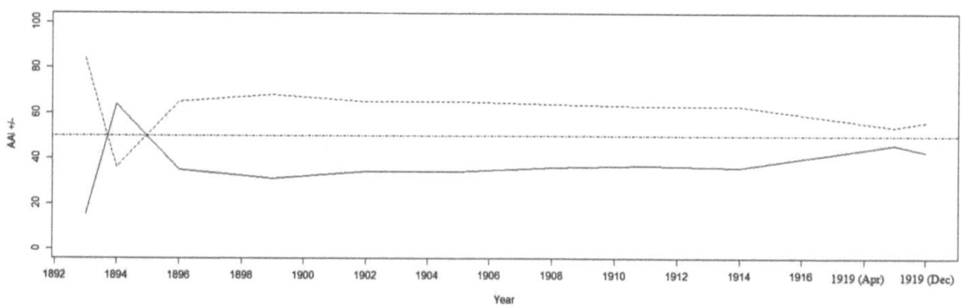

Fig. 4 Comparison of $A^{+}_{0.05g}$ (*solid line*) and $A^{-}_{0.05g}$ (*dotted line*) for each of the 11 NZ elections

association indices for each election. A graphical view of these changes can be seen by considering Fig. 4; the solid shows the variation in $A^{+}_{0.05g}$ over time, while the dotted line shows the variation in $A^{-}_{0.05g}$. Overall, the association tends to be more negative than positive suggesting that men were more likely to turn out and vote than women. This is very apparent in the first election of 1893 where $A^{+}_{0.05g} = 15$ and $A^{-}_{0.05g} = 84$. However, this finding is not so when considering the election held in 1894; recall that Fig. 3 presents an AAI curve for the 1894 election that is not consistent with the remaining ten elections held in New Zealand. In this case, $A^{+}_{0..05g} = 64$ and $A^{-}_{0.05g} = 36$, suggesting that the strong association between gender and voting behaviour (AAI = 100) is more likely to be positive than negative. Hence, the aggregate data suggests that for the 1894 election a higher proportion of women turned out to vote than men.

One can see that as time went on, the proportion of male to female voter turnout levelled off. From 1896 onwards the aggregate negative association index, using $\alpha = 0.05$, was consistently around 64. However, in 1919 the gender differences in voting behaviour were less evident. The aggregate positive and negative association indices suggest that the differences in turnout between male and female voters became more equitable. This equitability is very noticeable by comparing the difference between the solid and dotted lines of Fig. 4.

4 Discussion

We have briefly described the application of the AAI for early New Zealand voting data, although its applicability ranges across a diverse number of disciplines where aggregated data from stratified 2×2 tables are collected. Based only on the marginal information, it is clear that there is an association between gender and voter turnout at these elections. We have also revealed that, with the exception of the 1894 election, voter turnout for each gender was fairly homogeneous, although we are able to reflect electorate level variation as seen in Fig. 2. However, further research needs to be undertaken to elicit more information about voting turnout from the data and for the continual development of the AAI.

References

Aitkin, M., & Hinde, J. P. (1984). Comments to "Tests of significance for 2x2 contingency tables". *Journal of the Royal Statistical Society, Series A, 47*, 453–545.

Barnard, G. A. (1984). Comments to "Tests of significance for 2x2 contingency tables". *Journal of the Royal Statistical Society, Series A, 47*, 449–450.

Beh, E. J. (2008). Correspondence analysis of aggregate data. *Journal of Statistical Planning and Inference, 138*, 2941–2952.

Beh, E. J. (2010). The aggregate association index. *Computational Statistics and Data Analysis, 54*, 1570–1580.

Brown, P., & Payne, C. (1986). Aggregate data, ecological regression and voting transitions. *Journal of the American Statistical Association, 81*, 453–460.

Chambers, R. L., & Steel, D. G. (2001). Simple methods for ecological inference in 2×2 tables. *Journal of the Royal Statistical Society, Series A, 164*, 175–192.

Duncan, O. D., & Davis, B. (1953). An alternative to ecological inference. *American Social Review, 18*, 665–666.

Fisher, R. A. (1935). The logic of inductive inference (with discussion). *Journal of the Royal Statistical Society, Series A, 98*, 39–82.

Forcina, A., Gnaldi, M., & Bracalente, B. (2012). A revised Brown and Payne model of voting behaviour applied to the 2009 elections in Italy. *Statistical Methods and Applications, 21*, 109–119.

Goodman, L. A. (1953). Ecological regressions and behavior of individuals. *American Social Review, 18*, 663–664.

Hudson, I. L., Moore, L., Beh, E. J., & Steel, D. G. (2010). Ecological inference techniques: an empirical evaluation using data describing gender and voter turnout at New Zealand elections, 1893–1919. *Journal of the Royal Statistical Society, Series A, 173*, 185–213.

King, G. (1997). *A Solution to the Ecological Inference Problem*. Princeton University Press: Princeton.

Lattin, J., Carroll, J. D., & Green, P. E. (2003). *Analyzing multivariate data*. VIC: Thomson.

Plackett, R. L. (1977). The marginal totals of a 2x2 table. *Biometrics, 64*, 37–42.

Steel, D. G., Beh, E. J., & Chambers, R. L. (2004). The information in aggregate data. In G. King, O. Rosen, & M. Tanner (Eds.), *Ecological Inference: New Methodological Strategies* (pp. 51–68). Cambridge: Cambridge University Press.

Tran, D., Beh, E. J., & Hudson, I. L. (2012). The aggregate association index and its application in the 1893 New Zealand election. In: *Proceedings of the 5th ASEARC Conference* (pp. 22–25).

Wakefield, J. (2004). Ecological inference for 2 × 2 tables (with discussion). *Journal of the Royal Statistical Society, Series A, 167*, 385–424.

Estimating a Rasch Model via Fuzzy Empirical Probability Functions

Lucio Bertoli-Barsotti, Tommaso Lando, and Antonio Punzo

Abstract The joint maximum likelihood estimation of the parameters of the Rasch model is hampered by several drawbacks, the most relevant of which are that: (1) the estimates are not available for item or person with perfect scores; (2) the item parameter estimates are severely biased, especially for short tests. To overcome both these problems, in this paper a new method is proposed, based on a fuzzy extension of the empirical probability function and the minimum Kullback–Leibler divergence estimation approach. The new method warrants the existence of finite estimates for both person and item parameters and results very effective in reducing the bias of joint maximum likelihood estimates.

Keywords Bias • Maximum likelihood estimation • Modified likelihood • Rasch model

1 Introduction

In this paper we focus on the fixed-person and fixed-item version of the simple Rasch model (RM) for binary responses—that is the joint maximum likelihood (JML) version of the RM. According to the RM, the logit of the probability of a

L. Bertoli-Barsotti (✉)
University of Bergamo, Bergamo, Italy
e-mail: lucio.bertoli-barsotti@unibg.it

T. Lando
VŠB-Techn., University of Ostrava, Ostrava, Czech Republic
e-mail: tommaso.lando@vsb.cz

A. Punzo
University of Catania, Catania, Italy
e-mail: antonio.punzo@unict.it

D. Vicari et al. (eds.), *Analysis and Modeling of Complex Data in Behavioral and Social Sciences*, Studies in Classification, Data Analysis, and Knowledge Organization,
DOI 10.1007/978-3-319-06692-9_4,

1-response is $\ln[P(X_{vi}=1)/P(X_{vi}=0)]=\theta_v-\beta_i$, where $X_{vi}=1$ denotes a 1-response of person v ($v=1, \ldots, n$) to item i ($i=1, \ldots, k$) and where the parameters θ_v and β_i represent, respectively, the ability of person v and the difficulty of item i.

The joint maximum likelihood (JML) is an estimation method in which item parameters and ability parameters are estimated simultaneously. The log-likelihood function that is to be maximized is equal to $l=\sum_{v,i}\ln p_{vi}$, where $p_{vi}=P(X_{vi}=x_{vi})$ with x_{vi} taking values from the set $\{0, 1\}$. The JML method is actually adopted by the software WINSTEPS (Linacre 2009)—that may be considered the most popular program for estimating RMs (Cohen et al. 2008). Unfortunately, the JML approach is hampered by several drawbacks, the most relevant of which are:

1. (*Existence problem*) In a number of cases, finite estimates for item and/or person parameters are not available;
2. (*Bias problem*) Item parameter estimates are biased especially for short tests, independently from the sample size.

Problem 1 is due to the fact that JML estimators have always a positive probability of yielding infinite parameter values. This happens: (a) in the presence of zero and/or perfect person totals; (b) in the presence of zero and/or perfect items totals (i.e. null categories); (c) for other configurations of the dataset that, following Molenaar (1995), we shall call here “strange configurations”. Necessary and sufficient conditions for the existence of finite JML estimates are given by Fischer (1981) (see also Bertoli-Barsotti 2005). More specifically, Bertoli-Barsotti and Bacci (2014) showed that there are five different types of “JML-anomaly”; they refer to a dataset satisfying one or more of the specific conditions (a), (b) or (c) to as “JML-anomaluos”. Now, in order to prevent the estimates of the person parameters from diverging (case a) one could remove the persons with zero and/or perfect totals (of course, a similar argument would apply to items). Another more sophisticated trick consists in modifying the total scores for persons who achieve the minimum total score of zero or the maximum total score of k, by adding (in the former case) or subtracting (in latter case) an arbitrary constant r between 0 and 1. By default, the value $r=0.3$ is used in WINSTEPS, as well as in ConQuest (Wu et al. 2007). Besides, as far as we know, there still does not exist a remedy for the case of strange configurations.

As concerns the problem 2, Andersen (1980, p. 244) showed that for moderate values of k the JML estimate $\widehat{\beta}_{JML}$ of the item difficulty parameter β has an approximate bias that is a function of the ratio $(k-1)/k$. In particular, for the limit case of a two-item test, JML estimates of the item parameters are twice their theoretical values. For this reason, WINSTEPS uses $(k-1)/k$ as a simple test-length correction factor, for item estimates. This yields a “corrected” JML estimator of the difficulty parameter, which will be further denoted by C-JML: $\widehat{\beta}_{C\text{-}JML}=[(k-1)/k]\cdot\widehat{\beta}_{JML}$. Basing on an empirical study, Wright (1988) proved that for the most usual test lengths ($k\geq 20$) this correction factor works well.

Firth (1993) stated that there are two approaches that may reduce the MLE bias, one is a corrective approach and the other is a preventive approach. The C-JML method is corrective in nature, in that the JML estimate $\widehat{\beta}_{JML}$ is first calculated, and then corrected. In particular, it is apparent that this method requires, as a prerequisite, the existence of finite JML estimates.

Interestingly enough, the bias problem is connected to the problem of the existence of finite estimates for the parameters. In a sense, the reason that JML approach leads to biased estimates of the item parameters is that it includes the possibility of extreme scores in the estimation space, but cannot actually estimate them (Linacre 2009). Indeed, this leads to an overestimated dispersion of the item parameters. This problem is strictly related to the more general question of the bias of the MLE of the logit parameters. This issue has a long history and several adjustments have been proposed in the literature to reduce the bias and to make the MLE defined over the entire sample space. To be concrete, consider a simple random sample of size k from a Bernoulli distribution with parameter p. Then, the observed count Y is distributed as a binomial with parameters k and p. Haldane (1956) noted that the exact bias of the MLE of the logit of p, $\widehat{\eta} = \ln [Y/ (k - Y)]$, is undefined (in fact, none of its moments exists) because, for every fixed sample size, there is a non-zero probability that $\widehat{\eta}$ is infinite. Then, he suggested an additive modification of the binomial count by 0.5, leading to the modified MLE, say $\widehat{\eta}_H = \ln [(Y + 0.5) / (k - Y + 0.5)]$. The "Haldane correction" has the effect of solving the problem of infinite estimates and, more generally, of shrinking all the MLEs toward zero, resulting in an approximately unbiased estimator. Unfortunately, this formula is not suitable for the case in question because we are faced with a more general sampling scheme. We need a more general approach.

The present paper proposes a new method for overcoming both problems 1 and 2 simultaneously. The method presented is preventive, because is based on the maximization of a "modified" version of the log-likelihood function, and it does not need the existence of the JML estimate. A simulation study suggests that this method may be very effective in reducing the estimation bias, at least compared with the traditional C-JML estimator.

2 Fuzzy Empirical Probability Functions

For every v and i, let us define the empirical probability function $f_{vi} = f_{vi}(z)$ over the set $\{0, 1\}$ as follows: $f_{vi}(0) = 1$ and $f_{vi}(1) = 0$, if $x_{vi} = 0$; $f_{vi}(0) = 0$ and $f_{vi}(1) = 1$, if $x_{vi} = 1$. For each cell (v, i), the Kullback–Leibler (KL) divergence D_{vi} between the distributions of observed and expected counts (where the first of these is a degenerate distribution) reduces to $D_{vi} = f_{vi}(0) \cdot \ln \frac{f_{vi}(0)}{P(X_{vi}=0)} + f_{vi}(1) \cdot \ln \frac{f_{vi}(1)}{P(X_{vi}=1)} = -\ln p_{vi}$, where the quantity $0 \cdot \ln(0)$ is interpreted as zero. Thus, minimizing the sum $\sum_{v,i} D_{vi}$ is equivalent to maximizing log-likelihood function l. As mentioned above, the empirical distribution function is degenerate. The main problem with this design is that of reducing the bias and guaranteeing finiteness of log-odds estimation. Our approach can be explained intuitively basing on the following heuristic argument: if person v could have multiple independent attempts, say N, at item i, the difference $\theta_v - \beta_i$ could be estimated on the basis of the values $f_{vi}(0) = N_0/N$ and $f_{vi}(1) = N_1/N$, where N_1 is the number of times that person v responds correctly to item i, and $N_0 = N - N_1$. Now, both N_0/N and N_1/N can always be expected to be different from both 0 and 1, for N large enough (if the model is true). Then, there is some reason to modify the empirical probability function f_{vi} by replacing it with an ε-weighted version, say f_{vi}^{ε}, of the form: $f_{vi}^{\varepsilon}(0) = 1 - \varepsilon$ and $f_{vi}^{\varepsilon}(1) = \varepsilon$, if $x_{vi} = 0$; $f_{vi}^{\varepsilon}(0) = \varepsilon$ and $f_{vi}^{\varepsilon}(1) = 1 - \varepsilon$, if $x_{vi} = 1$, where ε is an arbitrarily small positive number. The function f_{vi}^{ε} may be considered a *fuzzy* extension of the usual empirical probability function. The ε-weighted minimum KL divergence method is equivalent to minimizing the function $-\sum_{v,i}[(1-\varepsilon)\cdot \ln p_{vi} + \varepsilon \cdot \ln(1-p_{vi})]$. It is worth noting that this is equivalent to maximizing the ε-adjusted log-likelihood function $l^{\varepsilon} = l + A_{\varepsilon}$, where l is the usual log-likelihood function and $A_{\varepsilon} = \varepsilon \cdot \sum_{v,i} \ln[(1-p_{vi})/p_{vi}]$. This yields an ε-adjusted JML estimator, which will be denoted in the sequel with the acronym ε-JML. It is worth noting that the ε-JML estimator belongs to a more general class of estimators—i.e. those obtained by maximizing a "modified" version of the log-likelihood function, specifically $l^* = l + A$, in which the function A is allowed to depend on both the data and the parameters. A special case of this class of estimators has been studied by Firth (1993). Firth's approach is defined in a rather general framework, that is for both exponential and nonexponential models. In particular, for the case of an exponential family in its canonical parameterization (as is the case of the RM), the method consists in maximizing a modified log-likelihood function $l^* = l + A$, where A is the logarithm of the square root of the determinant of the Fisher information matrix (that is, the logarithm of the Jeffreys prior; Jeffreys 1939, 1946). It may be noted that other modifications of a profile likelihood function are also possible; the interested reader is referred to Bartolucci et al. (2012) for an up-to-date review concerning techniques of bias reduction based on modified likelihood, or modified profile likelihood. Describing them here goes beyond the scope of the present study. Interestingly, Warm (1989) introduced a special case of Firth's bias reduction method, by applying a similar formula to the three-parameter logistic model (which includes the RM as a special case), but his approach is only devoted to the problem of the estimation of the ability parameter, under the assumption that the item parameters are *known*. This estimate is defined as the Weighted Likelihood Estimate (WLE). The WLE is currently adopted, as a default, by the software RUMM 2020 (Andrich et al. 2003) to obtain person parameter estimates. In a first step, this software uses the (Pairwise) Conditional Maximum Likelihood (CML) estimation method for obtaining the item parameters. Then, in a second step, the MLE (or WLE) approach is used to estimate ability parameters, treating the previously estimated item parameters as if they were the true quantities. Recently, Bertoli-Barsotti and Punzo (2012), by simulation, studied comparatively the Firth's and the C-JML approaches to JML estimation of the item parameters. They concluded that the C-JML method yields the less biased estimates (at least for test lengths ranging from 5 to 30, and sample sizes ranging from values between 100 and 1,000). Finally, it is interesting to note that, for the simple-sample case, the Haldane's estimator $\widehat{\eta}_H$ coincides with that yielded by the Firth's bias reduction formula (but that is not the case for a more general random sampling mechanism; see Firth 1993).

3 Simulation Study

This section describes design and results of a wide Monte Carlo simulation study. It was conducted to investigate the maximum possible gain, in terms of item parameters bias, which can be obtained by using the ε-JML estimation method with respect to the JML method with and without the standard $(k-1)/k$ bias-correction. The entire simulation was performed in the R computing environment (R Development Core Team, 2012). The R functions necessary to obtain the estimates, according to the considered methods, are available from the authors.

As it is evident, the core of the ε-JML estimation method is its ε-weighted empirical probability function, which in turn is function of the value of ε. Then, in the simulation study, three independent variables were manipulated: (a) the ε value; (b) the test-length k; (c) the sample size n. The complete simulation consisted in two parts: the first one for obtaining—empirically—an "optimal" value for ε, hereafter denoted as ideal-ε, say $\bar{\varepsilon}$, as the value yielding the greatest level of precision, for fixed dimensions of the dataset, and the second one for studying comparatively the effect of bias reduction of the $\bar{\varepsilon}$-JML estimator.

3.1 *Finding for an Ideal-Epsilon*

The test lengths considered were $k = 5$, 10, 15, 20, 25, and 30. The values of the item difficulties were equally spaced across the range from −2 to 2. The sample sizes considered were $n = 200$, 400, 1,000, and 4,000. The true ability parameters were generated from a standard normal distribution. We wrote an R code (always available from the authors) to generate each 0/1 response dataset. The generation procedure consisted of the following steps: (a) n ability parameters were randomly generated with the `rnorm` function of the `stats` package of R; (b) these ability parameters, and the defined item parameters, were used to compute the $n \times k$ probabilities of a correct response, according to an RM; (c) these probabilities were adopted in a Bernoulli distribution to generate the values 0 or 1. Thirty replications were made under each pair of n and k. For every replication j, in a first step the best value of ε, say ε_j^*, was obtained by minimizing, over a finite grid of 51 equidistant values in the range [0, 0.1], the sum of squares function $\sum_{i=1}^{k}(\beta_{ij}^{est} - \beta_{ij})^2$, where β_{ij}^{est} is the ε-JML estimate of item i in replication j, and β_{ij} represents the corresponding generating value of that item parameter and replication. Then, in a second step, the ideal-ε was obtained by averaging across the replications as

Table 1 Ideal-ε for different k and n. Empirical standard deviations in parentheses

k	$n=200$	$n=400$	$n=1{,}000$	$n=4{,}000$
5	0.05 (0.011)	0.05 (0.008)	0.05 (0.005)	0.05 (0.003)
10	0.03 (0.010)	0.03 (0.009)	0.03 (0.005)	0.03 (0.002)
15	0.02 (0.012)	0.02 (0.006)	0.02 (0.005)	0.02 (0.002)
20	0.02 (0.010)	0.02 (0.006)	0.02 (0.004)	0.01 (0.002)
25	0.02 (0.008)	0.02 (0.005)	0.01 (0.004)	0.01 (0.002)
30	0.02 (0.007)	0.01 (0.005)	0.01 (0.004)	0.01 (0.002)

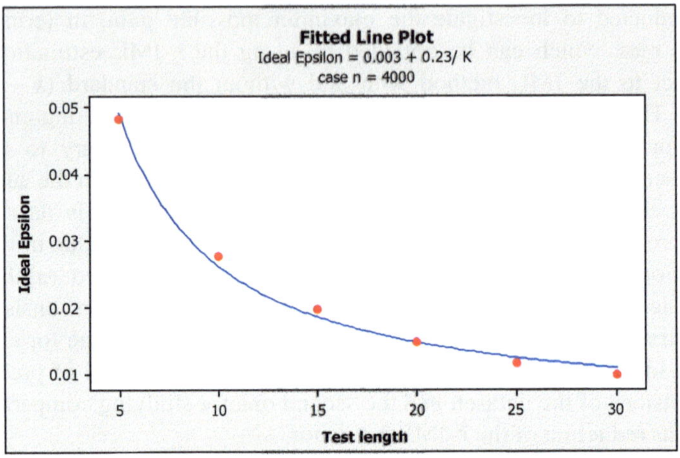

Fig. 1 Ideal-ε as a function of the test length (each *dot* is an average over 30 replications)

follows: $\bar{\varepsilon} = \sum_{j=1}^{30} \varepsilon_j^*/30$. The derived values of $\bar{\varepsilon}$ are given in Table 1, along with their empirical standard deviations, for each combination of n and k. They are also shown as dots and plotted in Fig. 1 against the test lengths, for the case of $n = 4,000$. Because of space constraints, only this sample size is shown, but the plots for the other sample sizes considered in this study are similar to that in Fig. 1. At a first glance, one can see that the relationship between $\bar{\varepsilon}$ and k is curvilinear. This function may be roughly approximated, for example, by the following family of functions: $\bar{\varepsilon} \approx a + b/k$, for a convenient choice of $a = a^*$ and $b = b^*$. For every fixed sample size n, different pairs a^* and b^* have been obtained by minimizing the sum of the squared residuals. We obtained: $a^* = 0.0120$, $b^* = 0.185$ for $n = 200$; $a^* = 0.0070$, $b^* = 0.211$ for $n = 400$; $a^* = 0.0047$, $b^* = 0.219$ for $n = 1,000$; $a^* = 0.003$, $b^* = 0.230$ for $n = 4,000$ (see Fig. 1). In their turn, a^* and b^* may be viewed as functions of the sample size, and may be roughly approximated, for example, by means of similar formulas $a^* = c + d/n$ and $b^* = e + f/n$, for convenient choices of c^*, d^*, e^* and f^*. In particular, by minimizing the sum of the squared residuals, we obtained $c^* = 0.0023$, $d^* = 1.82$, $e^* = 0.23$ and $f^* = -9.07$.

Table 2 MSE of item parameter estimates for the three different approaches and different k and n

k	n	$\overline{\varepsilon} = g(n,k)$	$\widehat{\beta}_{\overline{\varepsilon}\text{-}JML}$	$\widehat{\beta}_{C\text{-}JML}$	$\widehat{\beta}_{JML}$
5	200	0.05	*0.0269*	0.0527	0.3168
5	500	0.05	*0.0125*	0.0272	0.2613
5	1000	0.05	*0.0072*	0.0162	0.2349
10	200	0.03	*0.0264*	0.0453	0.0700
10	500	0.03	*0.0113*	0.0248	0.0467
10	1000	0.03	*0.0060*	0.0170	0.0373
20	200	0.02	*0.0277*	0.0302	0.0397
20	500	0.02	*0.0116*	0.0123	0.0190
20	1000	0.02	*0.0059*	0.0061	0.0121
30	200	0.02	*0.0291*	0.0314	0.0365
30	500	0.01	*0.0119*	0.0122	0.0159
30	1000	0.01	*0.0057*	0.0058	0.0086

Then, a formula $\overline{\varepsilon} = a^* + b^*/k$ for the ideal-ε can now be rewritten by replacing the estimated values of c^*, d^*, e^* and f^*. This leads to:

$$\overline{\varepsilon} \approx g(n,k) = 0.003 + 1.82 \cdot n^{-1} + 0.23 \cdot k^{-1} - 9.07 \cdot (nk)^{-1}, \qquad (1)$$

a formula that may be regarded as rough approximation of our simulation results concerning the ideal-ε.

3.2 Analysis of Bias Reduction

To investigate the performance of the proposed $\overline{\varepsilon}$-JML estimation method, a simulation study was carried out under various conditions. Three estimation methods were considered: JML, C-JML and $\overline{\varepsilon}$-JML, where $\overline{\varepsilon}$ was calculated by means of the formula (1). One hundred $n \times k$ datasets were randomly generated, according to a RM, for different values of k, $k = 5, 10, 20, 30$, and n, $n = 200, 500, 1{,}000$. More precisely, for each dataset, n person parameters were randomly generated from a normal distribution with mean zero and standard deviation 0.5 for the case $k = 5$ and 0.7 otherwise, while the item parameters were equally spaced in the interval $[-2, 2]$. In a variant of this analysis, for each dataset the item parameters were randomly selected from a uniform distribution on the same interval (the results were quite similar and are not reported here). The precision of the parameter estimates was assessed with the empirical mean squared error $\text{MSE} = 100^{-1} \cdot \sum_{j=1}^{100} \sum_{i=1}^{k} (\beta_{ij}^{est} - \beta_{ij})^2$, here used as an index (less is better) of the total error of estimation, which could reflect the amount of bias, the amount of standard error, or both.

Table 2 gives the results. Subscripts were added to distinguish between the different estimation methods. To facilitate comparisons, the values are highlighted in italic when the MSE is smaller. This simulation study shows that, setting ε conveniently, the new method may be very effective in reducing the bias of the JML.

Acknowledgments This research was partially funded in the framework of the project "Opportunity for young researchers", reg. no. CZ.1.07/2.3.00/30.0016, supported by Operational Programme Education for Competitiveness and co-financed by the European Social Fund and the state budget of the Czech Republic.

References

Andersen, E. B. (1980). *Discrete statistical models with social sciences applications*. Amsterdam: North-Holland.

Andrich, D., Lyne, A., Sheridan, B., & Luo, G. (2003). *RUMM 2020 [Computer Software]*. Perth, Australia: RUMM Laboratory.

Bartolucci, F., Bellio, R., Salvan, A., & Sartori, N. (2012). *Modified profile likelihood for panel data models*. Available at SSRN repository: http://ssrn.com/abstract=2000666.

Bertoli-Barsotti, L. (2005). On the existence and uniqueness of JML estimates for the partial credit model. *Psychometrika, 70*, 517–531.

Bertoli-Barsotti, L., & Punzo, A. (2012). Comparison of two bias reduction techniques for the Rasch model. *Electronic Journal of Applied Statistical Analysis, 5*(3), 360–366.

Bertoli-Barsotti, L., & Bacci, S. (2014). Identifying Guttman structures in incomplete Rasch datasets. *Communications in Statistics – Theory and Methods, 43*(3), 470–497. doi:10.1080/03610926.2012.66555.

Cohen, J., Chan, T., Jiang, T., & Seburn, M. (2008). Consistent estimation of Rasch item parameters and their standard errors under complex sample designs. *Applied Psychological Measurement, 32*, 289–310.

Fischer, G. H. (1981). On the existence and uniqueness of maximum-likelihood estimates in the Rasch model. *Psychometrika, 46*, 59–77.

Firth, D. (1993). Bias reduction of maximum likelihood estimates. *Biometrika, 80*, 27–38.

Haldane, J. B. S. (1956). The estimation and significance of the logarithm of a ratio of frequencies. *Annals of Human Genetics, 20*, 309–311.

Jeffreys, H. (1939). *Theory of probability*. Oxford: Oxford University Press.

Jeffreys, H. (1946). An invariant form for the prior probability in estimation problems. *Proceedings of the Royal Society of London. Series A: Mathematics and Physical Sciences, 186*, 453–461.

Linacre, J. M. (2009). *WINSTEPS®. Rasch measurement computer program*. Beaverton, OR: Winsteps.com.

Molenaar, I. W. (1995). Estimation of item parameters. In G. H. Fischer & I. W. Molenaar (Eds.), *Rasch models: Foundations, recent developments, and applications* (pp. 39–51). New York: Springer.

R Development Core Team. (2012). *R: A language and environment for statistical computing*. Vienna, Austria: R Foundation for Statistical Computing.

Warm, T. A. (1989). Weighted likelihood estimation of in item response theory. *Psychometrika, 54*, 427–450.

Wright, B. D. (1988). The efficacy of unconditional maximum likelihood bias correction: Comment on Jansen, van den Wollenberg, and Wierda. *Applied Psychological Measurement, 12*, 315–318.

Wu, M. L., Adams, R. J., Wilson, M. R., & Haldane, S. A. (2007). *ACER ConQuest: Generalised item response modeling software - version 2.0*. ACER Press edition

Scale Reliability Evaluation for A-Priori Clustered Data

Giuseppe Boari, Gabriele Cantaluppi, and Marta Nai Ruscone

Abstract According to the classical measurement theory, the reliability of a set of indicators related to a latent variable describing a true measure can be assessed through the Cronbach's α index. The Cronbach's α index can be used for τ-equivalent measures and for parallel measures and represents a lower bound for the reliability value in presence of congeneric measures, for which the assessment can properly be made only ex post, once the loading coefficients have been estimated, e.g. by means of a structural equation model with latent variables.

Once assumed the existence of an a-priori segmentation based upon a categorical variable Z, we test whether the construct is reliable all over the groups. In this case the measurement model is the same across groups, which means that loadings are equal within each group as well as they do not vary across groups. A formulation of the Cronbach's α coefficient is considered according to the decomposition of pairwise covariances in a clustered framework, and a test procedure assessing the possible presence of congeneric measures in a multigroup framework is proposed.

Keywords Cronbach's alpha • Multigroup reliability

1 Introduction

According to the classical measurement theory (Bollen 1989), the relationship between a latent variable τ_i describing a true measure and its corresponding q manifest proxies X_{ji}, for the generic i subject in a n-dimensional sample random variable, is:

G. Boari (✉) • G. Cantaluppi • M.N. Ruscone
Dipartimento di Scienze Statistiche, Università Cattolica del Sacro Cuore, Milano, Italy
e-mail: giuseppe.boari@unicatt.it; gabriele.cantaluppi@unicatt.it; marta.nairuscone@unicatt.it

D. Vicari et al. (eds.), *Analysis and Modeling of Complex Data in Behavioral and Social Sciences*, Studies in Classification, Data Analysis, and Knowledge Organization,
DOI 10.1007/978-3-319-06692-9_5,

$$X_{ji} = \lambda_{ji}\tau_i + E_{ji}, \quad j = 1,\ldots,q \quad i = 1,\ldots,n \tag{1}$$

where $E_{ji}, j = 1,\ldots,q, i = 1,\ldots,n$ are independent random error components, also independent of the latents τ_i which are assumed to be independently and identically distributed across the subjects, i.e. $\tau_i \sim \tau$.

In presence of congeneric measures the coefficients λ_{ji} equal λ_j, while for τ-equivalent measures are all equal to λ (parallel measures, requiring τ-equivalency and homoscedasticity of error components, are not considered here). So the i.i.d. random variables X_{ji} are distributed as X_j.

Let us now assume the existence of an a-priori segmentation, based upon a variable Z, with categories $z_1, z_2, \ldots, z_G$, where G is the number of groups.

See Wedel and Kamakura (1998) for a detailed presentation of a-priori/post-hoc segmentation techniques.

The latent construct in (1) is usually assumed to be valid for all the G groups, corresponding to the null joint hypothesis that λ_{ji} does not vary across subjects, within groups as well as among groups, and $\tau_i \sim \tau$, that is:

$$H_0 : \tau_i \; IID \; \tau \text{ and } \lambda_{ji} = \lambda_j. \tag{2}$$

As alternative hypotheses different measurement models can be defined for groups:

$$X_{ji} = \lambda_{jg}\tau + E_{ji}, \quad j = 1,\ldots,q, \quad i = 1,\ldots,n_g, \quad g = 1,\ldots,G, \tag{3}$$

when the true measures random variables τ_g are assumed i.i.d. across the groups and the λ_{ji} are the same within each group, or

$$X_{ji} = \lambda_j\tau_g + E_{ji}, \quad j = 1,\ldots,q, \quad i = 1,\ldots,n_g, \quad g = 1,\ldots,G, \tag{4}$$

when the measures may be congeneric and the random variables τ_i are i.i.d. τ_g within each group, with τ_g possibly not i.i.d. across the groups, or

$$X_{ji} = \lambda_{jg}\tau_g + E_{ji}, \quad j = 1,\ldots,q, \quad i = 1,\ldots,n_g, \quad g = 1,\ldots,G, \tag{5}$$

when the measures may be congeneric with λ_{jg} coefficients possibly different among groups, and random variables τ_g may not be i.i.d. across the groups. The random variables E_{ji} are assumed to be independent of Z. Only the latter alternative hypothesis will be considered below.

Relationship (5) can be related with measurement model (1) as follows:

$$\begin{bmatrix} \boldsymbol{X}_{j1} \\ \vdots \\ \boldsymbol{X}_{jG} \end{bmatrix} = \begin{bmatrix} \lambda_{j1}\mathbf{I}_{n_1} & \cdots & \mathbf{O} \\ \vdots & \ddots & \vdots \\ \mathbf{O} & \cdots & \lambda_{jG}\mathbf{I}_{n_G} \end{bmatrix} \begin{bmatrix} \tau_1\mathbf{1}_{n_1} \\ \vdots \\ \tau_G\mathbf{1}_{n_G} \end{bmatrix} + \begin{bmatrix} \boldsymbol{E}_{j1} \\ \vdots \\ \boldsymbol{E}_{jG} \end{bmatrix}, \quad j = 1,\ldots,q \tag{6}$$

where $\boldsymbol{X}_{jg}$ and $\boldsymbol{E}_{jg}$, $g = 1, \ldots, G$ are $(n_g \times 1)$ vectors, with generic elements X_{ji} and E_{ji} respectively; $\mathbf{1}_{n_g}$ is the $(n_g \times 1)$ unitary vector and $\mathbf{I}_{n_g}$ is the $(n_g \times n_g)$ identity matrix, being $\sum_{g=1}^{G} n_g = n$; $\mathbf{O}$ are null matrices.

The corresponding null hypothesis may be reformulated as:

$$H_0 : \tau_g \; IID \; \tau \text{ and } \lambda_{jg} = \lambda_j, \quad g = 1, \ldots, G, \tag{7}$$

which states, for a generic alternative at least one of the following issues: (a) the true measures have the same feature within groups but possibly different features among groups; (b) the latent construct may be possibly not linked to all the manifest variables, or may be linked to its manifest indicators with different intensities across the groups.

The Cronbach's α reliability coefficient is defined, in case of a unique group, as

$$\alpha = \frac{q}{q-1}\left(1 - \frac{\sum_{j=1}^{q} \sigma_j^2}{\sum_{j=1}^{q} \sum_{k=1}^{q} \sigma_{jk}}\right) \tag{8}$$

where σ_j^2, σ_{jk} are the elements of the covariance matrix of the vector random variable $(X_1, \ldots, X_q)$; so α can be estimated by replacing σ_j^2 and σ_{jk} with their sample counterparts (Zanella and Cantaluppi 2006).

The Cronbach's α index for the gth group is obtained by means of (8), making reference, for the involved sample variances and covariances, to the statistical units belonging to group g. Observe that Cronbach's α shows low values when at least one of the following situations occurs: (a) measures are highly congeneric, and in this case Cronbach's α is a lower bound for true reliability (Green and Yang 2009; Raykov 2002); (b) the variances of the errors E_{ji} have high values, denoting low 'signal to noise' ratios.

Now, according to the decomposition of pairwise covariances in a clustered framework we can redefine σ_j^2 and σ_{jk} as:

$$\begin{aligned} \sigma_j^2 &= \sigma_j^{*2} + \bar{\sigma}_j^2 \\ \sigma_{jk} &= \sigma_{jk}^{*} + \bar{\sigma}_{jk}, \end{aligned} \tag{9}$$

where $\bar{\sigma}_j^2$ and $\bar{\sigma}_{jk}$ are respectively the portions of the variances σ_j^2 and of the covariances σ_{jk} that can be explained conditional on the values of the grouping variable Z. See the Appendix for the derivation of relationships (9).

So

$$\alpha = \frac{q}{q-1}\left(1 - \frac{\sum_{j=1}^{q} (\sigma_j^{*2} + \bar{\sigma}_j^2)}{\sum_{j=1}^{q} \sum_{k=1}^{q} (\sigma_{jk}^{*} + \bar{\sigma}_{jk})}\right) \tag{10}$$

and the null hypothesis (7) finally becomes

$$H_0 : \bar{\sigma}_j^2 = 0, \bar{\sigma}_{jk} = 0, \quad j, k = 1, \ldots, q. \tag{11}$$

In this way one can assess the presence of reliability in a multigroup framework.

The rejection of H_0 may be interpreted, when Cronbach's α is not significantly high for all groups, as the possible presence of congeneric measures at the group level (different λ's among groups). Namely, should τ_g change in one group, its effect would be confused with a proportional change in all λ_{jg}, being the measures still τ-equivalent. In presence of congeneric measures one should use appropriate measures in order to assess reliability, see e.g. Green and Yang (2009) and Raykov (2002), whose procedures are based on a preliminary estimation of a structural equation model with latent variables.

Observe that the presence of high values of the error variances common to all groups can represent a confounding factor which cannot be detected by means of the analysis of variance and covariances given in (10) as we also observed in the simulation results presented below.

The hypothesis system (11) corresponds to the equality of variances and covariances of $X_{1g}, X_{2g}, \ldots, X_{qg}$ across groups, and can be tested by considering the comparison of the group covariance matrices.

The Box's M statistic (Anderson 1958)

$$(n - G) \log |\mathbf{S}| - \sum_{g=1}^{G} (n_g - 1) \log |\mathbf{S}_g|$$

where $\mathbf{S}$ and $\mathbf{S}_g$ are the pooled covariance and the group covariance matrix respectively, is used for testing the homogeneity of covariance matrices and can thus help identifying the presence of congeneric measures, i.e. different loadings with also the possible presence of null loadings, in some group.

The comparison of the group covariance matrices avoids the preliminary estimation of a structural equation model with latent variables.

2 Simulation Results and Conclusion

A Monte Carlo simulation procedure was set up in order to study the behaviour of the test in discovering whether all the groups are characterized by τ-equivalent measures. The true scores τ were independently generated according to standard Normal random variables (the same scores were thus considered for all the groups); $X_j, j = 1, \ldots, q$ (with $q = 2, \ldots, 5$) measures were obtained by linear transforming the true scores and adding Normal random errors with different variances for each item, the standard deviations of the errors ranging randomly in the

Table 1 Simulation results ($\tilde{\sigma}_E = 0.7$)

		$G = 2$			$G = 5$			$G = 10$		
q	n_g	10	20	50	10	20	50	10	20	50
2		0.052	0.042	0.058	0.047	0.059	0.041	0.058	0.049	0.054
3		0.049	0.048	0.050	0.054	0.053	0.055	0.041	0.050	0.050
4		0.052	0.049	0.055	0.045	0.056	0.034	0.046	0.052	0.042
5		0.050	0.046	0.057	0.055	0.048	0.037	0.067	0.055	0.049

Table 2 Simulation results ($\tilde{\sigma}_E = 1$)

		$G = 2$			$G = 5$			$G = 10$		
q	n_g	10	20	50	10	20	50	10	20	50
2		0.064	0.053	0.051	0.050	0.061	0.047	0.052	0.049	0.052
3		0.052	0.044	0.048	0.051	0.052	0.059	0.061	0.058	0.051
4		0.061	0.051	0.041	0.050	0.050	0.041	0.052	0.052	0.058
5		0.048	0.044	0.043	0.056	0.053	0.044	0.059	0.045	0.059

Table 3 Congeneric simulation plan

	$G = 2$ ($cong = 1$)	$G = 5$ ($cong = 2$)		$G = 10$ ($cong = 4$)			
A	0.8	0.8	1.2	0.7	0.8	1.2	1.4
B	0.7	0.7	1.3	0.5	0.7	1.3	1.5
C[a]	0.7	0.7	1.3	0.5	0.7	1.3	1.5
D[b]	0.7	0.7	1.3	0.5	0.7	1.3	1.5

[a] ± 0.075
[b] ± 0.1

interval $\tilde{\sigma}_E \pm 0.1$. Simulations were performed in presence of $G = 2, 5, 10$ groups, each with $10, 20, 50$ subjects. Tables 1 and 2 show results for the particular case of all τ-equivalent measures ($\lambda_{ij} = 1$). Simulations were replicated 1,000 times.

Tables 1 and 2 show the proportion of cases corresponding to a significant Box's M statistic (p-value less than 0.05): as expected figures are very close to the 5 % α level. Table 1 reports simulation results for the case $\tilde{\sigma}_E = 0.7$ while Table 2 for the case $\tilde{\sigma}_E = 1$ respectively.

The presence of congeneric measures was simulated in the following way. Some groups were given τ-equivalent measures while for the remaining groups a congeneric settlement was generated. In particular, $1, 2$ and 4 congeneric groups were created for $G = 2, 5$ and 10 groups respectively.

All λ_{ij} coefficients pertaining to the τ-equivalent measures were set to 1. The λ_{ij} coefficients pertaining the congeneric groups were created according to Table 3. For situations C and D, starting from each central value, say λ, reported in the last two rows of the table, the q congeneric loadings λ_{ij} are fixed equally spaced in the range $[\lambda \pm 0.075]$ in case C and or $[\lambda \pm 0.1]$ in case D.

Tables 4, 5, 6, 7, 8, 9, 10, and 11 show the proportion of cases corresponding to a significant Box's M statistic (p-value less than 0.05), expected to be larger than

Table 4 Simulation results ($\tilde{\sigma}_E = 0.7$) A

		$G = 2$ $(cong = 1)$			$G = 5$ $(cong = 2)$			$G = 10$ $(cong = 4)$		
q	n_g	10	20	50	10	20	50	10	20	50
2		0.053	0.065	0.133	0.078	0.115	0.220	0.113	0.259	0.705
3		0.060	0.072	0.133	0.066	0.081	0.199	0.108	0.188	0.590
4		0.056	0.055	0.124	0.070	0.071	0.174	0.082	0.157	0.510
5		0.063	0.074	0.098	0.067	0.065	0.132	0.090	0.131	0.385

Table 5 Simulation results ($\tilde{\sigma}_E = 1$) A

		$G = 2$ $(cong = 1)$			$G = 5$ $(cong = 2)$			$G = 10$ $(cong = 4)$		
q	n_g	10	20	50	10	20	50	10	20	50
2		0.047	0.066	0.099	0.065	0.084	0.166	0.102	0.182	0.496
3		0.060	0.071	0.097	0.068	0.079	0.147	0.078	0.153	0.428
4		0.054	0.069	0.107	0.047	0.072	0.122	0.082	0.129	0.418
5		0.057	0.065	0.085	0.063	0.077	0.118	0.081	0.120	0.351

Table 6 Simulation results ($\tilde{\sigma}_E = 0.7$) B

		$G = 2$ $(cong = 1)$			$G = 5$ $(cong = 2)$			$G = 10$ $(cong = 4)$		
q	n_g	10	20	50	10	20	50	10	20	50
2		0.081	0.128	0.305	0.103	0.183	0.530	0.215	0.563	0.989
3		0.072	0.103	0.258	0.099	0.171	0.497	0.163	0.464	0.965
4		0.074	0.100	0.226	0.093	0.140	0.396	0.173	0.356	0.924
5		0.056	0.088	0.182	0.088	0.129	0.327	0.111	0.284	0.887

Table 7 Simulation results ($\tilde{\sigma}_E = 1$) B

		$G = 2$ $(cong = 1)$			$G = 5$ $(cong = 2)$			$G = 10$ $(cong = 4)$		
q	n_g	10	20	50	10	20	50	10	20	50
2		0.063	0.097	0.180	0.083	0.153	0.373	0.141	0.374	0.861
3		0.058	0.092	0.191	0.088	0.094	0.336	0.145	0.326	0.870
4		0.061	0.074	0.162	0.086	0.113	0.322	0.105	0.255	0.827
5		0.056	0.080	0.151	0.081	0.108	0.263	0.133	0.210	0.759

Table 8 Simulation results ($\tilde{\sigma}_E = 0.7$) C

		$G = 2$ $(cong = 1)$			$G = 5$ $(cong = 2)$			$G = 10$ $(cong = 4)$		
q	n_g	10	20	50	10	20	50	10	20	50
2		0.070	0.147	0.334	0.111	0.224	0.595	0.228	0.586	0.988
3		0.068	0.124	0.309	0.083	0.167	0.549	0.193	0.481	0.974
4		0.071	0.100	0.271	0.076	0.143	0.437	0.144	0.371	0.949
5		0.071	0.088	0.239	0.067	0.127	0.349	0.131	0.297	0.901

95 % for an usual significance test level. The performance of the test improves when the number of groups increases. The level of the error standard deviation (0.7 vs 1) seems to represent a confounding factor. Similarly, an increase of the number of indicators appears to act as a confounding factor, since the loadings are defined in

Table 9 Simulation results ($\tilde{\sigma}_E = 1$) C

		$G = 2$ ($cong = 1$)			$G = 5$ ($cong = 2$)			$G = 10$ ($cong = 4$)		
q	n_g	10	20	50	10	20	50	10	20	50
2		0.067	0.108	0.228	0.077	0.134	0.418	0.156	0.363	0.899
3		0.075	0.090	0.185	0.091	0.137	0.337	0.130	0.314	0.883
4		0.069	0.101	0.181	0.071	0.114	0.316	0.132	0.252	0.822
5		0.067	0.066	0.178	0.085	0.096	0.269	0.124	0.240	0.777

Table 10 Simulation results ($\tilde{\sigma}_E = 0.7$) D

		$G = 2$ ($cong = 1$)			$G = 5$ ($cong = 2$)			$G = 10$ ($cong = 4$)		
q	n_g	10	20	50	10	20	50	10	20	50
2		0.075	0.150	0.343	0.121	0.226	0.614	0.248	0.625	0.985
3		0.089	0.124	0.308	0.087	0.187	0.555	0.199	0.491	0.973
4		0.078	0.111	0.250	0.089	0.156	0.461	0.145	0.417	0.951
5		0.080	0.085	0.245	0.089	0.138	0.400	0.143	0.311	0.918

Table 11 Simulation results ($\tilde{\sigma}_E = 1$) D

		$G = 2$ ($cong = 1$)			$G = 5$ ($cong = 2$)			$G = 10$ ($cong = 4$)		
q	n_g	10	20	50	10	20	50	10	20	50
2		0.057	0.099	0.237	0.083	0.154	0.422	0.151	0.401	0.879
3		0.057	0.077	0.200	0.064	0.154	0.394	0.139	0.324	0.888
4		0.066	0.084	0.194	0.060	0.116	0.343	0.137	0.287	0.868
5		0.044	0.091	0.169	0.090	0.100	0.291	0.106	0.243	0.798

a limited range: when their number increases they get closer. It is clearly difficult for the proposed test to find the presence of congeneric measures in presence of few group with smaller sizes. In these cases the test performance is very poor.

Appendix

Relationship (9) establishing the decomposition of variances and covariances with reference to the considered measurement model is here derived.

Let (X_j, X_k) be a two dimensional random variable whose values are observed conditional on the values of a grouping variable Z. In this case the covariance σ_{jk} between the marginal variables can be written as the sum of two components representing respectively the so-called residual covariance σ^*_{jk} (a sort of within groups measure) and the ecological covariance $\bar{\sigma}_{jk}$ (a sort of between groups measure) (Guseo 2010).

Let $\lambda_{j(Z)} = \lambda_{jg}$ for $Z = z_g$, $g = 1, 2, \ldots, G$. According to (1) we can obtain

$$
\begin{aligned}
Cov(X_j, X_k) = M_Z\left\{Cov(X_j, X_k|Z)\right\} + Cov_Z\left\{M(X_j|Z), M(X_k|Z)\right\} = \\
= \left[M_Z\left\{Cov(\lambda_{j(Z)}\tau + E_j, \lambda_{k(Z)}\tau + E_k|Z)\right\}\right] + \\
+ \left[Cov_Z\left\{M(\lambda_{j(Z)}\tau + E_j|Z), M(\lambda_{k(Z)}\tau + E_k|Z)\right\}\right],
\end{aligned}
$$

where M is used as the expectation operator and the subscript Z denotes that expectation is computed by using the relative frequencies of Z as weights; thus

$$
\begin{aligned}
Cov(X_j, X_k) = \left[M_Z\left\{\lambda_{j(Z)}\lambda_{k(Z)}Var(\tau|Z) + \lambda_{j(Z)}Cov(\tau, E_k|Z) + \right.\right. \\
\left.\left. + \lambda_{k(Z)}Cov(\tau, E_j|Z) + Cov(E_j, E_k|Z)\right\}\right] + \\
+ \left[Cov_Z\left\{\lambda_{j(Z)}M(\tau|Z), \lambda_{k(Z)}M(\tau|Z)\right\}\right],
\end{aligned}
$$

by remembering that E_j, $j = 1, \ldots, q$, are independent random error components, also independent of the latent τ which is assumed to be independently distributed across the subjects.

$M(E_j|Z) = 0$, $Cov(\tau, E_j|Z) = 0$, $Cov(E_j, E_k|Z) = 0$ and $Var(E_j|Z) = Var(E_j)$, since E_j is independent of Z. We have:

$$
Cov(X_j, X_k) = M_Z\left\{\lambda_{j(Z)}\lambda_{k(Z)}Var(\tau|Z)\right\} + Var_Z\left\{\lambda_{j(Z)}\lambda_{k(Z)}M(\tau|Z)\right\} = \sigma^*_{jk} + \bar{\sigma}_{jk}.
$$

An analogous decomposition holds for the variance of X_j which can be written as

$$
\begin{aligned}
Var(X_j) &= M_Z\left\{Var(X_j|Z)\right\} + Var_Z\left\{M(X_j|Z)\right\} = \\
&= M_Z\left\{Var(\lambda_{j(Z)}\tau + E_j|Z)\right\} + Var_Z\left\{M(\lambda_{j(Z)}\tau + E_j|Z)\right\} = \\
&= M_Z\left\{\lambda^2_{j(Z)}Var(\tau|Z) + Var(E_j|Z) + 2\lambda_{j(Z)}Cov(\tau, E_j|Z)\right\} + \\
&\quad + Var_Z\left\{\lambda_{j(Z)}M(\tau|Z) + M(E_j|Z)\right\} = \\
&= M_Z\left\{\lambda^2_{j(Z)}Var(\tau|Z)\right\} + M_Z\left\{Var(E_j|Z)\right\} + 2M_Z\left\{\lambda_{j(Z)}Cov(\tau, E_j|Z)\right\} + \\
&\quad + Var_Z\left\{\lambda_{j(Z)}M(\tau|Z)\right\} + Var_Z\left\{M(E_j|Z)\right\} + \\
&\quad + 2Cov_Z\left\{\lambda_{j(Z)}M(\tau|Z), M(E_j|Z)\right\}
\end{aligned}
$$

and by means of the assumptions stated above we have

$$
\begin{aligned}
Var(X_j) &= \left[M_Z\left\{\lambda^2_{j(Z)}Var(\tau|Z)\right\} + Var(E_j)\right] + Var_Z\left\{\lambda_{j(Z)}M(\tau|Z)\right\} = \\
&= \sigma^{*2}_j + \bar{\sigma}^2_j.
\end{aligned}
$$

Under the hypothesis (7) we have $M(\tau|Z) = M(\tau), Var(\tau|Z) = Var(\tau)$ and $\lambda_{j(Z)} = \lambda_j$, so $Cov(X_j, X_k) = \lambda_j\lambda_k Var(\tau) = \sigma^*_{jk}$ and $Var(X_j) = \lambda^2_j Var(\tau) + Var(E_j) = \sigma^{*2}_j$. Observe that previous results holds also with sample data.

References

Anderson, T. W. (1958). *Introduction to multivariate statistical analysis*. New York: Wiley.

Bollen, K. A. (1989). *Structural equations with latent variables*. New York: Wiley.

Green, S. B., & Yang, Y. (2009). Reliability of summed item scores using structural equation modeling: an alternative to coefficient alpha. *Psychometrika, 74*, 155–167.

Guseo, R. (2010). Partial and ecological correlation: a common three-term covariance decomposition. *Statistical Methods and Applications, 19*, 31–46.

Raykov, T. (2002). Examining group differences in reliability of multiple-component instruments. *British Journal of Mathematical and Statistical Psychology, 55*, 145–158.

Wedel, M., & Kamakura, W. A. (1998). *Market segmentation: conceptual and methodological foundations*. Boston: Kluwer Academic.

Zanella, A., & Cantaluppi, G. (2006). Some remarks on a test for assessing whether Cronbach's coefficient α exceeds a given theoretical value. *Statistica Applicata, 18*, 251–275.

An Evaluation of Performance of Territorial Services Center (TSC) by a Nonparametric Combination Ranking Method

The IQuEL Italian Project

Mario Bolzan, Livio Corain, Valeria De Giuli, and Luigi Salmaso

Abstract The work presents some results about a national project IQuEL-2010, aimed to solve some problems associated to the digital divide by Territorial Services Center (TSC). A specific survey was carried out by sample of local operator in the three Italian provinces (Padova, Parma Pesaro-Urbino). We applied a nonparametric combination (NPC) ranking method on a set of nine dimensions related the public services supplied. The results show important differences among the provinces, at least for six out of nine TSC abilities or performances and producing a Global satisfaction ranking.

Keywords Global Satisfaction Ranking • Nonparametric combination ranking method • TSC

1 Introduction

During 2010, the most frequent type of interaction of enterprises with public administration (PA) in the EU27 using the Internet was downloading electronic forms (76 %), followed by obtaining information (74 %) and submitting completed forms (69 %). More than 90 % of all enterprises in Slovakia, Lithuania and Finland and Sweden reported that they used the Internet to obtain information from public

M. Bolzan (✉)
Department of Statistical Sciences, University of Padua, Padua, Italy
e-mail: mario.bolzan@unipd.it

L. Corain • L. Salmaso
Department of Management and Engineering, University of Padua, Padua, Italy

V. De Giuli
Department of Industrial Engineering, University of Padua, Padua, Italy

D. Vicari et al. (eds.), *Analysis and Modeling of Complex Data in Behavioral and Social Sciences*, Studies in Classification, Data Analysis, and Knowledge Organization,
DOI 10.1007/978-3-319-06692-9_6,

authorities' websites in 2010, while it was less than half of enterprises in Romania and the Netherlands. On the other hand, more than 90 % of enterprises in the Netherlands, Lithuania, Greece, Poland and Finland reported that they used the Internet in 2010 to submit completed forms electronically to public authorities. In Italy and Romania it was less common (both 39 %, Eurostat 2011). In Italy, after 10 years since the start of the eGovernment Italian project, only 28.9 % of population contact electronically the public administration (PA) services to find information and not more than 10 % to fill and send documents and submitting complete forms online (Rapporto eGov Italia 2010). IQuEL, Project (Innovation and Quality supplied and perceived from Local Units) in 2009 involved large local units (commons, provinces and regions) of around 15 millions of people. This is a trial to be expanded to all the local units in the nation. The goal was to monitor and evaluate the quality of the services supplied from PA by Territorial Services Centers (TSC). The TSCs are the government' executive branch of IQuEL project aimed to solve some problems associated to the digital divide existing in fringe areas. IQuEL uses a quali-quantitative approach on services supplied, based on a series of statistical indicators shared with the stakeholders. The indicators are about the level of access, of specific performance for each channel of distribution, and of Customer Satisfaction about Citizen Relationship Management services. A particular experiments was carried out in three provinces in North and Centre of Italy: Padova (PD), Parma (PR), and Pesaro-Urbino (PU) involved in the project has been evaluated by a sample of local operators ($nPd = 25$, $nPr = 42$; $nPU = 13$) on each of nine abilities to communicate and to transfer information from TSC: (1) Precision on Communication at start point of Service; (2) Politeness of TSC's office worker; (3) Clarity in the application forms; (4) Performances of TSC's office worker; (5) Friendly use and quality of the service by the TSC; (6) Timeliness to assistance; (7) Quality of assistance for a clarification; (8) Level of competitiveness with respect to market needs; (9) Level of global satisfaction towards the service provided. The answer for each dimension ranged in the interval 0–5.

We applied a nonparametric combination (NPC) ranking method on the three provinces (Pesarin et al. 2010).

2 An Overview of the Methodology

The NPC Ranking method has been shown to generally perform better than the 'usual' arithmetic mean. Let us assume that a sample is drawn from each of the C populations of interest in order to make inference on the true ranking of that populations. In order to formalize the problem let us refer to so-called one way MANOVA layout: let Y be the ordered categorical multivariate response variable representing a p-vector of the observed data and let us assume, without loss of generality, that high values of each univariate aspect Y correspond to a better performance and therefore to a higher ranking position; in other words, we are assuming the criterion "the higher the better". We recall that our inferential goal is to rank C multivariate populations (i.e. items/groups/treatments) with respect to p different variables where n replicates from each populations are available. Note that for sake of simplicity we are referring to the balanced design but the extension to unbalanced designs is straightforward.

Under the hypothesis of equality of the C populations, all true ranks would necessarily be equal to one, hence we would be in a full ex-equo situation. This situation of equal ranking may be formally represented in an hypothesis testing framework where the hypotheses of interest are:

$$\begin{cases} H_0 : \mathbf{Y}_1 \overset{d}{=} \mathbf{Y}_2 \overset{d}{=} \dots \overset{d}{=} \mathbf{Y}_C \\ H_1 : \exists \mathbf{Y}_j \overset{d}{\neq} \mathbf{Y}_h, \ j,h = 1,\dots,C, \ j \neq h. \end{cases}$$

In case of rejection of the global multivariate hypothesis H_0, it is of interest to perform inference on pairwise comparisons between populations, i.e.

$$\begin{cases} H_{0(jh)} : \mathbf{Y}_j \overset{d}{=} \mathbf{Y}_h \\ H_{1(jh)} : \mathbf{Y}_j \overset{d}{\neq} \mathbf{Y}_h, j,h = 1,\dots,C, j \neq h. \end{cases}$$

Note that a rejection of at least one hypothesis $H_{0(jh)}$ implies that we are not in an equal ranking situation, that is some multivariate population has a greater rank than some others.

Finally, it is useful to also consider the inferences on univariate pairwise comparisons between populations, which are defined by the following hypotheses:

$$\begin{cases} H_{0k(jh)} : Y_{jk} \overset{d}{=} Y_{hk} \\ H_{1k(jh)} : \left(Y_{jk} \overset{d}{<} Y_{hk}\right) \cap \left(Y_{jk} \overset{d}{>} Y_{hk}\right) \end{cases}, \quad j,h = 1,\dots,C, \ j \neq h, k = 1,..,p.$$

Looking at the univariate alternative hypothesis $H_{0k(jh)}$, note that since our goal is on the ranking of several multivariate populations, we are more interested in decide whether a population is greater of other one (not only different). In this connection, we can take account of the directional type alternatives, namely those that are suitable for testing both the one-sided alternatives.

We now present in detail the general algorithm underlying the proposed method aimed at obtaining a ranking procedure for multivariate populations. Let us take p response variables, C random samples from the relative reference populations, n sample size (balanced design), B independent permutations (CMCM method), and let $K = [C \cdot (C-1)]/2$ be the number of pair-wise comparisons between samples.

The algorithm consists of the following steps:

1. Let A be the K block matrix, each with p columns and $2{\cdot}n$ rows. The blocks represent all the possible pairs of the observed samples while the p columns refer to the different observed variables for each sample. Therefore matrix A has a total of $K{\cdot}p$ columns. Let us take, for example, X, Y and Z as three populations ($C = 3$). Matrix A will be as follows:

x11b	...	x1pb	x11b	...	x1pb	y11b	...	y1pb
...	...	...	...	...	...	...	...	...
xn1b	...	xnpb	xn1b	...	xnpb	yn1b	...	ynpb
y11b	...	y1pb	z11b	...	z1pb	z11b	...	z1pb
...	...	...	...	...	...	...	...	...
yn1b	...	ynpb	zn1b	...	znpb	zn1b	...	znpb

where b = 1, …, (B+1).
For each pair of samples, for each of the p relative variables and B permutations of the samples were carried out, obtaining a depth A^* for matrix A above.

With reference to the three populations, X, Y and Z, we have

$$\{x, y, z\}_{rj}$$

where $i, r = 1, \ldots, r$, $j = 1, \ldots, p$ and $b = 1, \ldots, (B+1)$ and where if $b = 1$ $x^*_{ijb} = x_{ijb} = x_{ij}$.

Note that we are dealing with balanced designs (each sample is the same size) but extension to unbalanced designs is easily achieved.

Essentially two types of permutations can be applied:

- *synchronized* permutations: the row indices (from 1 to $2{\cdot}n$) are randomly permuted and as a result the respective rows of matrix A are permuted;
- *unsynchronized* or independent permutations: permutation of the row indices is carried out in each block, therefore permutation of the respective rows of the matrix is carried out independently for each block.

Note that with unbalanced samples, only the latter type of permutation is applicable, given that each block has a different number of rows.

2. For each block k, $k = 1, \ldots, K$, for each variable j, $j = 1, \ldots, p$ and for each permutation b, $b = 1, \ldots, (B+1)$, calculate the difference of means (or other suitable statistic) of two samples. Note that such a statistic $T^b_{(ih)j}$ calculated on pairs of samples (i, h), for each variable j, refers to the observed statistics on the original samples for $b = 1$, whereas for $b = 2, \ldots, (B+1)$ the statistics are calculated on the permuted samples. This set of statistics calculated for each variable represents a sort of estimate of the null distribution of the statistic of interest for the given variable, i.e. under the hypothesis that data derives from the same population. From this distribution it is possible to estimate the p-value $\lambda^b_{(ih)j}$ related to the two pairs of directional hypotheses:

$$\begin{cases} {}_{(ih)j}H_0^+ : T^b_{ij} = T^b_{hj} \\ {}_{(ih)j}H_1^+ : T^b_{ij} > T^b_{hj} \end{cases}, \qquad i, h=1, \ldots, C, i<h, j=1, \ldots, p, b=1, \ldots, (B+1)$$

and

$$\begin{cases} {}_{(ih)j}H_0^- : T_{ij}^b = T_{hj}^b \\ {}_{(ih)j}H_1^- : T_{ij}^b < T_{hj}^b \end{cases}, \quad i,h=1,\ldots,C, i<h, j=1,\ldots,p, b=1,\ldots,(B+1)$$

respectively with $\widehat{\lambda}_{(ih)j}^b = \frac{\#T_{(ih)j}^* > T_{(ih)j}^b}{B}$ and $\widehat{\lambda}_{(hi)j}^b = \frac{\#T_{(hi)j}^* > T_{(hi)j}^b}{B}$.

These p-values will form another p-block matrix. In other words, by comparing each pair of samples, we obtain K comparisons (and consequently K p-values) and given that the comparison is one-directional in both the directions, we have a total of $2 \cdot K$ p-values per block. These will be calculated for all p considered variables.

	$\lambda_{(12b)1}$	$\lambda_{(13b)1}$	...	$\lambda_{(1Cb)1}$
$\lambda_{(21b)1}$		$\lambda_{(23b)1}$	...	$\lambda_{(2Cb)1}$
$\lambda_{(31b)1}$	$\lambda_{(32b)1}$		...	$\lambda_{(3Cb)1}$
...	...	...		...
$\lambda_{(C1b)1}$	$\lambda_{(C2b)1}$	$\lambda_{(C3b)1}$	...	

	$\lambda_{(12b)2}$	$\lambda_{(13b)2}$	...	$\lambda_{(1Cb)2}$
$\lambda_{(21b)2}$		$\lambda_{(23b)2}$	...	$\lambda_{(2Cb)2}$
$\lambda_{(31b)2}$	$\lambda_{(32b)2}$		...	$\lambda_{(3Cb)2}$
...	...	...		...
$\lambda_{(C1b)2}$	$\lambda_{(C2b)2}$	$\lambda_{(C3b)2}$	...	

...

	$\lambda_{(12b)p}$	$\lambda_{(13b)p}$	...	$\lambda_{(1Cb)p}$
$\lambda_{(21b)p}$		$\lambda_{(23b)p}$	...	$\lambda_{(2Cb)p}$
$\lambda_{(31b)p}$	$\lambda_{(32b)p}$		...	$\lambda_{(3Cb)p}$
...	...	...		...
$\lambda_{(C1b)p}$	$\lambda_{(C2b)p}$	$\lambda_{(C3b)p}$	...	

Where $\lambda_{(ihb)j} = \widehat{\lambda}_{(ih)j}^b$ $i,h = 1,\ldots,C, i < h, j = 1,\ldots,p, b = 1,\ldots,(B+1)$ as defined above.

3. From this matrix we can obtain the combinations of the p-values related to the same comparisons for the different variables. Consider, for example, the pair of samples (1, 2) and combine (using a suitable combining function) the p-values related to the comparison (1>2) obtained for each of the p variables, and so on for all K pairs of samples and for both directions, obtaining a total of $2 \cdot K$ combinations. These combinations are also repeated on the p-values related to the comparisons of the permuted samples.

   Let ϑ_{ih}^b (or alternatively ϑ_{hi}^b) where $i,h = 1,\ldots,C, i < h, b = 1,\ldots,(B+1)$ represent the above combinations. A normalization is applied to these combinations in respect of the number of samples and variables. Such normalization is to divide ϑ_{ih}^b by the quantity $(C-1) \cdot p$. For simplicity of notation, from this moment on ϑ_{ih}^b represents these normalized combinations.
4. From step 3 obtain an estimate of the null distribution for each of the $2 \cdot K$ combinations, $\vartheta_{ih}^* = \left(\vartheta_{ih}^2, \vartheta_{ih}^3, \ldots, \vartheta_{ih}^{(B+1)}\right)$ and alternatively $\vartheta_{hi}^* = \left(\vartheta_{hi}^2, \vartheta_{hi}^3, \ldots, \vartheta_{hi}^{(B+1)}\right)$ where $i,h = 1,\ldots,C, i < h$, and compute the quantities: $\lambda_{ih}'' = \frac{\#\vartheta_{ih}^* > \vartheta_{ih}}{B}$ or alternatively $\lambda_{hi}'' = \frac{\#\vartheta_{hi}^* > \vartheta_{hi}}{B}$.
5. From the table of p-values in 2 above, also compute C combinations ϑ_i, $i = 1,\ldots,C$, of all p-values of each row. The above normalization is applied to these combinations as well.
6. Sort in descending order the normalized ϑ_i obtaining an initial indication of the ranking of the C samples.

7. Consider the values $\lambda''_{(ih)}$ where $i, h = 1, \ldots, C, i < h$ and where the parentheses indicate the order of the samples.

   Assume for example that we have to compare $C = 5$ samples and have obtained the following order in step 6: $(\vartheta_5, \vartheta_3, \vartheta_4, \vartheta_1, \vartheta_2)$. Therefore we consider only the K of the $2{\cdot}K$ $\lambda''_{(ih)}$ and $\lambda''_{(hi)}$ in the order $\lambda''_{53}, \lambda''_{54}, \lambda''_{51}, \lambda''_{52}, \lambda''_{34}, \lambda''_{31}, \lambda''_{32}, \lambda''_{41}, \lambda''_{42}, \lambda''_{12}$.

   These will form a triangular matrix from which we obtain the final ranking. With reference to the previous example, the above matrix will be

	1°=5	2°=3	3°=4	4°=1	5°=2
1°=5		λ''(12)	λ''(13)	λ''(14)	λ''(15)
2°=3			λ''(23)	λ''(24)	λ''(25)
3°=4				λ''(34)	λ''(35)
4°=1					λ''(45)
5°=2					

   Therefore λ_{ih} refers to the p-values related to the comparison between the group in position i and the group in position h, where $i, h = 1, \ldots, C, i < h$. In this example, therefore, we have $\lambda''_{(12)} = \lambda''_{53}, \lambda''_{(13)} = \lambda''_{54}$, and so on.
8. The Bonferroni–Holm–Shaffer correction applies to the matrix in step 7. For simplicity of notation we will continue to refer to $\lambda''_{(ih)}$ as the general corrected p-value. Each p-value in the table from step 7 and corrected in step 8, is compared with $\alpha/2$ where α is the confidence level we are considering.
9. The previous matrix is converted into a matrix of 0 and 1 depending on whether $\lambda''_{ih} < \alpha/2$ or $\lambda''_{ih} > \alpha/2$. In particular each cell will be

$$\begin{cases} 1 \; \lambda''_{ih} < \alpha/2 \\ 0 \; \lambda''_{ih} \geq \alpha/2 \end{cases}.$$

10. This matrix passes for a further algorithm which reinstates the global ranking. This algorithm consists of the following steps:

    – Transpose the matrix of 0 and 1 from the previous step;
    – Calculate the sum along the columns to obtain a *score*.
    – On these transposed scores, calculate the rank transformation to obtain the global ranking.

Table 1 Medians of the nine investigated abilities from TSC by Province

Abilities from TSC	Province		
	PD	PR	PU
Precision on communication	4	3	4
Politeness of TSC's office worker	4	4	4
Clarity in the application forms	4	3	4
Performances of TSC's office worker	4	4	4
Friendly use and quality of the service	4	3	4
Timeliness to assistance	4	3	4
Quality of assistance	4	4	4
Level of competitiveness	4	3	4
Global satisfaction	4	3	4

Table 2 *P*-values of permutation C-sample test (Anderson–Darling) and of pairwise comparisons

	Precision on Communication (0.208)			Politeness to TSC's office worker (0.458)			Clarity in the application forms (**0.047**)		
Province	PD	PR	PU	PD	PR	PU	PD	PR	PU
PD		0.628	0.527		0.238	0.759		**0.008**	0.884
PR			0.284			0.247			**0.036**
PU									
	Performances of TSC's office worker (**0.016**)			Friendly use and quality of the service (**0.002**)			Timeliness to assistance (0.071)		
PD		**0.011**	0.471		**0.027**	0.578		**0.017**	0.799
PR			**0.028**			0.067			0.123
PU									
	Quality of assistance (0.262)			Level of competitiveness (**0.001**)			Global satisfaction (**0.001**)		
PD		0.278	0.560		**0.013**	0.293		**0.035**	0.158
PR			0.194			**0.014**			**0.006**
PU									

(*P*-values at 5% significance level are highlighted in bold)

Table 3 Global ranking

Province	PD	PR	PU
Ranking	1	3	1

3 Some Results

In Tables 1, 2, and 3 we present some main results of the application of the multivariate ranking method (at 5% significance level). The analysis by means of NPC Ranking method gives both PD and PU on the first place in the ranking. The results show important heterogeneity among the three Italian provinces, at least for six on nine TSC abilities or performances (see Table 2) producing a Global satisfaction ranking displayed in Table 3. The Performances of TSC office worker, the Level of competitiveness with respect to market needs and the Level of competitiveness looks to be important since they give a good 'image' of TSC to the local community in particular to attract new enterprises in the area. The main differences among provinces are due to the specificity expressed by areas but also to the different policies to training personnel. Before to start with IQuEL project, PD and PU had work strong to promote the competences and skills of the municipal managers since those are the keys to organize all the services by means of a vision of New Public Management aimed at changing perspectives: from an administration based on authority to one that provides services to citizens or enterprises (Bolzan 2010). In general the analysis showed a positive evaluation on TSC, emphasizing some critical situation (not sufficient information about the procedures to use it, the operative and the communicative interface device which is much formal and not of friendly use, and again feedback not expected). About the workers: low profile of competence about the most innovative aspects of the network. Such problems might be faced by updating and training (see also Bolzan 2010; Bolzan and Boccuzzo 2011).

References

Bolzan, M. (2010). L'identità professionale del personale dirigente degli enti locali del Veneto: competenze, fabbisogno formativo in relazione al territorio e missione dell'ente. In *Competenze e processi formativi per i dirigenti degli enti locali* (pp. 13–54). Atti a cura di Bolzan M. Ed Cleup. ISBN 978-88-6129-489-9.

Bolzan, M., & Boccuzzo, G. (2011). Skills, formative needs and criticalities of municipal managers in Italy. In *Improving the quality of public services: A multinational conference on public management international conference*: Moscow 27–29 June, Panel 12: Understanding Contemporary Public Administration.

Pesarin, F., & Salmaso, L. (2010). *Permutation tests for complex data: Theory, applications and software*. Chichester: Wiley.

Rapporto eGov Italia. (2010). Milano. http://epp.eurostat.ec.europa.eu/185/2011. 13 December 2011 Internet access and use of ICT in enterprises in 2011.

A New Index for the Comparison of Different Measurement Scales

Andrea Bonanomi

Abstract In psychometric sciences, a common problem is the choice of a good response scale. Every scale has, by its nature, a propensity to lead a respondent to mainly positive- or negative- ratings. This paper investigates possible causes of the discordance between two ordinal scales evaluating the same goods or services. In psychometric literature, Cohen's Kappa is one of the most important index to evaluate the strength of agreement, or disagreement, between two nominal variables, in particular in its weighted version. In this paper, a new index is proposed. A proper procedure to determine the lower and upper triangle in a non-square table is also implemented, as to generalize the index in order to compare two scales with a different number of categories. A test is set up with the aim to verify the tendency of a scale to have a different rating compared to a different one. A study with real data is conducted.

Keywords Agreement index • Measurement ordinal scale • Weighted kappa

1 Cohen's Kappa and Weighted Kappa

In psychometric sciences, a common problem is the choice of a good response scale. Several studies (see Bonanomi 2004) have shown that different measurement scales can lead to highly dissimilar evaluations of goods/services, in particular in the measurement of observable variables in latent variables models. Every scale has, in its proper nature, a tendency or propensity to lead a respondent to mainly positive (or negative) ratings. This paper investigates possible causes of the discordance between two ordinal scales evaluating the same good or service. The question we

A. Bonanomi (✉)
Dipartimento di Scienze Statistiche, Università Cattolica del Sacro Cuore di Milano, Milano, Italy
e-mail: andrea.bonanomi@unicatt.it

D. Vicari et al. (eds.), *Analysis and Modeling of Complex Data in Behavioral and Social Sciences*, Studies in Classification, Data Analysis, and Knowledge Organization,
DOI 10.1007/978-3-319-06692-9_7,

would like to study is the following: "Is the eventual discordance random, or, on the contrary, is it the result of a systematic tendency to assign mainly positive (or negative) evaluations?". In psychometric literature, Cohen's Kappa (Cohen 1960) is one of the most important index used to evaluate the strength of agreement (or disagreement) between two categorical variables. Let us consider two different measurement scales, with k response modalities, used to evaluate the same generic item by the same sample of respondents. In a $k \times k$ square table, with generic relative frequencies p_{ij}, Cohen's Kappa compares the probability of agreement $p_o = \sum_{i=1}^{k} p_{ii}$, where p_{ii} are the relative observed frequencies on the main diagonal, to the expected if the variables were independent, $p_e = \sum_{i=1}^{k} p_{i\cdot} \cdot p_{\cdot i}$, with $p_{i\cdot}$ and $p_{\cdot j}$ respectively the relative marginal frequency of the i-th row and the relative marginal one the j-th column, by $Kappa = (p_o - p_e)/(1 - p_e)$. The value of *Kappa* is 1 when perfect agreement between the two scales occurs, 0 when agreement is equal to what is expected under independence. If there is more disagreement than the zero case, then Kappa assumes a negative value, although this is generally not a probable situation in practice. Kappa does not take into account the degree of disagreement between different scales, and all disagreement is treated only as total disagreement. Therefore, when the categories are ordered, it is preferable to use *Weighted Kappa* (*WK*) (Agresti 2002). *WK* is appropriate when the relative seriousness of the different possible disagreements can be specified. Whereas *Kappa* does not distinguish among degrees of disagreement, *WK* incorporates the magnitude of each disagreement and provides partial credit for disagreements when agreement is not complete. The usual approach is to assign weights to each disagreement pair with larger weights indicating greater disagreement. Weight matrix cells on the main diagonal represent agreement and thus contain zero. Off-diagonal cells contain weights indicating the seriousness of that disagreement. The choice of the weights is arbitrary, and it represents an open question in literature (Vanbelle and Albert 2009). Standard weights are the linear (Cicchetti and Allison 1971) and the quadratic weights (Fleiss et al. 1969). Quadratic weights are the most popular choice because the quadratically weighted Kappa can be interpreted as an interclass correlation coefficients; Warrens (2012) shows several limits and paradoxical properties of this choice. On the contrary, the linearly weighted Kappa presents some suitable properties, in particular when ordinal categories are easily confused and can be collapsed (Warrens 2013). We refer to linear weights in the present study. For the generic cell (i, j) the corresponding weight is the absolute difference between i and j, and it measures the distances from the main diagonal (situation of perfect agreement). Let $w_{ij} = |i - j|$, $p_{ow} = \sum_{i=1}^{k} \sum_{j=1}^{k} w_{ij} p_{ij}$ and $p_{ew} = \sum_{i=1}^{k} \sum_{j=1}^{k} w_{ij} p_{i\cdot} \cdot p_{\cdot j}$, Linearly Weighted Kappa is defined by $WK = 1 - p_{ow}/p_{ew}$. *WK* is generally ranging in [0;+1]. The value of the index is 1 when perfect agreement between the two scales occurs, 0 when agreement is equal to what is expected under independence. Negative values are not very probable, and they indicate that agreement is less than expected by chance. In the comparison of two different scales, *Kappa* and *Weighted Kappa* do not completely answer the question whether a scale has a systematic tendency to assign more positive or more negative evaluations than other one with the same number of response modalities. They only indicate the magnitude of agreement.

2 A Proposal of a New Index

A new measure to investigate the eventual systematic propensity of a scale to assign more positive (or negative) evaluations is proposed. Although the literature on association coefficients is large, the idea is to set up an index that measures not only the magnitude of agreement and association, but eventually the direction of disagreement and the skewness of the evaluation of two scales, that is which scale has the propensity to assign more positive evaluations when evaluating the same good. The evaluation of an item with two response ordinal scales, S_1, with I categories and S_2, with J categories, $(I = J)$, is presented to the same sample of n subjects, randomizing the presentation sequence of the scales, in order to avoid distortion effects. Considering a generic item, let p_{ij} be the relative frequency of subjects which assigned the *i*-th category evaluating with S_1 and the *j*-th with S_2, in an $I \times J$ contingency table. Let $w_{ij} = |i - j|$ be the generic linear weight for the cell (i, j), $p_{ow}^{+} = \sum_{i=1}^{I} \sum_{j=i+1}^{J} w_{ij} p_{ij}$ be the weighted relative frequencies in the upper triangle of the contingency table (above the main diagonal) and $p_{ow}^{-} = \sum_{i=1}^{I} \sum_{j=1}^{i-1} w_{ij} p_{ij}$ be the weighted relative frequencies in the lower triangle, $p_{ew} = \sum_{i=1}^{I} \sum_{j=1}^{J} w_{ij} p_{i.} p_{.j}$, the weighted relative frequencies in the situation of independence. The new measure, named *Skewness Evaluation Scales Index* (*SESI*), is absolutely arbitrary and it is defined by

$$SESI = \frac{p_{ow}^{+} - p_{ow}^{-}}{p_{ew}}. \tag{1}$$

The proposed *SESI* measures the ratio among the difference between the weighted relative frequencies in the upper triangle of the matrix and in the lower triangle and the sum of the weighted relative expected frequencies. The utility of dividing the difference between p_{ow}^{+} and p_{ow}^{-} for the sum of the weighted relative expected frequencies is double: (1) the denominator is always positive that so the index is always defined; (2) as it is possible to prove that $|p_{ow}^{+} - p_{ow}^{-}| \leq p_{ew}$, the *SESI* is a normalized index, ranging from -1 to $+1$. For simplicity, let show the proof in case of a 2×2 contingency table. Let $a = p_{11}$, $b = p_{12}$, $c = p_{21}$ and $d = p_{22}$ be the conjoint relative frequencies. With reference to the definition of the terms in (1), let $p_{ow}^{+} = b$, $p_{ow}^{-} = c$ and $p_{ew} = (a + b)(b + d) + (a + c)(c + d)$. In order to prove that $|p_{ow}^{+} - p_{ow}^{-}| \leq p_{ew}$, and so that the proposed index is ranging from -1 to $+1$, it is necessary to show that $(a+b)(b+d)+(a+c)(c+d) \geq |b-c|$. Developing the products and gathering up common terms, you obtain $b(1-c)+c(1-b)+2ad \geq |b-c|$. If $b \geq c$ then, when developing the products and simplifying the opposite terms, you obtain $c(1-b) \geq -ad$, while if $b < c$ you obtain $b(1-c) \geq -ad$. In

Table 1 Example 1: contingency matrix of the observed frequencies

	Numerical scale				
Verbal scale	1	2	3	4	5
Insufficient	**2**	13	3	0	0
Mediocre	2	**14**	18	3	0
Sufficient	0	10	**61**	38	0
Good	0	2	3	**59**	7
Excellent	0	0	1	1	**13**

both cases, the first member of the inequality is always a non-negative term, while the second member is always a non-positive term. Thus the inequality is always verified, and consequently the proposed index is always ranges from -1 to 1. The interpretation of the index is:

- $SESI = +1$: all frequencies are in the upper triangle, corresponding to a systematic tendency to assign more positive evaluations using the scale with categories by column;
- $SESI = -1$: all frequencies are in the lower triangle, corresponding to a systematic tendency to assign more positive evaluations using the scale with categories by row;
- $-1 < SESI < 0$: propensity to assign more positive evaluations using the scale with categories by row;
- $0 < SESI < +1$: propensity to assign more positive evaluations using the scale with categories by column;
- $SESI = 0$: indifferent evaluation, result of two possible situations: (1) the weighted frequencies in the upper triangle balance the frequencies in the lower triangle causing no propensity for any scale; (2) all frequencies are in the main diagonal with perfect concordance between the two scales. A joint evaluation of *WK* and *SESI* brings a better understanding of the situation of $SESI = 0$: if *WK* tends to its maximum value 1, then the situation 2 above showed is verified (in this case *WK* permits to weigh the relative importance of the elements on the main diagonal). Also, the trace of the matrix of observed relative frequencies allows to better explain the situation of $SESI = 0$. Let $Trace = \sum_{i=1}^{I} p_{ii}$: if *Trace* tends to its maximum value 1, then the situation 2 above showed is verified.

Example 1. In a survey of customer satisfaction at an Italian university, the same item was presented to 250 students of different faculties, gender and age: "How do you evaluate the Didactic Services of the University?" Each respondent had to assign the evaluation by two different measurement ordinal scales, a numeric five-point scale (ranging from *1*, the minimum, to *5*, the maximum) and a verbal five-point Likert scale (insufficient, mediocre, sufficient, good or excellent). Table 1 reports the contingency matrix of the observed frequencies. The main diagonal (bold frequencies) indicates the perfect concordance between the numeric and verbal scale. The other frequencies indicate a no-concordance of rank between numeric and verbal scale.

The perfect situation should be with all the frequencies in the main diagonal. As there are many frequencies in the upper or lower triangle of the matrix, this indicates a non-perfect concordance between the two scales. In this example, *WK* is equal to 0.572. This value indicates, in the classical interpretation (Landis and Koch 1977), a fair/moderate agreement between the ordinal scales. In this particular context, it shows that a numerical scale and a verbal scale probably do not evaluate the item in the same way. *SESI* is equal to 0.257, which indicates a tendency to assign more positive evaluations using the numeric scale (with modalities by column). In Sect. 3, a test is set up to verify if this disagreement and skewness is significant.

A proper procedure to determine the lower and upper triangle in a non-square table is also implemented, as to generalize the index *SESI* in order to compare two scales with a different number of categories. Let consider the evaluation of an item with two response ordinal scales, S_1, with I categories and S_2, with J categories, $(I \neq J)$. *SESI* is based on a comparison between relative frequencies in the lower and in the upper triangle as in the case of a square contingency table. Given that the contingency table is rectangular $(I \neq J)$, a procedure to define the "main diagonal" is necessary. Let us suppose that the evaluation is a continuous latent variable defined in the close interval [0; 1]. The choice by a generic respondent of a specific response category is conduced by a latent operation that permits us to transform the continuous evaluation in a discrete rating. For example, if an evaluation in the latent trait [0; 1] is equal to 0.7 and the measurement scale is on four modalities, the choice would be for the third modality in the ordinal sequence or the fourth if the scale is on five modalities. By the determination of the main diagonal, if two different scales (also with different number or response categories) have the same "evaluation propensity", all frequencies are naturally in a "main diagonal", defined by an evaluation concordance. In this way, it is possible in the contingency table to define the cells corresponding to the "main diagonal", and consequently the lower and the upper triangle. In algebra literature, there is a definition of "main diagonal" of a non-square table (Horn and Johnson 1985), but it does not satisfy our requirement. Therefore an arbitrary new procedure is proposed and set up. Formally, in a $I \times J$ rectangular contingency table, a generic cell belongs to the main diagonal if its row and column indices (respectively i and j) satisfy conjointly the conditions

$$\begin{cases} i/I > (j-1)/J \\ j/J \geq (i-1)/I \end{cases} \qquad (2)$$

Cells satisfying the condition (2) define the "*quasi main diagonal*", and they ensure indifferent evaluations between different scales (the evaluation using S_1 or S_2 is invariant). In case of a rectangular contingency table, it is not possible to define cells on the main diagonal as cells of "perfect concordance", in particular when the number of rows and columns is very different. It is more appropriate, in this case, to speak of "quasi-perfect concordance". Cells above the main diagonal belong to the upper triangle, on the contrary cells below the main diagonal belong to the lower triangle. The formula (1) of *SESI* is unchanged, with the natural adjustment of cells

$$w_{ij} = \begin{vmatrix} 0 & 0 & 1 & 2 \\ 1 & 0 & 0 & 1 \\ 2 & 1 & 0 & 0 \end{vmatrix}.$$

Fig. 1 Matrix weights for a 3 × 4 contingency table

Table 2 Example 2: contingency matrix 5 × 4 of the observed frequencies

	Likert scale			
Semantic scale	*1*	*2*	*3*	*4*
Bad	**5**	5	0	0
Quite bad	**3**	**7**	2	0
Neither bad or good	1	**10**	**8**	1
Quite good	0	0	**4**	**8**
Good	0	0	1	**12**

in the upper and lower triangles. In the general situation of a $I \times J$ rectangular contingency table, we define the weights w_{ij} as the absolute horizontal distance from the main diagonal previously defined: for example, if $I = 3$ and $J = 4$ the main diagonal consists of the cells (1,1), (1,2), (2,2), (2,3), (3,3), (3,4). Both cells (2,1) and (1,3) are adjacent to the main diagonal. Their distance from the main diagonal is the same and so $w_{21} = w_{13} = 1$. Figure 1 reports the weights for a 3×4 rectangular contingency matrix. The frequencies in the upper and lover triangle are respectively equal to $p^+_{ow} = \sum_{j=1}^{J} \sum_{\forall i: \frac{i}{I} \leq \frac{j-1}{J}} w_{ij}\, p_{ij}$ and $p^-_{ow} = \sum_{j=1}^{J} \sum_{\forall i: \frac{i-1}{I} > \frac{j}{J}} w_{ij}\, p_{ij}$.

Example 2. A sample of 67 respondents evaluated the appreciation of a tv-show, using two different scales, a four-point Likert Scale (ranging from *1*, the minimum, to *4*, the maximum, with categories expressed in columns) and a five-point Semantic Scale, with categories expressed in rows. Table 2 reports the contingency matrix of the observed frequencies. The main diagonal (bold frequencies) indicates the "quasi-perfect" concordance between the Likert and the Semantic scale. The other frequencies indicate a no-concordance of rank between the two ordinal scales. In this example, *WK* is equal to 0.834. This value indicates a strong agreement between the ordinal variables. In this particular context, it shows that a Likert scale and a Semantic scale probably do not evaluate the item exactly in the same way. *SESI* is equal to 0.100, and it indicates a slight tendency to assign more positive evaluations using the Likert scale (with modalities by column).

3 A Parametric Test for *SESI*

In order to verify the tendency of a scale to have a mainly positive or negative rating compared to a different one, a parametric test is set up. The null hypothesis refers to the perfect concordance between the two considered scales. Acceptance or refusal

Table 3 Results of Examples 1 and 2

Indices	Example 1 (5 × 5) table	Example 2 (5 × 4) table
SESI	0.257	0.100
SE(*SESI*)	0.035	0.048
Statistic Z	4.775	2.424
p-value	0.000	0.007
Inferior limit	0.189	0.005
Superior limit	0.325	0.195

general norm of scales concordance null hypothesis is determined by the confidence interval: if 0 (situation of indifference between the two scales) falls between interval limits, the null hypothesis is accepted, refused otherwise. At a prefixed significance level $\alpha = 0,05$, the setting of a confidence interval is bounded to the estimation of the Asymptotic Standard Error. The Standard Error formula has long been debated in the psychometrical literature (Cohen 1960; Fleiss et al. 1969; Hubert 1978). In particular Fleiss et al. (1969) showed the exact formula of *Var*(*Kappa*) and *Var*(*WK*) using in the determination of the confidence interval for *Kappa* and *WK*: Everitt (1968) derived the exact formulas of *Var*(*Kappa*) and *Var*(*WK*) assuming a generalized hypergeometric distribution, but they are far too complicated for routine use. Assuming binomial distributions and for large samples n, it is possible to use alternative and approximate formulas, they are slightly incorrect (the bias is in the direction of a slight overestimation), but very simple for a routine use (Cohen 1968). The formula of *Var*(*SESI*) for the determination of confidence intervals is the natural arrangement of the Asymptotic Variance of Weighted Kappa, adapted for the particular case of the new proposed index:

$$Var(SESI) \cong \frac{\sum_{i=1}^{I}\sum_{j=1}^{J} w_{ij}^2 p_{ij} - (p_{ow}^{+} + p_{ow}^{-})^2}{n(\sum_{i=1}^{I}\sum_{j=1}^{J} w_{ij} p_{eij})^2}. \quad (3)$$

Since with large samples in general kappa statistics are asymptotically distributed as $N(0,1)$ under H_0 (Cohen 1968), (3) can be used for the determination of the confidence interval of *SESI*. Under the hypothesis of normality, a statistic test and a p-value are also calculated. To test an obtained *SESI* for significance, the *Var*(*SESI*) when the population *SESI* equals zero is needed, and it is obtained by substituting observed relative weighted frequencies with expected relative weighted ones. In this way, (3) assumes the following formulation:

$$Var(SESI_0) \cong \frac{\sum_{i=1}^{I}\sum_{j=1}^{J} w_{ij}^2 p_{eij} - (p_{ew}^{+} + p_{ew}^{-})^2}{n(\sum_{i=1}^{I}\sum_{j=1}^{J} w_{ij} p_{eij})^2}. \quad (4)$$

Table 3 reports the index *SESI*, the *Var*(*SESI*), the statistic z, the p-value and the interval confidence for Examples 1 and 2. In both the examples, confidence intervals and p-values suggest that the two considered scales do not show concordance of

Table 4 University customer satisfaction: comparison between a verbal and numerical scale

Item	SESI	95 % CI's	Item	SESI	95 % CI's	Item	SESI	95 % CI's
Item 1	0.262	0.17–0.35	*Item 5*	0.067	0.00–0.13	*Item 9*	0.294	0.22–0.37
Item 2	0.219	0.13–0.31	*Item 6*	0.220	0.14–0.30	*Item 10*	0.252	0.18–0.33
Item 3	0.065	−0.01–0.14	*Item 7*	0.151	0.08–0.23	*Item 11*	0.154	0.08–0.23
Item 4	0.151	0.08–0.23	*Item 8*	0.104	0.04–0.17	*Item 12*	0.096	0.02–0.15

evaluation or agreement. The different propensity of the scales is significative, the choice of a scale becomes very important in the measurement process of a service or a good. The index SESI and the associated test validity have been verified for several items of a university customer satisfaction survey, evaluated with two different measurement ordinal scales (a five point Verbal Scale by row in the contingency matrix and a five point Numerical Scales by column) by 250 respondents. In Table 4, the results of the index *SESI* and 95 % CI's for all the items of the survey are reported. For each item, the index is positive, indicating a generalized propensity to assign more positive evaluation using the Numerical Scale (modalities by column in the contingency matrix). The index is quite stable and robust. The associated test, for most items, confirms that this propensity is significative. Only for two items (Items 3 and 5) does the test suggests to not reject the null hypothesis of concordance between the scales. In the other cases, CI's suggest that the statistic we observed in the sample is unlikely to have occurred when the null hypothesis is true. It is confirmed the idea that the evaluation with a numerical and with a verbal scale does not bring to the same results.

References

Agresti, A. (2002). *Categorical data analysis*. Ney York: Wiley.

Bonanomi, A. (2004) *Variabili ordinali e trasformazioni di scala, con particolare riferimento alla stima dei parametri dei modelli interpretativi con variabili latenti*. Methodological and Applied Statistical, University of Milan Bicocca.

Cicchetti, D. V., & Allison, T. (1971). A new procedure for assessing reliability of scoring EEG sleep recording. *The American Journal of EEG Technology, 11*, 101–109.

Cohen, J. (1960). A coefficient of agreement for nominal scales. *Educational and Psychological Measurement, 20*, 37–46.

Cohen, J. (1968). Weighted kappa: nominal scale agreement with provision for scaled disagreement or partial credit. *Psychological Bullettin, 70*, 213–220.

Everitt, B. S. (1968). Moments of the statistics kappa and weighted kappa. *British Journal of Mathematical and Statistical Psycology, 21*, 97–103.

Fleiss, J. L., Cohen, J., & Everitt, B. S. (1969). Large sample standard errors of kappa and weighted kappa. *Psychological Bullettin, 72*, 323–327.

Horn, R. A., & Johnson, C. R. (1985). *Matrix analysis*. Cambridge: Cambridge University Press.

Hubert, L. J. (1978). A general formula for the variance of Cohen's weighted kappa. *Psychological Bullettin, 85*(1), 183–184.

Landis, J. R., & Koch, G. G. (1977). The measurement of observer agreement for categorical data. *Biometrics, 33*(1), 159–174–327.

Vanbelle, S., & Albert, A. (2009). A note on the linearly weighted kappa coefficient for ordinal scales. *Statistical Methodology, 6*, 157–163.

Warrens, M. J. (2012). Some paradoxical results for the quadratically weighted kappa. *Psychometrika, 77*, 315–323.

Warrens, M. J. (2013). Cohen's weighted kappa with additive weights. *Advances Data Analysis Classification, 7*, 41–55.

Asymmetries in Organizational Structures

Giuseppe Bove

Abstract Relationships in organizational structures are frequently asymmetric (e.g., the number of e-mail messages that an employee sends to a colleague is usually different from the number of e-mail messages he received from that colleague). So organizational data are usually represented by asymmetric square matrices that cannot be analyzed by standard symmetric approaches. For this reason methods based on Singular Value Decomposition and Asymmetric Multidimensional Scaling were proposed to analyze these types of matrices. In many situations information concerning hierarchies or aggregations in the organizational structure is available and can be used in the analysis of the data (e.g., professional levels or departments belonging). In this paper three-way unfolding is proposed to take into consideration this additional information and applied to Krackhardt (Social Networks 9:109–134, 1987) data on advice-giving and getting in an organization.

Keywords Asymmetry • Multidimensional scaling • Visualization

1 Introduction

Within the area of network analysis data frequently concern relationships between all pairs of members in an organizational structure (e.g., employees in a firm, students at school, friends in the web, etc.). Different types of relations can be considered, e.g., 'A is a friend of B' or 'A approaches B for help and advice', and each person is asked to judge if person i is in relation or not with person j, for all pairs (i, j). Thus, if there are n persons, each of them provides an $n \times n$ matrix 0/1

G. Bove (✉)
Dipartimento di Scienze dell'Educazione, Università degli Studi *Roma Tre*, Via Milazzo 11/b, Roma, Italy
e-mail: bove@uniroma3.it

D. Vicari et al. (eds.), *Analysis and Modeling of Complex Data in Behavioral and Social Sciences*, Studies in Classification, Data Analysis, and Knowledge Organization, DOI 10.1007/978-3-319-06692-9_8,

(a slice), and we can collect all this information in a three-way $n \times n \times n$ dichotomous data array $\boldsymbol{\Omega} = (\omega_{ijk})$ consisting of the n slices.

Different approaches were considered to analyse the previous three-way array, in this paper we focalize on graphical representation, following an explorative approach. When n is not small, slices are usually aggregated. So, in the next section we briefly recall different way to aggregate the n slices, including the case when external information on persons is available to make aggregations. In Sect. 3 the main aspects of the proposed unfolding model will be provided. Then an example of application of three-way unfolding to Krackhardt (1987) data on advice-giving and getting in an organization will be presented.

2 Aggregations of Network Data

The n dichotomous slices can be aggregated in many different ways, according to the aim of the analysis. We can distinguish three different approaches: locally aggregated structures, consensus structures, three-way structures.

2.1 Locally Aggregated Structures

An aggregated two-way matrix $\mathbf{A} = (a_{ij})$ can be obtained by the methods proposed in Krackhardt (1987, pp. 116–117). These methods take into consideration only parts of the three-way array $\boldsymbol{\Omega}$. For instance, the "union-intersection" methods consider only the entries ω_{iji} and ω_{ijj}, that are compared to determine the entry a_{ij} for each pair (i, j). The relation exists ($a_{ij} = 1$) if at least one of the two persons say it exists (*Union rule*) or if both persons say it exists (*Intersection rule*), otherwise $a_{ij} = 0$.

2.2 Consensus Structures

In this approach for each pair (i, j) an opportune synthesis of the entire vector of perceptions is defined by the corresponding entries in each slice:

$$a_{ij} = f(\omega_{ij1}, \omega_{ij2}, \ldots, \omega_{ijn}). \tag{1}$$

The aggregation function f can be defined in different ways (e.g., Krackhardt 1987). Frequently, for each pair (i, j) it is considered the sum of the corresponding entries in each slice (i.e., the number of persons who perceived that person i is in relation with person j), and the resulting two-way matrix $\mathbf{A} = (a_{ij})$ (possibly symmetrized) is analysed. Okada (2010) proposed a method based on the singular value decomposition of matrix $\mathbf{A}$ to represent graphically centrality in asymmetric social network, and he compared his proposal to conjoint measurement and asymmetric multidimensional scaling (Okada 2011). Freeman (1997) dichotomized matrix **A** applying the rule that for any pair (i, j) fixes the corresponding value equal to one if $a_{ij} > a_{ji}$ and equal to zero otherwise. Then he performed the singular value decomposition of the skew-symmetric component (originally studied by Gower 1977) of the dichotomized matrix, in order to detect a dominance ordering between persons. Matrix **A** can also be dichotomized by the "threshold" rule proposed in Krackhardt (1987). In this case, the mean of the vector $(\omega_{ij1}, \omega_{ij2}, \ldots, \omega_{ijn})$ is compared with a fixed threshold value taking a fractional value from 0 to 1 (e.g., a threshold of 0.5 would be interpreted as meaning that a relation exists from i to j if and only if the mean of the previous vector is greater than 0.5, that is a majority of the persons of the network perceive that it exists). We note that for the "union-intersection" rule and the "threshold" rule the sum of the two entries (i, j) and (j, i) in the dichotomized matrix can assume the values zero, one or two, while in the approach proposed by Freeman (1997) is always equal to zero (presence of a tie) or one.

2.3 Three-Way Structures

In many organizational studies external information concerning members can be available (e.g., professional levels, departments belonging). Researchers could be interested in ascertaining if different patterns of perceived relationships exist in the groups determined by the categories of external variables. In these cases, an aggregation function f_k is separately applied to each group of slices determined by the categories of the external variables, and we obtain a reduced three-way array with entries

$$a_{ijk} = f_k(\omega_{ijk_1}, \omega_{ijk_2}, \ldots, \omega_{ijk_{h_k}}) \tag{2}$$

where persons $(k_1, k_2, \ldots, k_{h_k})$ belong to group k, and we obtain an aggregated $n \times n$ matrix $\mathbf{A}_k = (a_{ijk})$ for each group. We can consider the aggregation function given in (1) as the particular case of the aggregation function given in (2) in the case of only one group of n persons.

Several approaches can be considered to compare the slices $\mathbf{A}_k$ of the three-way array. Given the intrinsically asymmetric structure of each slice, three-way unfolding models seem particularly appropriate to detect and to depict in a diagram a common pattern of perceived relationships between the n persons. In the next section we briefly recall the main features of the unfolding model.

3 Three-Way Unfolding Model

The application of the unfolding model to asymmetric proximity data was proposed by Gower (1977) and Constantine and Gower (1978) for the two-way case, and Bove and Rocci (1999) presented a review of three-way methods for asymmetric three-way scaling, including an application of three-way unfolding to import-export data. In scalar notation the model can be expressed as

$$\gamma_{ijk} = g_k(a_{ijk}) = \sqrt{(\mathbf{x}_i - \mathbf{y}_j)'\mathbf{W}_k(\mathbf{x}_i - \mathbf{y}_j)} + e_{ijk} \tag{3}$$

where: γ_{ijk}'s (also called pseudo-distances) are obtained as monotonic increasing (or decreasing) transformations (by the g_k's) of the entries a_{ijk}; $\mathbf{x}_i, \mathbf{y}_j$ are vectors of coordinates respectively for row i and column j; $\mathbf{W}_k$ is a positive semi-definite diagonal matrix of weights for dimensions associated with slice $\mathbf{A}_k$, and e_{ijk} is an error term. A diagram for the common pattern of relationships (usually named *Common Space*) is obtained by the coordinate vectors, and each person is represented as a row-point (vector $\mathbf{x}_i$) and as a column-point (vector $\mathbf{y}_i$). In this way we can easily analyse asymmetry by comparing distances in both directions (i, j) and (j, i). So interpretation of the diagram focus mainly on distances between the set of rows and the set of columns, and the within-set distances are less relevant. Additionally, the system of dimensional weights provided by the matrices $\mathbf{W}_k$ allows to represent the specific pattern of each slice $\mathbf{A}_k$ in diagrams named *individual spaces*. A disadvantage of the model is that we have a double number of points in the diagrams.

An algorithm for model (3) based on the minimization of a penalized Stress loss function was proposed by Busing et al. (2005), and it was implemented in a computer program called PREFSCAL (now also included in Spss software package). The algorithm allows us to avoid degenerate solutions typical in ordinal and interval unfolding, incorporating in the loss function a penalty based on the coefficient of variation. In the next section an application of the model to data on advice-giving and getting in an organization will be presented.

4 Application

Krackhardt (1987) reported an empirical study concerning relationships among 21 managers occupied in a small manufacturing organization. The analysed relationship (of the type 'A approaches B for help and advice') was submitted to all 21 managers. Each manager was presented with a questionnaire of 21 questions. Each question was contained in a separate page and it asked, "Who would manager X go to for help or advice at work?". The rest of the page consisted of the list of the names of the other 20 managers, and the responded was instructed to put a check mark near the names of the managers to whom manager X was likely to go for advice. So the data consist of 21 square (21×21) matrices, each produced by one manager. Diagonal elements in each matrix were fixed equal to one. External information concerning position level (1 president, 4 vice-presidents, 16 supervisors) and Department belonging (4 Departments) was also available. The

Table 1 Department composition

Department	Vice president	Supervisors
1	21	6,8,12,17
2	14	3,5,9,13,15,19,20
3	18	10,11
4	2	1,4,16

following Table 1 (reproduced by Okada 2011, Table 1, p. 221) provides belonging of vice presidents and supervisors.

The unfolding model (3) was applied to the original 21 matrices and to the 4 matrices obtained by aggregating respect to the Departments, in order to detect differences in the perceptions of the four groups. A matrix conditional interval transformation (without intercept) for similarity data was chosen in PREFSCAL algorithm. Distances are inversely proportional to entries a_{ijk}. Each manager is represented by two points in the diagrams, so graphical interpretation need some caution to distinguish row-points (labels 1,2,3, ...) and column-points (labels m1,m2,m3, ...). The three-way unfolding of the original 21 matrices and the three-way unfolding of the 4 department matrices provide very similar common patterns of perceived relationships between managers (*common spaces*). For reasons of space here we present only the results of the unfolding of the departments.

PREFSCAL algorithm was applied with different number of dimensions. As usually in unfolding analyses we choose the two-dimensional solution, a good balance between the need to analyse relationships in only one diagram and an acceptable data approximation (we also notice that the three-dimensional solution had only a small reduction in the stress functions). The values of the Stress functions (e.g., *Stress-I*=0,30) and the others variation indices, along with the analysis of dimensional weights, allow us to analyse the relationships between the managers in the common space depicted in Fig. 1. In fact, the analysis in the individual spaces (that is the diagrams for the four departments) obtained by model (3) is quite similar to the analysis in the common space. The weights of the first and second dimension are very similar for each source, as it is possible to see in Fig. 2, so distance ratios remain approximately the same in the common and the individual spaces. Thus the comparison between departments allow us to say that the perception of the pattern of relationships between managers is almost the same when analysed aggregated in each department.

Now we focus on distance analysis between row points and column points in the diagram depicted in Fig. 1. First we remark that supervisors column points are located far from the centre of the configuration, where are positioned most of the row points. This means that supervisors are not frequently approached for help and advice. Besides some supervisors appear to be quite isolated because they go to very few people for advice (e.g. manager 12 row point is also far from the centre of the configuration). Row and column points corresponding to the President (label 7) and to three Vice presidents (labels 2, 14, 18) are in a central position in the diagram, that means they are centres of focus for advice. Vice president 21 is depicted more in the direction of the first quadrant. The analysis of the positions of these five column

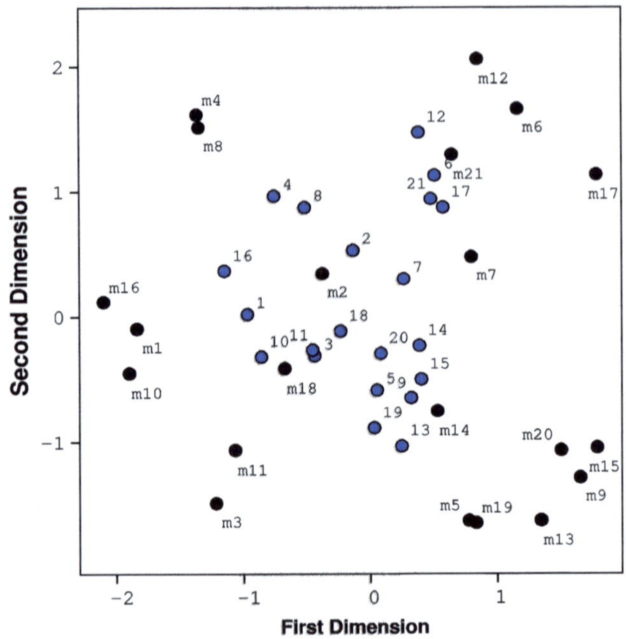

Fig. 1 Three-way unfolding of the four departments (Common space)

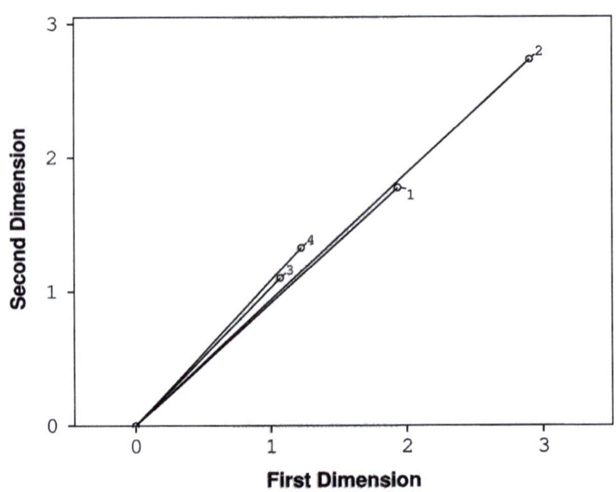

Fig. 2 Three-way unfolding of the four departments (Weights representation)

points allow us to detect a certain degree of hierarchy, where vice president 2 seems approached more frequently for advice respect to the others vice presidents, and he is more linked to the president (manager 7). The asymmetry in the relationship is always favourable to vice president 2. Besides vice presidents 14 and 18 appear more central than vice president 21. We notice that a similar result is reported in Krackhardt (1987, p. 119).

The distances between the managers reflect quite well the pattern of Department belonging, (especially for the large Departments: Dep. 1 in the first quadrant and Dep. 2 in the fourth quadrant), with just a few exceptions (managers 3 and 8). So the position of Vice president 21, close to the managers of Department 1 that he heads up, seems to suggest that his advice is almost exclusively oriented to the managers of Department 1 (a similar result was obtained by Okada 2011).

5 Conclusion

In this paper we showed how external information can be incorporated in the analysis of asymmetric relationships in organizational structures by using three-way unfolding. In the application presented in the previous section external information allowed us to reorganize the original data array in an aggregated three-way array to be submitted to the unfolding procedure. In particular, a department aggregation was considered but other possibilities could be exploited (e.g., professional levels aggregations).

Other approaches incorporating directly external information in the adopted model like asymmetric two-way MDS with external information (e.g., De Leeuw and Heiser 1980; Bove 2006) or three-way scalar product models (e.g., Bove and Rocci 1999) could be considered. An advantage of the unfolding approach is the possibility to represent the common and individual spaces by distances, that are usually easier to analyse respect to scalar products.

References

Bove, G. (2006). Approaches to asymmetric multidimensional scaling with external information. In S. Zani, A. Cerioli, et al. (Eds.), *Data analysis, classification and the forward search* (pp. 69–76). Berlin: Springer.

Bove, G., & Rocci, R. (1999). Methods for asymmetric three-way scaling. In M. Vichi & O. Opitz (Eds.), *Classification and data analysis. Theory and application* (pp. 131–138). Berlin: Springer.

Busing, F. M. T. A., Groenen, P. J. F., & Heiser, W. J. (2005). Avoiding degeneracy in multidimensional unfolding by penalizing on the coefficient of variation. *Psychometrika, 70*, 71–98.

Constantine, A. G., & Gower, J. C. (1978). Graphical representation of asymmetric matrices. *Applied Statistics, 27*, 297–304.

De Leeuw, J., & Heiser, W. J. (1980). Multidimensional scaling with restrictions on the configuration. In P. R. Krishnaiah, et al. (Eds.), *Multivariate analysis V* (pp. 501–522). Amsterdam: North Holland.

Freeman, L. C. (1997). Uncovering organizational hierarchies. *Computational and Mathematical Organization Theory, 3*, 5–18.

Gower, J. C. (1977). The analysis of asymmetry and orthogonality. In J. R. Barra, et al. (Eds.), *Recent developments in statistics* (pp. 109–123). Amsterdam: North Holland.
Krackhardt, D. (1987). Cognitive social structures. *Social Networks, 9*, 109–134.
Okada, A. (2010). Two-dimensional centrality of asymmetric social network. In N. C. Lauro, et al. (Eds.), *Data analysis and classification* (pp. 93–100). Heidelberg: Springer.
Okada, A. (2011). Centrality of asymmetric social network: singular value decomposition, conjoint measurement, and asymmetric multidimensional scaling. In S. Ingrassia, et al. (Eds.), *New perspectives in statistical modeling and data analysis* (pp. 219–227). Heidelberg: Springer.

A Generalized Additive Model for Binary Rare Events Data: An Application to Credit Defaults

Raffaella Calabrese and Silvia Angela Osmetti

Abstract We aim at proposing a new model for binary rare events, i.e. binary dependent variable with a very small number of ones. We extend the Generalized Extreme Value (GEV) regression model proposed by Calabrese and Osmetti (Journal of Applied Statistics 40(6):1172–1188, 2013) to a Generalized Additive Model (GAM). We suggest to consider the quantile function of the GEV distribution as a link function in a GAM, so we propose the Generalized Extreme Value Additive (GEVA) model. In order to estimate the GEVA model, a modified version of the local scoring algorithm of GAM is proposed. Finally, to model default probability, we apply our proposal to empirical data on Italian Small and Medium Enterprises (SMEs). The results show that the GEVA model has a higher predictive accuracy to identify the rare event than the logistic additive model.

Keywords Generalized additive model • Generalized extreme value distribution • Rare event

1 Introduction

Let Y be a binary random variable of parameter π that describes a binary rare event (small number of one than zero). A Generalized Linear Model (GLM) (see Agresti 2002) considers a monotonic and twice differentiable function $g(\cdot)$, called *link function*, such that

R. Calabrese
Essex Business School, University of Essex, Wivenhoe Park, Colchester CO4 3SQ, UK
e-mail: rcalab@essex.ac.uk

S.A. Osmetti (✉)
Università Cattolica del Sacro Cuore di Milano, Largo Gemelli 1, 20123, Milano, Italy
e-mail: silvia.osmetti@unicatt.it

D. Vicari et al. (eds.), *Analysis and Modeling of Complex Data in Behavioral and Social Sciences*, Studies in Classification, Data Analysis, and Knowledge Organization,
DOI 10.1007/978-3-319-06692-9_9,

$$g(\pi) = \eta = \alpha + \sum_{j=1}^{p} \beta_j x_j.$$

By applying the inverse function of $g(\cdot)$, we obtain the response curve

$$\pi = g^{-1}\left(\alpha + \sum_{j=1}^{p} \beta_j x_j\right).$$

Several models for binary response variable have been proposed by considering different link functions $g(\cdot)$: logit, probit, log–log and complementary log–log models. When the binary dependent variable Y is a rare event, the logistic and probit models show relevant drawbacks because a symmetric link function is used (see Calabrese and Osmetti 2013; King and Zeng 2001). In particular, if the link function is symmetric, the response curve approaches zero at the same rate it is approaches one. Instead, the characteristics of the rare events ($Y = 1$) are more informative than those of the others ($Y = 0$), so the probability that the rare event occurs is underestimated.

In order to overcome this drawback and to focus our attention on the tail of the response curve for values close to 1, we choose an asymmetric link function. Since the Generalized Extreme Value (GEV) random variables is used for modelling the tail of a distribution (see Kotz and Nadarajah 2000; Falk et al. 2010), Calabrese and Osmetti (2013) suggested its quantile function as a link function of a GLM

$$\frac{[-ln\pi(x)]^{-\tau} - 1}{\tau} = \alpha + \sum_{j=1}^{p} \beta_j x_j, \tag{1}$$

so the GEV regression model was proposed. The log–log and the complementary log–log link functions are asymmetric link functions used in GLMs (Agresti 2002) and they represent particular cases of the GEV model.

In a GLM the predictor η is assumed to be linear in the covariates. This assumption is not satisfied by many empirical studies, e.g. credit scoring models (Thomas et al. 2002). For this reason in this work we extend the GEV model to a Generalized Additive Model (GAM). We call this model Generalized Extreme Value Additive (GEVA) model.

The present paper is organised as follows. In the next section we describe the characteristic of a GAM. In Sect. 3 we propose the GEVA model and an iterative estimation procedure. Since defaults in credit risk analysis are rare events, in Sect. 4 we apply the GEVA model to empirical data on Italian Small and Medium Enterprises (SMEs) to model their default probability. In particular, the dataset is described and we show that the default probability does not depend on a linear predictor of the covariates. In the Sect. 4.1 the predictive accuracies of the additive logistic regression model and the GEVA model are compared for different percentages of the defaults in the sample.

2 Generalized Additive Model

Hastie and Tibshirami (1990) proposed the Generalized Additive Models (GAMs). These models assume that the mean of the dependent variable depends on an additive predictor through a non-linear link function. GAM extends the GLM by replacing the linear form $\alpha + \sum_{j=1}^{p} \beta_j x_j$, where x_j with $j = 1, 2, \ldots, p$ are the covariates, with the additive form $\alpha + \sum_{j=1}^{p} s_j(x_j)$, where s_j with $j = 1, 2, \ldots, p$ are arbitrary smooth functions. This means that a GAM model is defined as

$$g(\pi) = \eta = \alpha + \sum_{j=1}^{p} s_j(x_j), \quad (2)$$

where $\pi = P(Y = 1|\mathbf{x})$. In the GLMs the relationship between the independent variable and the predictor η is constrained to be linear. Instead, the GAMs do not involve strong assumptions about this relationship, which is merely constrained to be smooth. Furthermore, GAM overcomes the problem of rapidly increasing variance of the non-parametric estimator for increasing dimensionality, called curse of dimensionality (Hastie and Tibshirami 1990).

To estimate the functions $s_1, s_2, \ldots, s_p$, Hastie and Tibshirami (1990) proposed the local scoring algorithm. Estimation of $s_1, s_2, \ldots, s_p$ is accomplished by replacing the weighted linear regression in the adjusted dependent variable regression by an appropriate algorithm for fitting a weighted additive model. The name local scoring derives from the fact that local averaging is used to generalized the Fisher scoring procedure.

3 A New Model for Rare Events: Generalized Extreme Value Additive Model

In a GLM the predictor is assumed to be linear in the covariates. This assumption is not satisfied by many empirical studies, e.g. credit scoring models (Thomas et al. 2002). For this reason we extend the GEV model to a GAM model. In particular, the linear form of the systematic component defined in (1) is replaced by the following additive form

$$\frac{[-ln\pi(x)]^{-\tau} - 1}{\tau} = \alpha + \sum_{j=1}^{p} s_j(x_j). \quad (3)$$

We call this model Generalized Extreme Value Additive (GEVA) model.

The response curve of the GEVA model is so

$$\pi(\mathbf{x}) = exp\left\{-\left[1 + \tau(\alpha + \sum_{j=1}^{p} s_j(x_j))\right]^{-1/\tau}\right\}. \tag{4}$$

To estimate the GEVA model (3), we need to estimate both the functions $s_j(j = 1, 2, \ldots, p)$ and the τ parameter. This means that we need to modify the local scoring algorithm in order to introduce the estimation of the τ parameter in the estimation procedure. We propose the following m-steps iterative procedure:

Modified Local Scoring Algorithm

1. (Initialization) Step $m = 0$:
   Set $\tau^0 \simeq 0$, $\alpha^0 = \ln[-\ln(\overline{y})]$ and $s_1^0(\cdot) = s_2^0(\cdot) = \ldots = s_p^0(\cdot) = 0$
2. (Iterate) Step $m = m + 1$:
   (a) Construct the adjusted dependent variable $Z = \eta^{m-1} + (Y - \pi^{m-1})\dfrac{\partial\eta}{\partial\pi^{m-1}}$ where
   $\eta^{m-1} = \alpha^{m-1} + \sum_{j=1}^{p} s_j^{m-1}(X_j)$
   $\pi^{m-1} = \exp\left\{-[1 + \tau^{m-1}\eta^{m-1}]^{-1/\tau^{m-1}}\right\}$
   $\dfrac{\partial\eta}{\partial\pi^{m-1}} = \dfrac{[-\ln(\pi^{m-1})]^{-\tau^{m-1}-1}}{\pi^{m-1}}.$
   (b) Construct the weights $W^{-1} = \left(\dfrac{\partial\eta}{\partial\pi^{m-1}}\right)^2 V^{m-1}$ where V is the variance of Y.
   (c) Fit an additive model to Z using the backfitting algorithm (see Hastie and Tibshirami 1990) with weights W, get estimated functions $s_j^m(\cdot)$ and α^m.
   (d) To estimate π^m, maximize the conditional log-likelihood function with respect to τ^m with $\tau^m > -0.5$

$$l(\tau^m|\alpha^m, s_1^m, \ldots, s_p^m) = \sum_{i=1}^{n} \left\{ y_i \ln\left[\pi^m(\alpha^m, s_1^m, \ldots, s_p^m; \mathbf{x}_i)\right] + \right.$$
$$\left. + (1 - y_i)\ln\left[1 - \pi^m(\alpha^m, s_1^m, \ldots, s_p^m; \mathbf{x}_i)\right]\right\}$$

   where $\pi^m(\alpha^m, s_1^m, \ldots, s_p^m; \mathbf{x}_i)$ is defined in (4).
3. Iterate m until the change from π^{m-1} to π^m is sufficiently small.

4 An Application to Credit Defaults

SMEs play a very important role in the economic system of many countries (see Altman and Sabato 2006; Ansell et al. 2009) and particularly in Italy (see Fantazzini et al. 2009; Fantazzini and Figini 2009; Vozzella and Gabbi 2010). Since defaults in credit risk analysis are rare events (Basel Committee on Banking Supervision 2005) and the default probability could not depend on a linear predictor of the covariates (Thomas et al. 2002), we apply the GEVA to empirical data on Italian SMEs to model the default probability (PD). Compliant to Basel II, the PD is 1 year forecasted (Basel Committee on Banking Supervision 2005). Therefore, let Y_t be a binary r.v. such that $Y_t = 1$ if a firm is defaulted at time t and let $\mathbf{x}_{t-1}$ be the covariate vector at time $t-1$. In this application we aim at estimating the conditional PD,

$$\pi(\mathbf{x}_{t-1}) = P(Y_t = 1|\mathbf{x}_{t-1}), \tag{5}$$

by comparing the GEVA, the additive logistic and the logistic regression models.

We consider defaulted and non-defaulted Italian SMEs over the years 2005–2009. In particular, since the default probability is 1 year forecasted, the covariates concern the period of time 2004–2008. The database contains accounting data of around 210,000 Italian firms with total asset below 10 millions euro (Vozzella and Gabbi 2010).

In accordance with Altman and Sabato (2006), we apply a choice-based or endogenous stratified sampling to this dataset. In this sampling scheme data are stratified by the values of the response variable. We draw the observations randomly within each stratum defined by the two categories of the dependent variable (1 = default, 0 = non-default) and we consider all the defaulted firms.

In order to model the default event, we choose 16 independent variables that represent financial and economic characteristics of firms according to the recent literature (Altman and Sabato 2006; Ciampi and Gordini 2008; Vozzella and Gabbi 2010). These covariates cover the most relevant aspects of the firm's operations: leverage, liquidity and profitability. We apply the multicollinearity analysis and then we choose the significant variables at the level of 5 % for the PD forecast in the GEVA model.

The significant covariates are:

- *Solvency ratio*: the ratio of a company's income over the firm's total debt obligations;
- *Return on equity*: the amount of net income returned as a percentage of shareholders equity;
- *Turnover per employee*: the ratio of sales divided by the number of employees;
- *Added value per employee*: the enhancement added to a product or service by a company divided by the number of employees;
- *Cash flow*: the amount of cash generated and used by a company in a given period;

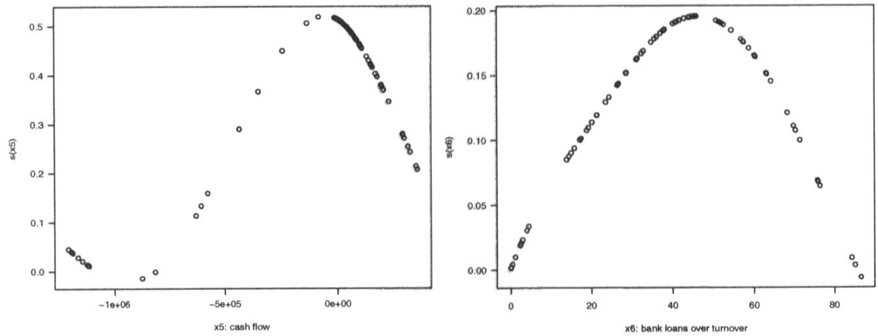

Fig. 1 Estimates of the smooth functions of the covariates "Cash flow" and "Bank loans over turnover"

- *Bank loans over turnover*: short and long term debts with banks over sales volume net of all discounts and sales taxes;
- *Total personnel costs over added value*: the ratio of a company's labour costs divided by the enhancement added to a product or service by a company.

Figure 1 shows that the predictor $\eta = g(\pi)$ has a nonlinear relationship with the covariates "Cash flow" and "Bank loans over turnover". This means that the linearity assumption of a GLM is not satisfied. For this reason we remove this assumption and we choose an additive predictor in the model (3).

4.1 Forecasting Accuracy

We compare the predictive accuracy of the GEVA regression model with the ones of the additive logistic and the logistic regression models. The predictive accuracy is assessed using the Mean Square Error (MSE) and the Mean Absolute Error (MAE). It is more costly to classify an SME as non-defaulter when it is a defaulter than to classify an SME as defaulter when it is a non-defaulter. In particular, when a defaulted firm is classified as non-defaulter by the scoring model, banks will give it a loan. If the borrower becomes defaulter, the bank may lose the whole or a part of the credit exposure. On the contrary, when a non-defaulter is classified as defaulter, the bank only loses interest on loans. For this reason, the identification of defaulters is a pivotal aim for banks internal scoring models. For all these reasons, we compute the MAE and the MSE only for defaulters, denoted MAE^+ and MSE^+.

Another measure used to validate a scoring model is the Area Under the Curve (AUC) (Giudici 2003). This measure is equivalent to averaging the misclassification loss over a cost ratio distribution which depends on the score distributions, so different classifiers are incoherently evaluated using different metrics (Hand 2009).

Table 1 Forecasting accuracy measures for different PDs on the sample (denoted by the subscript "s") and the out-of-time sample (denoted by the subscript "cs")

Sample percentage of defaults	Error	Models		
		GEVA	Additive logistic	Logistic
5 %	MAE_s^+	0.6712	0.8601	0.8829
	MSE_s^+	0.4509	0.7834	0.8171
	MAE_{cs}^+	0.6480	0.9129	0.9302
	MSE_{cs}^+	0.4202	0.8351	0.8467
1 %	MAE_s^+	0.6468	0.9320	0.9502
	MSE_s^+	0.4184	0.9033	0.9270
	MAE_{cs}^+	0.6482	0.9791	0.9820
	MSE_{cs}^+	0.4202	0.9589	0.9684

Table 2 Forecasting accuracy measures for different PDs on the sample (denoted by the subscript "s") and the out-of-sample (denoted by the subscript "cs")

Sample percentage of defaults	Error	Models		
		GEVA	Additive logistic	Logistic
5 %	MAE_s^+	0.8413	0.8609	0.8815
	MSE_s^+	0.7079	0.7833	0.8161
	MAE_{cs}^+	0.8364	0.8462	0.9100
	MSE_{cs}^+	0.6996	0.7729	0.8489
1 %	MAE_s^+	0.6295	0.8609	0.9478
	MSE_s^+	0.3963	0.9052	0.9246
	MAE_{cs}^+	0.6296	0.9156	0.9797
	MSE_{cs}^+	0.3964	0.8880	0.9601

For this reason we do not compute the AUC for analysing the performance of our proposal.

Models were validated on observations that were not included in the sample used to estimate the model. Specifically, we used out-of-sample and out-of-time tests. In the out-of-time approach, we estimate the model on the sample of observations from the years 2005–2008 and we test the model on 1 year default horizon, corresponding to 2009 (out-of-time sample). Table 1 reports the values of the MAE^+ and MSE^+ on the sample (subscript "s") and on the out-of-time sample (subscript "cs"). Instead, in the out-of-sample approach we test the models on a randomly draw 10 % of the observations from the years 2005–2008 (out-of-sample). The models are estimated on the residual 90 % (the sample). Table 2 reports the values of the MAE^+ and MSE^+ on the sample (subscript "s") and on the out of sample (subscript "cs").

In both cases we select a random sample of non-defaulted SMEs over the same year of defaults in order to obtain a percentage of defaults in our sample as close as possible to the default percentage (5 %) for Italian SMEs (Cerved Group 2011). In order to analyze the properties of our model for different percentages of rare events, we consider also a default percentage of 1 %.

The predictive accuracy of the additive logistic model is higher than the one of the logistic model. This means that the systematic component of the model for credit defaults is not a linear predictor.

Since for banks the underestimation of the PD could be very risky, this application shows that the GEVA model overcomes the drawback of the additive logistic regression model in the underestimation of rare events. The GEVA model shows errors lower than those of the logistic model and the additive logistic model for both the sample and the control sample and for both the sample percentages of the rare event. In particular, our model improves its accuracy by reducing the probability of the rare event.

In order to analyse the robustness of the GEVA model we estimate the coefficients of both the regression models on a sample with a given percentage of defaulters and we evaluate the accuracy on a control sample with a different percentage of defaulters. Because the errors do not change, our model is robust for different sample percentages of defaulters. On the contrary, for the additive logistic regression models these errors are significantly different.

Compliant with the expectation, the GEVA model is suitable for rare event analysis because its performance improves by decreasing the probability of the rare event.

5 Conclusions

We propose the GEVA model that removes the linearity assumption of a GLM and overcomes the drawbacks of the logistic additive regression in rare event studies. To estimate the model, we suggest a modification of the local scorig procedure. We investigate the performance of our proposal on data on Italian SMEs for modeling the probability of default. The application shows that the additive logistic regression model underestimates the default probability. On the contrary, the GEVA model overcomes this drawback. A possible extension of this paper could be to improve the estimation procedure and to implement it in a R package. Moreover, we could apply GEVA model to compare the characteristics of SMEs in two different countries.

References

Agresti, A. (2002). *Categorical data analysis*. New York: Wiley.

Altman, E., & Sabato, G. (2006). Modeling credit risk for SMEs: evidence from the US market. *Abacus, 19*(6), 716–723.

Ansell, J., Lin, S., Ma, Y., & Andreeva, G. (2009, August). Experimenting with modeling default of small and medium sized enterprises (SMEs). In *Credit Scoring and Credit Control XI Conference*.

Basel Committee on Banking Supervision. (2005). *International convergence of capital measurement and capital standards: A revised framework*. Basel: Bank for International Settlements.

Calabrese, R., & Osmetti, S. A. (2013). Modelling small and medium enterprise loan defaults as rare events: the generalized extreme value regression model. *Journal of Applied Statistics, 40*(6), 1172–1188.

Cerved Group. (2011, February). Caratteristiche delle imprese, governance e probabilità di insolvenza. Report. Milan.

Ciampi, F., & Gordini, N. (2008). *Using economic-financial ratios for small enterprize default prediction modeling: an empirical analysis*. In Oxford Business & Economics Conference, Oxford.

Falk, M., Haler, J., & Reiss, R. (2010). *Laws of small numbers: extremes and rare events* (3rd ed.). Basel: Springer.

Fantazzini, D., & Figini, S. (2009). Random survival forests models for SME credit risk measurement. *Methodology and Computing in Applied Probability, 11*, 29–45.

Fantazzini, D., Figini, S., De Giuli, E., & Giudici P. (2009). Enhanced credit default models for heterogeneous SME segments. *Journal of Financial Transformation, 25*(N.1), 31–39.

Giudici, P. (2003). *Applied data mining: statistical methods for business and industry*. London: Wiley.

Hand, D. J. (2009). Measuring classifier performance: a coherent alternative to the area under the ROC curve. *Machine Learning, 77*, 103–123.

Hastie, T. J., & Tibshirami, R. J. (1990). *Generalized additive models*. Boca Raton: Chapman & Hall.

King, G., & Zeng, L. (2001). Logistic regression in rare events data. *Political Analysis, 9*, 321–354.

Kotz, S., & Nadarajah, S. (2000). *Extreme value distributions. Theory and applications*. London: Imperial Colleg Press.

Thomas, L., Edelman, D., & Crook, J. C. (2002). *Credit scoring and its applications*. Philadelphia: Society for Industrial and Applied Mathematics.

Vozzella, P., & Gabbi, G. (2010). *Default and asset correlation: An empirical study for Italian SMEs*. Working Paper.

The Meaning of *forma* in Thomas Aquinas: Hierarchical Clustering from the *Index Thomisticus* Treebank

Gabriele Cantaluppi and Marco Passarotti

Abstract We apply word hierarchical clustering techniques to collect the occurrences of the lemma *forma* that show a similar contextual behaviour in the works of Thomas Aquinas into the same or closely related groups. Our results will support the lexicographers of a data-driven new lexicon of Thomas Aquinas in their task of writing the lexical entry of *forma*. We use two datasets: the *Index Thomisticus* (IT), a corpus containing the opera omnia of Thomas Aquinas, and the *Index Thomisticus* Treebank, a syntactically annotated subset of the IT.

Results are evaluated against a manually labeled subset of the occurrences of *forma*.

Keywords Divisive hierarchical clustering analysis • Index Thomisticus

1 Background and Motivation

Started in 1949 by father Roberto Busa (1913–2011), the *Index Thomisticus* (IT; Busa 1974–1980) represents the first digital corpus of Latin and has been a groundbreaking project in computational linguistics and literary computing. The IT contains the opera omnia of Thomas Aquinas (118 texts) as well as 61 texts by other authors related to Thomas, for a total of around 11 million tokens. The corpus

G. Cantaluppi (✉)
Dipartimento di Scienze Statistiche, Università Cattolica del Sacro Cuore, Milano, Italy
e-mail: gabriele.cantaluppi@unicatt.it

M. Passarotti
Centro Interdisciplinare di Ricerche per la Computerizzazione dei Segni dell'Espressione, Università Cattolica del Sacro Cuore, Milano, Italy
e-mail: marco.passarotti@unicatt.it

D. Vicari et al. (eds.), *Analysis and Modeling of Complex Data in Behavioral and Social Sciences*, Studies in Classification, Data Analysis, and Knowledge Organization, DOI 10.1007/978-3-319-06692-9_10,

is morphologically tagged and lemmatised, and it is available on paper, CD-ROM and on-line (www.corpusthomisticum.org).[1]

The *Index Thomisticus* Treebank (IT-TB: http://itreebank.marginalia.it) is an ongoing project that aims at performing the syntactic annotation of the entire IT corpus (McGillivray et al. 2009). The IT-TB is a dependency-based treebank consisting of around 150,000 annotated tokens for a total of approximately 8,000 sentences excerpted from three works of Thomas[2]: *Scriptum super Sententiis Magistri Petri Lombardi, Summa contra Gentiles* and *Summa Theologiae*. This means that the syntactic structure of the sentences is graphically represented by dependency trees, i.e. trees whose nodes are labeled with words, which are connected by hierarchical relations called 'dependencies' labeled with syntactic tags. The IT-TB shares the same annotation guidelines with the Latin Dependency Treebank (LDT), developed by the Perseus Digital Library (Boston, MA) on text of the Classical era. These guidelines resemble those of the Prague Dependency Treebank of Czech (PDT) and are similar to those of the PROIEL corpus of the New Testament in a number of Indo-european languages (Oslo, Norway).

The IT-TB is part of a bigger project named 'Lessico Tomistico Biculturale' (LTB). LTB aims at building a new lexicon of Thomas Aquinas by empirical confrontation with the evidence provided by the IT. Indeed, the entries of the available lexica of Thomas are systematically biased by the criteria for the selection of the examples adopted to describe the different meanings of lemmas. This limitation can now be overcome by exploiting the data of the IT and of the IT-TB.

The first lemma we want to analyse for the purposes of LTB is *forma*. This lemma has 18,357 occurrences in the IT corpus, 16,525 of which in Thomas' works and 1,832 in the texts of other authors. Thus, we devoted the first years of the project to annotate those sentences that feature at least one occurrence of *forma*. Five thousand one hundred and ninety one occurrences of *forma* have been annotated so far in the IT-TB, corresponding to around one third of all the occurrences of *forma* in Thomas' works.

Forma is a 'technical' word in Thomas' writings, showing high polysemy. In the lexicon of Thomas Aquinas (Deferrari and Barry 1948–1949), *forma* has five meanings: (a) 'form, shape', synonym of *figura*; (b) 'form', the configuration of an artificial thing as distinct from 'figure' (which is the configuration of natural things); (c) 'form', the actualizing principle that makes a thing to be what it is; (d) 'mode, manner'; (e) 'formula'. In Latin Wordnet (Minozzi 2008), *forma* has 21 senses, which do not include all those present in Thomas.

[1]The IT was lemmatised manually. Participles were always reduced to verbs unless they feature a separate lexical entry in the Latin dictionary provided by Forcellini (1771; extended in 1896 by R. Klotz, G. Freund & L. Doderlein); for instance, the word *falsus* is always lemmatised as a form of the adjective *falsus* and not of the verb *fallo*. Disambiguation of the homographs is partly available in the IT and it is completed in the *Index Thomisticus* Treebank.

[2]Sentences in the IT-TB were splitted automatically by strong punctuation marks (period, colon, semicolon, question mark, exclamation mark). At times, manual modifications of automatic sentence splitting were made by annotators.

2 Contribution

We apply word hierarchical clustering techniques to cluster the occurrences of *forma* in both the IT and the IT-TB, so that occurrences showing similar behaviour fall in the same cluster(s) (Pedersen 2006). The results of our work will support the LTB lexicographers in their task of writing the lexical entry of *forma*. Indeed, collecting the occurrences of *forma* into contextually homogeneous groups will provide lexicographers with an efficient managing of the occurrences based on all data available. This will allow both to verify previous intuition-based assumptions about the meanings of *forma* on the evidence provided by data and to bring to light further meanings (or refinements of already known ones) that may be overlooked by previous intuition-based studies.

Our theoretical starting point is the notion of 'context of situation' (Firth 1957), pointing out the context-dependent nature of meaning, as reported in Firth's famous quotation: 'You shall know a word by the company it keeps'.

We produced two matrices of data, one from the IT-TB and one from the IT:

- a matrix (A) consisting of 5,191 observations. For each occurrence of *forma* in the IT-TB, one observation (organised into 14 columns) report the lemmas of: (a) its parent and grandparent in the dependency tree: columns 1–2, respectively including 105 and 104 categories; (b) up to 4 attributives (dependent nodes with syntactic label 'Atr' in the tree): columns 3–6, including 128 categories in total; (c) up to 4 coordinated nodes in the tree: columns 7–10, including 302 categories in total; (d) up to 2 words preceding and 2 words following the occurrence of *forma* concerned in the observation: columns 11–14, including 31 categories in total.
  While (a), (b) and (c) report information extracted from the IT-TB (i.e. syntactic information), (d) features information concerning the linear word order of the text (taken from the IT);
- a matrix (B) consisting of 18,357 observations organized into 6 columns. For each occurrence of *forma* in the IT, observations report the lemmas of up to 3 words preceding (columns 1–3: 271 categories) and 3 words following (columns 4–6: 227 categories) the occurrence of *forma* concerned.

We carried out a DIvisive hierarchical clustering ANAlysis (Kaufman and Rousseeuw 1990) by using the function DIANA (Maechler et al. 2012), available in the package 'cluster' of R (R Core Team 2012), starting from a dissimilarity matrix generated by considering a modification of the *simple matching distance* (Sokal and Michener 1958): such a modification is needed because we deal with groups of variables instead of simple ones. For each pair of observations, r and s, with categories $(x_{r1}, x_{r2}, \ldots, x_{rk})$ and $(x_{s1}, x_{s2}, \ldots, x_{sk})$ over k variables we computed their similarity rate in groups of variables, e.g. for the 2 words preceding and 2 words following the occurrence of *forma*. This implies the comparison of the group of values $(x_{r11}, \ldots, x_{r14})$ with the group of values $(x_{s11}, \ldots, x_{s14})$ (and not for each value as specified by the simple matching distance, involving comparisons e.g. separately from x_{r7} with x_{s7} to x_{r10} with x_{s10}). Thus, we defined:

$$\text{diss}(r,s) = \begin{cases} 1 - \frac{1}{\text{sim}_{\max}} \min\left(\sum_{g=1}^{G} \text{sim}(x_{rg}, x_{sg}), \sum_{g=1}^{G} \text{sim}(x_{sg}, x_{rg})\right), & r \neq s \\ 0, & r = s \end{cases} \quad (1)$$

where G is the number of groups of variables; $\text{sim}(x_{rg}, x_{sg})$ is an asymmetric measure for the number of elements in the s observation matching with the elements in the r observation for group g (since multiple occurrences of the same term may occur into one observation in a group); and

$$\text{sim}_{\max} = \max_{r,s}\left(\min\left(\sum_{g=1}^{G} \text{sim}(x_{rg}, x_{sg}), \sum_{g=1}^{G} \text{sim}(x_{sg}, x_{rg})\right)\right)$$

is the overall observed maximum number of matches.

The dissimilarity function (1) has the following properties: $\text{diss}(r,s) \geq 0$ and $x_r = x_s \rightarrow \text{diss}(r,s) = 0$, i.e. if the groups of the two observations r and s feature the same elements, they have null dissimilarity. The dissimilarity function is symmetric $\text{diss}(r,s) = \text{diss}(s,r)$. According to the first part of (1), the distance between element r and itself is 0 only if two observations feature exactly the same elements in each corresponding group. This happens in case of two identical sequences of words. So, we defined $\text{diss}(r,r) = 0$.

We performed several experiments on the matrices produced from the IT-TB and the IT, by changing the settings for grouping the variables used by the clustering algorithm. In particular, two groups and four groups of variables were considered. Moreover, we studied the effects of excluding specific kinds of words (like function words, pronouns and some verbs) when computing the dissimilarity between observations.

3 Evaluation and Results

In order to evaluate our results, we built two different gold standards.

- Gold standard A (GsA): we manually annotated the meaning of 672 randomly chosen occurrences of *forma* (approx. 13 % of the total in the IT-TB). We used a tagset featuring ten different values ('semantic tags'; see Table 1) that were defined according to Deferrari and Barry (1948–1949) and Minozzi (2008), Latin Wordnet and lexico-syntactic information from the IT-TB;
- Gold standard B (GsB): among the observations of GsA, we selected a subset of 357 featuring a clear (i.e. not ambiguous and easy-to-detect) meaning of *forma*.

Table 1 Evaluation tagset

Tag	Description	No. occurrences	Clear
1	*forma* as *substantia, principium, essentia*; *forma inhaerens, substantialis, subsistens*	231	97
2	*forma corporis, mentis, hominis*	165	88
3	*forma artis, artificiati*	17	10
4	*forma naturalis, speciei*	45	25
5	*forma praedicati*	17	1
6	*forma materialis*	106	75
7	*forma* as '*shape/figure*' (*forma domus*)	16	9
8	*forma accidentalis*	23	17
9	*forma* (*formula*) *baptismi/sacramenti*	50	34
10	*forma coniuncta/participata/participabilis*	2	1

We evaluated our results by using the following evaluation metrics (Van Rijsbergen 1979): precision, recall and f-score, respectively defining the proportion of retrieved material that is actually relevant (that is the distribution of the tags in each cluster), the proportion of relevant material actually retrieved in answer to a search request (in our case the distribution of clusters pertaining each tag), and the harmonic mean of precision and recall (referring thus to each cluster and tag combination). Precision and recall are here defined according to each semantic tag and each cluster. So, for instance, let's consider one cluster consisting of 100 observations (i.e. 100 occurrences of *forma*), 80 out of which have the same semantic tag in the gold standard. This means that the precision of that tag in that cluster is 0.80; conversely, if in the gold standard the total number of occurrences labeled with that specific semantic tag is 80, that tag in that cluster has recall 1.

The best performing setting on 5,191 observations is the following:

- function words, pronouns and verb *sum* excluded from computing dissimilarity;
- grouping setting: two separate groups, namely syntactic information and textual information.

With this setting we reached the best f-score of 0.93 (precision 0.95; recall 0.90) for the GsB observations tagged with label 6 (*forma materialis*; *forma* connected with *materia*). The GsB observations tagged with other labels show lower f-scores (ranging from 0.86 to 0.5). If GsA observations are concerned, the f-score ranges from 0.8 (precision: 0.93; recall: 0.7) for label 6, to 0.28 for label 2. As the analyses made on the IT-TB show better results than on the IT, we can conclude that the syntactic annotation of the corpus provides a positive contribution to the clustering process. Further, we observed that the presence of specific categories of words like function words, pronouns and verbs represents a confounding element.

Table 2 reports the best results achieved with the above settings.

In the columns featuring precision and recall, we report in round parentheses a number corresponding to one label used in the evaluation tagset. For instance, the best precision rate on GsA (matrix A) concerns those observations that are tagged with label 2.

Table 2 Best results

	Precision	Recall
GsA (matrix A)	0.9545 (2)	0.8235 (5)
	PM: 0.9787 (2)	PM: 1 (5/8)
GsA (matrix B)	0.9444 (6)	0.6522 (8)
GsB (matrix A)	0.9687 (6)	0.8933 (6)
		PM: 0.9067 (6)
GsB (matrix B)	0.9375 (8)	0.8267 (6)

In a number of experiments, we also allowed partial matches (PM) between variables. When full matches are allowed, objects are considered equal to themselves only (for instance, the lemma *baptisma* is equal to the value *baptisma* only). Instead, when partial matches are allowed, objects are considered equal to a set of values (for instance, the lemma *baptisma* is equal not only to *baptisma*, but also to *baptismalis*, *baptizo* etc.). For each lemma reported in the input matrices, we built (semi-automatically) a set of lemmas sharing the same morphological root, which are considered equal if partial matches are allowed.

Plotting the results of hierarchical clustering in dendrograms allows the distribution of clusters to be visually checked. In particular, this is helpful to verify if those observations annotated with the same label in the gold standard do indeed appear close to each other in the dendrogram. Figure 1 reports the dendrogram showing the results on GsB (matrix A).

Most of the observations labeled with tag 6 occur within the bracketed cluster. Those elements (leaves) that share a common ancestor-branch at distance 0 have the maximum similarity, i.e. the maximum number of common words (6) across groups of variables. Elements that share a common ancestor-branch at distance 1/6 differ for one word; elements at distance 2/6 for two words, and so on.

As a number of observations shows no (or very low) similarity with others, this implies the presence of orphan branches in the dendrogram. Such observations are informative, as the frequency of some semantic tags (for instance $3, 5, 7$ and 10) is very low in our data.

One problematic issue is to decide the height where the dendrogram has to be cut for evaluation purposes. Indeed, this affects the number of clusters considered in the evaluation process and, thus, the precision and recall rates.

In our experiments we pursued an heuristic approach, aimed at best balancing precision and recall. Cutting the dendrogram in higher position implies a lower number of clusters and higher recall rates. This happens because observations are not spread over a higher number of clusters. Instead, if the dendrogram is cut at lower position, more clusters and higher precision rates are achieved. We performed the evaluation process by cutting the dendrograms at height 0.95, which looks like the most suitable solution to balance precision and recall rates.

Results pertaining the distribution of recall can be visualised graphically for each semantic tag by means of Lorenz curves, which summarise the concentration of one semantic tag across all clusters (Fig. 2).

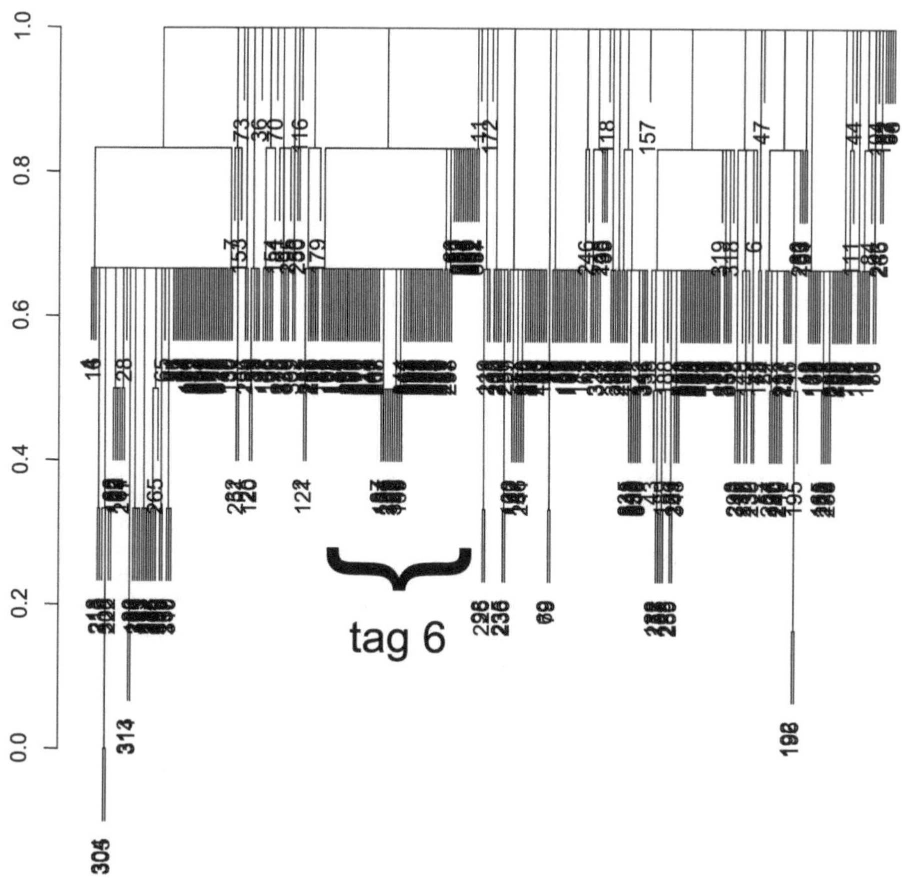

Fig. 1 Hierarchical clustering dendrogram for GsB on matrix A (357 elements featuring an easy-to-detect meaning of *forma*)

4 Conclusion and Future Work

Looking at results, we can conclude the following: (a) while calculating the dissimilarity between observations, excluding function words and pronouns leads to higher results; (b) excluding the verb *sum* leads to higher results on GsA, while excluding all verbs leads to higher results on GsB; (c) recall rates on matrix B are much lower than on matrix A; precision rates on matrix B are fairly lower than on matrix A; (d) results on matrix A are always higher than on matrix B. Our general conclusion is that syntactic metadata (available in matrix A, but not in matrix B) are indeed helpful for clustering purposes, particularly if recall is concerned. We must consider this aspect in our future work, when we will have to decide which metadata from annotation we should use while applying other clustering techniques, as for

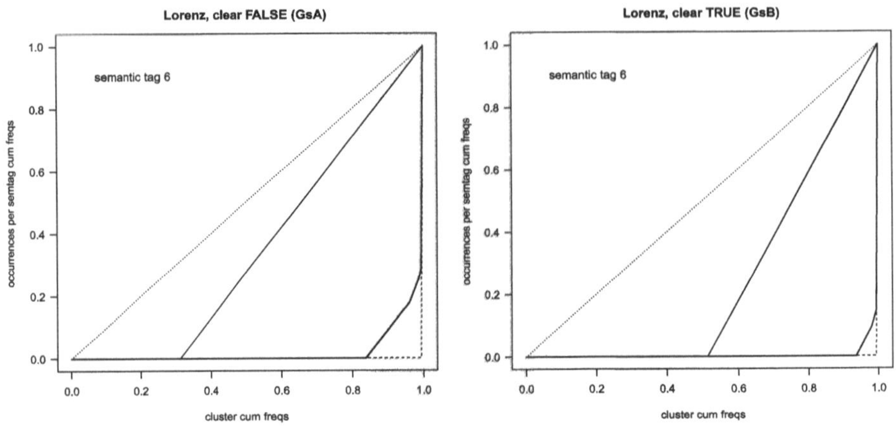

Fig. 2 Lorenz curves for semantic tag 6. Concentration structure of recall in both the observed situation (*thick line*) and the two extreme situations [i.e. no concentration (*thin line*) and maximum concentration (*dashed line*)]. Gini concentration indices are 0.9432 on GsA and 0.9827 on GsB. Low values of recall hold across all clusters when the occurrences of a semantic tag are uniformly distributed over clusters, that is in case of no concentration of the occurrences. Only in case of an integer value, say μ, for the average number of occurrences per cluster, the Lorenz curve representing the situation of no concentration is a *straight line*. Otherwise, it consists of two joint segments, whose slopes are $[\mu]/\mu$ and $([\mu] + 1)/\mu$, where $[\cdot]$ is the floor function returning the integer part of a number

instance Latent Semantic Analysis. We also would like to exploit Latin Wordnet in order to perform word sense disambiguation and associate sense labels to clusters.

In the near future we plan to search for automatic selection criteria aimed at best cutting the dendrogram, in order to manage the variability of precision and recall rates across clusters and semantic tags.

References

Busa, R. (1974–1980). *Index Thomisticus*. Stuttgart-Bad Cannstatt: Frommann-Holzboog

Deferrari, R. J., & Barry, M. I. (1948–1949). *A Lexicon of St. Thomas Aquinas: based on the Summa Theologica and selected passages of his other works*. Washington, DC: Catholic University of America Press

Firth, J. R. (1957). *Papers in linguistics 1934–1951*. London: London University Press.

Forcellini, A. (1771). Totius Latinitatis lexicon, consilio et cura Jacobi Facciolati opera et studio Aegidii Forcellini, lucubratum, typis Seminarii, Patavii.

Kaufman, L., & Rousseeuw, P. J. (1990). *Finding groups in data: an introduction to cluster analysis*. New York: Wiley.

Maechler, M., Rousseeuw, P. J., Struyf, A., Hubert, M., & Hornik, K. (2012). *Cluster: Cluster analysis basics and extensions. R package version 1.14.3*. http://CRAN.R-project.org/package=cluster.

McGillivray, B., Passarotti, M., & Ruffolo, P. (2009). The *Index Thomisticus* treebank project: Annotation, parsing and valency lexicon. *Traitement Automatique des Langues, 50*(2), 103–127.

Minozzi, S. (2008). La costruzione di una base di conoscenza lessicale per la lingua latina: Latinwordnet. In G. Sandrini (Ed.), *Studi in onore di Gilberto Lonardi* (pp. 243–258). Verona: Fiorini.

Pedersen, T. (2006). Unsupervised corpus-based methods for WSD. In E. Agirre & P. Edmonds (Eds.), *Word sense disambiguation: algorithms and applications* (pp. 133–166). New York: Springer.

R Core Team (2012). *R: a language and environment for statistical computing*. Vienna: R Foundation for Statistical Computing. ISBN: 3-900051-07-0. http://www.R-project.org/.

Sokal, R. R., & Michener, C. D. (1958). A statistical method for evaluating systematic relationships. *University of Kansas Science Bulletin, 38*, 1409–1438.

Van Rijsbergen, 'Keith' C. J. (1979) *Information retrieval*. London: Butterworths

The Estimation of the Parameters in Multi-Criteria Classification Problem: The Case of the Electre Tri Method

Renato De Leone and Valentina Minnetti

Abstract In this work we will address the estimation of the parameters of the well-known Electre Tri method, used to model the ordinal sorting problem. This is a multi-criteria classification problem, in which classes are in the strict preference relation. The parameters are profiles, thresholds, weights, and cutting level; they are linked each other either directly or indirectly with mathematical relations. We propose a new procedure composed of two phases, taking into account that the core of the analysis is the profiles estimation made by linear programming problem.

Keywords Multicriteria Classification • Electre Tri Method • Artificial Intelligence

1 Introduction

Given a finite set of units and a finite set of classes, the aim of a classification problem, in general, is to assign units to a class, according to their known characteristics. Note that the problem, as stated, is not necessarily of statistic nature. Many real problems can be formulated by defining a set of criteria, which evaluate alternatives (actions, units, objects, projects) performances. When modelling a real world decision problem using multiple criteria decision aid, several problems need to be taken into account (Roy 1996). They differ in the way alternatives are considered and in the type of results expected from the analysis. Roy (1996)

R. De Leone (✉)
School of Science and Technology, University of Camerino, Camerino, Italy
e-mail: renato.deleone@unicam.it

V. Minnetti
Department of Statistics, Sapienza University of Rome, Rome, Italy
e-mail: valentina.minnetti@uniroma1.it

D. Vicari et al. (eds.), *Analysis and Modeling of Complex Data in Behavioral and Social Sciences*, Studies in Classification, Data Analysis, and Knowledge Organization,
DOI 10.1007/978-3-319-06692-9_11,

distinguishes among four issues: choice, sorting, ranking, and describing, in order to guide the analyst in structuring the decision problem. In this work we will address to the sorting issue, which consists in assigning the alternatives into pre-defined preference categories (i.e. classes). It can be formulated as a classification problem that utilizes multi-criteria *preference information (p.i.)* provided by a Decision Maker (DM). We focus our attention on methods using the outranking relation, since they allow to deal with incompatibilities arising between actions. Several methods using the outranking approach have been proposed in literature to solve the Multiple Criteria Sorting Problem (MCSP): Trichotomic Segmentation (Moscarola and Roy 1977; Roy 1981), N-Tomic (Massaglia and Ostanello 1991), ORClass (Larichev and Moskovich 1994; Larichev et al. 1988), Electre Tri (Roy and Bouyssou 1993; Yu 1992a, 1992b). Moreover, filtering methods based on concordance and non-discordance principles have been studied in Perny (1998). Finally, rough sets theory (Greco et al. 1998; Pawlak and Slowinski 1994; Slowinski 1992) has also allowed significant progress in this field.

In the following paragraph, the Electre Tri method is presented and then the proposed new two-phase procedure is described in short.

2 The Electre Tri Method

Given p categories and a finite set of n alternatives $A = \{a_1, a_2, \dots, a_n\}$ evaluated on m criteria $g_1, g_2, \dots, g_m$, the sorting issue consists in assigning each alternative to one of the predefined ordered categories $C_1, C_2, \dots, C_p$, limited by $p-1$ profiles $b_1, b_2, \dots, b_{p-1}$. Without any loss of generality, we assume that preferences increase with the value on each criterion (gain criterion). Therefore, C_1 and C_p are the worst and the best categories. The generic profile b_h is the upper limit of category C_h and lower limit of category C_{h+1}. The profiles b_0 and b_p are often considered constant for all criteria as shown in the figure below (Fig. 1); we assume $b_0 = -\infty$ and $b_p = +\infty$.

The assignment of an alternative to a specific category follows from the comparison of it with the profiles values on all criteria. The comparison is realized using the outranking relation, indicated with S, as explained below.

Referring to a generic alternative a and to a profile b_h, the outranking relation validates or invalidates the assertion "*a outranks* b_h" that is "a is at least as good as b_h" (aSb_h). The validation of this assertion is made through the computation of four indices (Mousseau and Slowinski 1998):

1. the partial concordance indices $c_j(a, b_h), \forall j = 1, \dots, m$;
2. the global concordance index $c(a, b_h)$;
3. the partial discordance indices $d_j(a, b_h), \forall j = 1, \dots, m$;
4. the credibility indices $\sigma(a, b_h)$.

For the computation of the partial concordance indices, it is necessary to known the profiles, preference and indifference thresholds values. The global concordance

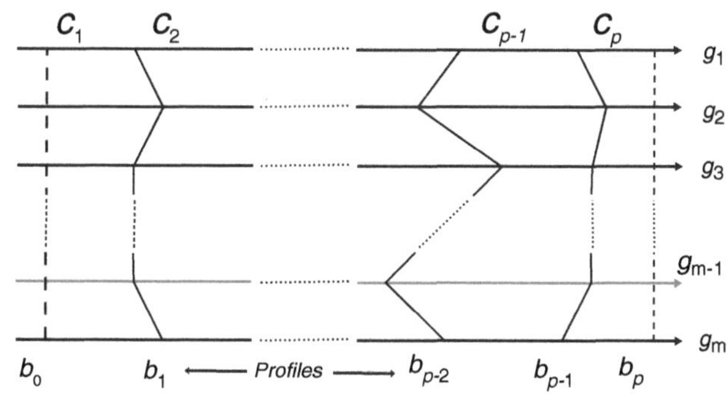

Fig. 1 Scheme of an ordinal sorting problem with m criteria and p categories

index is computed as the weighted average of the partial concordance indices multiplied by the weights, representing the importance of the criteria. If veto thresholds are not considered, the partial discordance indices cannot to be computed. Since the credibility index corresponds to the global concordance index weakened by eventual veto effects, if veto thresholds do not enter in the model, then the credibility index is equal to the global concordance index.

In order to define the outranking relation S between alternative a and profile b_h, it is necessary to define the minimum credibility index value, indicated with λ: $\sigma(a, b_h) \geq \lambda \Rightarrow aSb_h$. Finally, the assignment will be based on one of the two exploitation procedures, which are:

- the pessimistic procedure: compare the alternative a with the highest profile b_{p-1}, then with the next one b_{p-2}, etc. until the relation aSb_h is validated; then the alternative a is assigned to the category on the right side of profile b_h;
- the optimistic procedure: compare the alternative a with the lowest profile b_1, then with the next one b_2, etc. until the relation $b_h \succ a$ (i.e. b_hSa and $\neg aSb_h$) is validated; then the alternative a is assigned to the category on the left side of the profile b_h.

It may happen that divergence exists among the results of the two assignment procedures due to incomparability between the examined alternative and one or several profiles. In particular, in such case the pessimistic assignment rule assigns the alternative to a lower category than the assignment made by the optimistic one.

3 The New Two-Phase Procedure

3.1 The Motivations

One of the main difficulties that an analyst must face when interacting with a DM is the elicitation of the parameters that in the Electre Tri method are: profiles, weights, cutting level, preference, indifference and veto thresholds. In the general case, it could be difficult to fix directly their values and to have a clear global understanding of the implications of their values on the output (Mousseau and Slowinski 1998).

However, elicitation of *p.i.* and technical parameters is a crucial stage in a MCDA models and it may be rather difficult, requiring significant cognitive effort. Difficulties in eliciting *p.i.* may arise from imprecise, partial, or incomplete data, uncertainty of the DM on the correct values of some parameters, etc. Therefore, it is more realistic to assume that the DM provides some Assignment Examples (A.E.s) rather than to fix directly the value of the parameters. A generic A.E. will be denoted with $a \rightarrow C_h$ that indicates the DM assigns the alternative a to the category C_h. The A.E.s set defines the training set and must be sufficiently large and representative of the entire "universe", i.e., it must contain "enough" information. The empirical behaviour of the inference procedure was studied in Mousseau et al. (2001) whose experiments showed that the number of A.E.s equal to the double of the number of criteria is sufficient. The A.E. set represents a part of the holistic information, provided by DM, which constitutes the *p.i.* Inferring a form of knowledge from *p.i.* provided by the DM is a typical approach in *artificial intelligence*.

In order to infer the values of the parameters, Mousseau and Slowinski (1998) formulated an interesting methodology, in which the core is the solution of a Non Linear Programming (NLP) problem, based on a certain form of regression on A.E.s. The problem has $3mp + m + 2n + 1$ variables and $4n + 3mp + 2$ constraints, where m is the number of criteria, p the number of profiles, n the number of A.E.s. The main advantage of this approach is that all the parameters are estimated at the same time; the disadvantage is that a nonlinear, non convex and nondifferentiable programming problem must be solved. Both the non convexity and the no differentiability are caused by the partial concordance index function. The implication is that the classical gradient techniques cannot be used. To overcome these difficulties, the authors suggested to approximate the partial concordance indices by a sigmoidal functions. Moreover, standard solution techniques only determine a local optimal solution while, instead, a global optimal solution is needed. In this direction the ides is to formulate a simpler problem than the proposed NLP one.

To better understand the how the new proposed two-phase procedure was born, we start to comment the results of the illustrative example proposed in Mousseau and Slowinski (1998).

Given a set of six alternatives $A = \{a_1, a_2, a_3, a_4, a_5, a_6\}$ evaluated on three gain criteria $G = \{g_1, g_2, g_3\}$, and three categories C_1, C_2, C_3, the matrix of the performances is:

$$\begin{aligned}
a_1 &= [\ 70 \quad 64.75 \quad 46.25\ ]' \\
a_2 &= [\ 61 \quad 62 \quad 60\ ]' \\
a_3 &= [\ 40 \quad 50 \quad 37\ ]' \\
a_4 &= [\ 66 \quad 40 \quad 23.125\ ]' \\
a_5 &= [\ 20 \quad 20 \quad 20\ ]' \\
a_6 &= [\ 15 \quad 15 \quad 30\ ]'
\end{aligned}$$

The A.E.s set is composed by all the alternatives, constituting the training set necessary to build the assignment model:

$$A^* = \{a_1 \to C_3,\ a_2 \to C_3,\ a_3 \to C_2,\ a_4 \to C_2,\ a_5 \to C_1,\ a_6 \to C_1\}$$

The NLP problem results (Mousseau and Slowinski 1998) lead to a model with 0 misclassified alternatives with respect to the A.E.s; that is the model best fits the A.E.s, provided by DM. In order to solve the NLP problem, the authors Mousseau and Slowinski (1998) suggested to start with the following *starting point*:

- profiles:

$$b_1^{SP} = [35.25\ \ 31.25\ \ 27.5]',$$
$$b_2^{SP} = [59.25\ \ 54.19\ \ 41.6]'.$$

- indifference thresholds:

$$q_1^{SP} = [1.762\ \ 1.563\ \ 1.375]',$$
$$q_2^{SP} = [2.960\ \ 2.71\ \ 2.08]'.$$

- preference thresholds:

$$p_1^{SP} = [3.524\ \ 3.126\ \ 2.750]',$$
$$p_2^{SP} = [5.92\ \ 5.42\ \ 4.16]'.$$

- weights: $w^{SP} = [1\ \ 1\ \ 1]'$.
- cutting level: $\lambda^{SP} = 0.75$.

These *starting point* values correspond to profiles obtained by what the authors called the *heuristic rule*: the profiles are the average points of the centroids of contiguous categories; the thresholds are fixed arbitrarily (5 % for indifference thresholds, 10 % for preference thresholds); the weights are the same for all criteria and the cutting level λ is fixed to 0.75. With these values, Electre Tri leads to a model with 1 misclassified alternative: the model assigns a_4 to the category C_1 instead of category C_2, as preferred by the DM. This result is a logic consequence of a wrong choice of the fixing values (De Leone and Minnetti 2011). In order to obtain a model with 0 misclassified alternatives, with respect to the A.E.s set, if the DM requires equal weights for all criteria, then the cutting level value must be chosen in the interval $(0.33, 0.67]$. As a consequence, the values outside the interval lead to a model with one or more misclassified alternatives. This interval $(0.33, 0.67]$ is obtained by solving a system of nonlinear inequalities (De Leone and Minnetti 2011), developed after the computation of global concordance indices and before the exploitation procedure.

3.2 The New Two-Phase Procedure

In Mousseau and Slowinski (1998) a very interesting linear relation between thresholds and profiles is used that will be included in the procedure proposed in this work.

However, there is another interesting relation, which derives automatically from the mathematical structure of the Electre Tri method. The knowledge of the cutting level allows moving from the credibility index to the outranking relation S using nonlinear inequalities. For all alternatives of training set, in order to find all the possible outranking relations, a system of nonlinear inequalities is defined and solved. In absence of the veto thresholds, the nonlinear relation between cutting level and weights is determined by solving a system of nonlinear inequalities (De Leone and Minnetti 2011).

The linear relation between thresholds and profiles and the nonlinear relation between cutting level and weights suggested us to formulate the new procedure composed of two phases. The first is dedicated to profiles and thresholds estimations and the second one to the weights and cutting level estimations.

The core of the new two-phase procedure is the profiles estimation achieved by a Linear Programming (LP) problem that utilized the information from the training set (A.E. set) (De Leone and Minnetti 2011)

Suppose that the alternative a_k is assigned to the category C_h by DM ($a_k \to C_h$).

Let $g_j(a_k)$ be the value of the alternative a_k on the jth criterion and let $g_j(b_h)$ be the hth profile on the jth criterion. If

$$g_j(b_{h-1}) \leq g_j(a_k) \leq g_j(b_h)$$

the alternative is correctly classified. Otherwise, $\theta_j(a_k)$ is the error in satisfying the above inequalities (the error in classification).

The objective function (to be minimized) is the sum of these errors on all alternative and all the criteria. The LP problem is:

$$\begin{array}{lll}
\min & z = \sum_{j=1}^{m} \sum_{a_k \to C_h} \theta_j(a_k) & \\
\text{subject to} & \theta_j(a_k) \geq g_j(a_k) - g_j(b_h) & \forall j = 1, \ldots, m, \forall a_k \to C_h \\
& \theta_j(a_k) \geq g_j(b_{h-1}) - g_j(a_k) & \forall j = 1, \ldots, m, \forall a_k \to C_h \\
& g_j(b_h) \geq g_j(b_{h-1}) & \forall j = 1, \ldots, m, \forall h = 2, \ldots, p-1 \\
& \theta_j(a_k) \geq 0 & \forall j = 1, \ldots, m, \forall a_k \to C_h
\end{array}$$

The first group of constraints do not exist for $h = p$ since we have assumed $b_p = +\infty$; similarly the second group of constraints do not exist for $h = 1$ since $b_0 = -\infty$. Let $|C_p|$ be the cardinality of category p and $|A^*|$ the cardinality of the training set. For the above LP, the total number of variables is $m|A^*| + m(p-1)$ and the number of constraints is equal to $(|C_1| + 2|C_2| + \ldots + 2|C_p - 1| + |C_p|)m + (p - 2)m + |A^*|m$. They linearly increase with the cardinality of the A.E.s set.

Note that the feasible region is always not empty, and the LP problem has always an optimal solution. Without going into too much detail, to find feasible solutions corresponds to find variation intervals (v.i.) for each profile on each criterion as:

$$\max_{a_v \to C_h} \{g_j(a_v) - \theta_j(a_v)\} \leq g_j(b_h) \leq \min_{a_s \to C_{h+1}} \{g_j(a_s) + \theta_j(a_s)\} \quad \forall j, \forall h = 1, \ldots, p-1$$

Turning now to the objective function, the optimal objective function value is 0 when all errors are equal to zero, i.e., when all alternatives of the A.E.s set (the training set) belong to their own category as indicated above:

$$g_j(a_k) \in [g_j(b_{h-1}), g_j(b_h)], \quad \forall j = 1, \ldots, m, \forall\, a_k \to C_h.$$

The second phase of the proposed procedure corresponds to the definition and solution of a system of nonlinear inequalities in order to estimate both the cutting level and the weights (De Leone and Minnetti 2011). In the case the system is inconsistent, we may always find a model which assigns the alternatives, because the constraints which causes the inconsistency would be deleted. But in this case the model will contain a number of misclassified alternatives equal to the number of deleted constraints in the system (De Leone and Minnetti 2011, 2012). As regard as to inconsistencies, many authors (Mousseau et al. 2004, 2003) proposed a Mixed Integer Linear Programming and LP problems to delete or revise the A.E.s, which causes the inconsistencies. Both weights and cutting level values can be estimated using the algorithm recently proposed by Giampaolo Liuzzi and Stefano Lucidi (2008) with the variant that in this case the derivatives are available.

4 Conclusion

The Electre Tri method uses parameters to built a classification model which assigns all the alternatives to one of the predefined categories. However, a direct elicitation of the parameters is not easy to obtain; therefore, several authors (Dias et al. 2000; Mousseau et al. 2001; Mousseau and Slowinski 1998) formulated mathematical programming problems in order to estimate these parameters. More specifically, in Mousseau and Slowinski (1998) a NLP problem is formulated in order to estimate all the parameters; the disadvantage of this procedures is that the values are linked by intrinsic mathematical relations, creating an "intrinsic chain". If the DM wants to add a constraint on one parameter and/or to fix a parameters value, then the NLP problem will provide solutions, always different from the initial. As a consequence the DM is constrained to accept the results, even when he may have precise ideas on the values to be assigned to some parameters. To dissolve this chain means to adopt a methodology composed of multiple phases, which can guarantee more precise results on each single parameter. In addition, each phase of the procedure has to plan for managing the worst case in which the DM have not any ideas about the parameters values. Moreover, the analyst has to suggest how to do in the case a phase fails. We proposed a two-phase procedure to estimate all the parameters (veto thresholds included): the first phase is dedicated to both profiles and thresholds estimations and the second to the weights and cutting levels estimations (De Leone and Minnetti 2011). With two phases the DM keeps better under control the parameters about their variations and he is able to choose some parameters—as weights—, taking into account their implications in the model. Only the profiles have to be inferred in an objective way; they are estimated by means of LP problem, which is the core of our procedure. Moreover, if the LP problem provides multiple optimal solutions, we suggest to use a differential evolutionary algorithm (De Leone and Minnetti 2012). While in the case the LP "fails", then we suggest a local search algorithm (De Leone and Minnetti 2012). In other words, since the LP problem cannot fail but the second phase can fail due to the inconsistence of the system of nonlinear inequalities, the LP solution is not valid. Whereas the inconsistency of the system is maintained by DM in the case he wants to fix necessary some parameters between weights and cutting level. Future works will focus on the estimation of thresholds, in order to suggest a procedure to compute the best percentages. Another important aspect is to improve the differential algorithm performances by changing the different schemes proposed and suited to classification problems. Finally we will focus on the research of the training set; we are testing a model formulated by means of a Mixed Integer Non Linear Programming (MINLP) problem.

References

De Leone, R., & Minnetti, V. (2011). New approach to estimate the parameters of electre tri model in the ordinal sorting problem. In *Proceedings of the 42nd Annual Conference of AIRO*, Brescia.

De Leone, R., & Minnetti, V. (2012). The estimation of the parameters of the electre tri method: A proposal of a new two step procedure. In *Proceedings of the 43nd Annual Conference of AIRO*, Vietri sul Mare (SA).

Dias, L. C., Mousseau, V., Figueira, J., & Clímaco, J. N. (2000). *An aggregation/disaggregation approach to obtain robust conclusions with ELECTRE TRI* (Cahier du LAMSADE no. 174, Université de Paris-Dauphine).

Greco, S., Matarazzo, B., & Slowinski, R. (1998). A new rough set approach to evaluation of bankruptcy risk. In C. Zopounidis (Ed.), *Operational tools in the management of financial risks* (pp. 121–136). Dodrecht: Kluwer Dordrecht.

Larichev, O. I., & Moskovich, H. M. (1994). An approach to ordinal classification problems. *International Transactions in Operational Research, 1*(3), 375–385.

Larichev, O. I., Moskovich, H. M., & Furems, E. M. (1988). Decision support system CLASS. In B. Brehmer, H. Jungermann, P. Lourens, & G. Sevon (Eds.), *New Directions in Research on Decision Making* (pp. 303–315). North Holland: Elsevier Science.

Liuzzi, G., & Lucidi, S. (2008). A derivative-free algorithm for systems of nonlinear inequalities. *Optimization Letters, 2*(4), 521–534.

Massaglia, R., & Ostanello, A. (1991). N-tomic: A support system for multicriteria segmentation problems. In P. Korhonen, A. Lewandowski, & J. Wallenius (Eds.), *Multiple criteria decision support*, LNEMS (Vol. 356, pp. 167–174). Berlin: Springer.

Moscarola, J., & Roy, B. (1977). Procédure automatique d'examen de dossiers fondée sur une segmentation trichotomique en présence de critères multiples. *RAIRO Recherche Opérationnelle, 11*(2), 145–173.

Mousseau, V., & Slowinski, R. (1998). Inferring an ELECTRE TRI model from assignment examples. *Journal of Global Optimization, 12*, 157–174.

Mousseau, V., Dias, L. C., & Figueira, J. (2004). *Dealing with inconsistent judgments in multiple criteria sorting models* (Research report no.17/2004, INESC Coimbra).

Mousseau, V., Figueira, J., Dias, L. C., Gomes da Silva, C., & Climaco, J. (2003). Resolving inconsistencies among constraints on the parameters of an mcda model. *European Journal of Operational Research, 147*(1), 72–93.

Mousseau, V., Figueira, J., & Naux, J.-Ph. (2001). Using assignment examples to infer weights for electre tri method: Some experimental results. *European Journal of Operational Research, 130*(2), 263–275.

Pawlak, Z., & Slowinski, R. (1994). Decision analysis using rough sets. *International Transactions on Operational Research, 1*, 107–114.

Perny, P. (1998). Multicriteria filtering methods based on concordance/non-discordance principles. *Annals of Operations Research, 80*, 137–167.

Roy, B. (1981). A multicriteria analysis for trichotomic segmentation problems. In P. Nijkamp, & J. Spronk (Eds.), *Multiple criteria analysis: Operational methods* (pp. 245–257). Aldershot: Gower Press.

Roy, B. (1996). *Multicriteria methodology for decision aiding*. Dordrecht: Kluwer Academic.

Roy, B., & Bouyssou, D. (1993). *Aide Multicritère à la Décision : Méthodes et Cas*. Paris: Economica.

Slowinski, R. (Ed.). (1992). *Intelligent decision support - Handbook of applications and advances of the Rough Sets theory*. Dordrecht: Kluwer Academic.

Yu, W. (1992). *Aide multicritère à la décision dans le cadre de la problématique du tri: méthodes et applications*. (Ph.D. thesis, LAMSADE, Université Paris Dauphine, Paris).

Yu, W. (1992). *ELECTRE TRI: Aspects méthodologiques et manuel d'utilisation* (Document du LAMSADE no 74, Université Paris-Dauphine).

Dynamic Clustering of Financial Assets

Giovanni De Luca and Paola Zuccolotto

Abstract In this work we propose a procedure for time-varying clustering of financial time series. We use a dissimilarity measure based on the lower tail dependence coefficient, so that the resulting groups are homogeneous in the sense that the joint bivariate distributions of two series belonging to the same group are highly associated in the lower tail. In order to obtain a dynamic clustering, tail dependence coefficients are estimated by means of copula functions with a time-varying parameter. The basic assumption for the dynamic pattern of the copula parameter is the existence of an association between tail dependence and the volatility of the market. A case study with real data is examined.

Keywords Copula function • Tail dependence • Time series clustering

1 Introduction

Time series clustering is a topic approached from different perspectives. In the literature we find a number of different proposals, ranging from basic settings, with dissimilarities merely derived by the comparison between observations or some simple statistics computed on the time series data, to several more complex solutions based on stochastic approaches and using tools as periodograms and density forecasts (see for example Piccolo 1990; Corduas and Piccolo 2008; Otranto

G. De Luca (✉)
University of Naples Parthenope, Naples, Italy
e-mail: giovanni.deluca@uniparthenope.it

P. Zuccolotto
University of Brescia, Brescia, Italy
e-mail: paola.zuccolotto@unibs.it

D. Vicari et al. (eds.), *Analysis and Modeling of Complex Data in Behavioral and Social Sciences*, Studies in Classification, Data Analysis, and Knowledge Organization,
DOI 10.1007/978-3-319-06692-9_12,

2008; Galeano and Peña 2006; Caiado et al. 2006; Alonso et al. 2006, but the list of citations could be even longer).

As a matter of fact, a key point concerns the preliminary identification of the clustering task, on which we have to found the choice of the dissimilarity measure. In the context of financial time series, for example, it can be reasonable to define a dissimilarity measure able to account for some interesting financial feature of the analysed time series. Following this line, in this work we resort to the procedure proposed by De Luca and Zuccolotto (2011) in order to cluster time series of returns of financial assets according to their association in the lower tail, that is in case of extremely negative returns. The dissimilarity measure used for such clustering is based on tail dependence coefficients estimated by means of copula functions. As shown by De Luca and Zuccolotto (2011, 2014), the results of the clustering can be used for a portfolio selection purpose, when the goal is to protect investments from the effects of adverse extreme events, for example during a financial crisis.

The aim of this paper is to propose a dynamic variant of the above described procedure, obtained by means of a copula function with a time-varying parameter, defined as a function of the volatility of the market. This assumption is connected with the idea of contagion, which is known as a situation when assets are characterized by a significant increase in cross-correlation during the crisis periods. The time-varying pattern of the tail dependence coefficients obtained in this way leads to a dynamic clustering of the time series.

The paper is organized as follows: in Sect. 2 the clustering procedure of De Luca and Zuccolotto (2011) is briefly recalled and an application to real data is presented in Sect. 3. The case study is divided in two parts: firstly we carry out a preliminary analysis aimed at exploring the reliability of the assumption of association between lower tail dependence and volatility, secondly we estimate the time-varying copula function and perform the dynamic clustering. Concluding remarks follow in Sect. 4.

2 The Lower Tail Dependence Based Clustering Procedure

The interest of researchers in modelling the occurring of extreme events has several empirical motivations, especially in contexts where it can be directly associated to risk measurement, such as, for example, financial markets. Recently, a great deal of attention has been devoted also to the study of association between extreme values of two or more variables. From a methodological point of view, the problem of quantifying this association has been addressed in different ways. One of the proposed approaches consists in analyzing the probability that one variable assumes an extreme value, given that an extreme value has occurred to the other variables (see Cherubini et al. 2004). This probability is known as lower or upper tail dependence and we will restrict its analysis to the bivariate case.

Let Y_1 and Y_2 be two random variables and let $U_1 = F_1(Y_1)$ and $U_2 = F_2(Y_2)$ be their distribution functions. The lower and upper tail dependence coefficients are defined respectively as $\lambda_L = \lim_{v \to 0^+} P[U_1 \leq v | U_2 \leq v]$ and $\lambda_U = \lim_{v \to 1^-} P[U_1 > v | U_2 > v]$.

In practice, the tail dependence coefficients have to be estimated from observed data. A very effective way of modeling financial returns is the use of a copula function thanks to which tail dependence estimation is both simple and flexible. A two-dimensional copula function for two random variables Y_1 and Y_2 is defined as a function $C : [0,1]^2 \to [0,1]$ such that $F(y_1, y_2; \theta) = C(F_1(y_1; \vartheta_1), F_2(y_2; \vartheta_2); \tau)$, for all y_1, y_2, where $F(y_1, y_2; \theta)$ is the joint distribution function of Y_1 and Y_2 (see Nelsen 2006) and $\theta = (\vartheta_1, \vartheta_2, \tau)$. It is straightforward to show that the tail dependence coefficients can be expressed in terms of the copula function. In particular, the lower tail dependence coefficient, which is the focus of the work, is given by $\lambda_L = \lim_{v \to 0^+} C(v, v)/v$.

In the analysis of the relationship between financial returns, the lower tail dependence coefficient gives an idea of the risk of investing on assets for which extremely negative returns could occur simultaneously. So, the lower tail dependence is strictly linked to the diversification of investments, especially in financial crisis periods. For this reason, De Luca and Zuccolotto (2011) proposed to cluster time series of financial returns according to a dissimilarity measure defined as $\delta(\{y_{it}\}, \{y_{jt}\}) = \delta_{ij} = -\log(\hat{\lambda}_{L,ij})$, where $\{y_{it}\}_{t=1,\ldots,T}$ and $\{y_{jt}\}_{t=1,\ldots,T}$ denote the time series of returns of two assets i and j, and $\hat{\lambda}_{L,ij}$ is their estimated tail dependence coefficient. In this way we obtain clusters of assets characterized by high tail dependence in the lower tail. From a portfolio selection perspective, it should then be avoided portfolios containing assets belonging to the same cluster.

Given p assets, the clustering procedure proposed by De Luca and Zuccolotto (2011) is composed by two steps: firstly, starting from the dissimilarity matrix $\boldsymbol{\Delta} = (\delta_{ij})_{i,j=1,\ldots,p}$, an *optimal* representation of the p time series $\{y_{1t}\}, \ldots, \{y_{pt}\}$ as p points $\mathbf{y}_1, \ldots, \mathbf{y}_p$ in R^q is found by means of a non-metric Multidimensional Scaling (MDS) taking into account the ordinal nature of the proposed dissimilarity measure; secondly, the k-means clustering algorithm is performed using the obtained geometric representation of the p time series. The above mentioned term *optimal* means that the Euclidean distance matrix $D = (d_{ij})_{i,j=1,\ldots,p}$, with $d_{ij} = \|\mathbf{y}_i - \mathbf{y}_j\|$, of the points to be defined in the first step has to fit as closely as possible the dissimilarity matrix $\boldsymbol{\Delta}$. The extent to which the interpoint distances d_{ij} "match" the dissimilarities δ_{ij} is measured by an index called *stress*, which should be as low as possible. MDS works for a given value of the dimension q, which has to be given in input. So, it is proposed to start with the dimension $q = 2$ and then to repeat the analysis by increasing q until the minimum stress of the corresponding optimal configuration is lower than a given threshold $\bar{s}$. The lower the value $\bar{s}$, the higher the dimension q of the resulting geometric representation in R^q of the p time series. This means a better representation but, at the same time, a higher computational burden, which has to be taken into account in the dynamic setting presented later. In general it should be fixed at least $\bar{s} = 0.05$.

3 Dynamic Clustering: A Case Study

We analyse the time series of the log-returns of the $p = 24$ stocks which have been included in FTSE MIB index during the whole period from January 3, 2006 to October 31, 2011. The total number of observation is $T = 1{,}481$. De Luca and Zuccolotto (2014) use these data to cluster the 24 stocks into 4 groups following the above described two-steps clustering procedure. In this paper we try to get a deeper insight on the same data by exploring the possibility to carry out clustering in a dynamic setting. We rely on the idea that the lower tail dependence between assets could be somehow associated to the volatility of the financial market. The empirical analysis is divided in two steps: firstly we carry out a preliminarily exploratory analysis in order to assess if the assumption of association between volatility and lower tail dependence has some empirical foundation, secondly we incorporate this assumption in the model by means of a time-varying definition of the copula parameter.

3.1 Preliminary Exploratory Analysis on the Assumption of Association Between Volatility and Lower Tail Dependence

In order to explore the existence of association between volatility and lower tail dependence, we repeatedly estimate

- the 276 lower tail dependence coefficients $\hat{\lambda}_{L,ij}$, $i, j = 1, \ldots, p$ corresponding to all the possible pairs of the 24 stocks, using the bivariate Joe-Clayton copula function,
- the average volatility $\bar{\sigma}$ of the FTSE MIB index, obtained by averaging conditional standard deviations estimated with a Student's t AR(1)-GARCH(1,1) model

over rolling time spans $(1, \ldots, t)$, $(2, \ldots, t+1)$, …, $(T - t + 1, \ldots, T)$, with $t = 1{,}000$. As a result, we have $T - t + 1 = 482$ estimates of the lower tail dependence coefficients and of the average volatilities, which will be denoted $\hat{\lambda}^{rolling}_{Lt,ij}$ and $\bar{\sigma}_t$, respectively, with $t = 1{,}000, \ldots, 1{,}481$.

For each stock i we plot the values of its lower tail dependence coefficients with the other stocks, $\hat{\lambda}^{rolling}_{Lt,ij}$, $j = 1, \ldots, p, j \neq i$, against the volatilities $\bar{\sigma}_t$, thus obtaining $p - 1$ scatterplots for each stock. For instance, Fig. 1 displays the 23 scatterplot drawn for the stock ENI, with the interpolation line indicating tendency. We detect a positive association between volatility and lower tail dependence of ENI with some stocks, such as for example ATLANTIA and FIAT. If confirmed, the existence of such a relation would have a meaningful interpretation from a financial point of view, as it would highlight the tendency of some pairs of stocks to increase their association in extreme events just during turbulence periods, thus recalling the idea of contagion. The other stocks present similar patterns, so we decide to proceed to a more formal tractation.

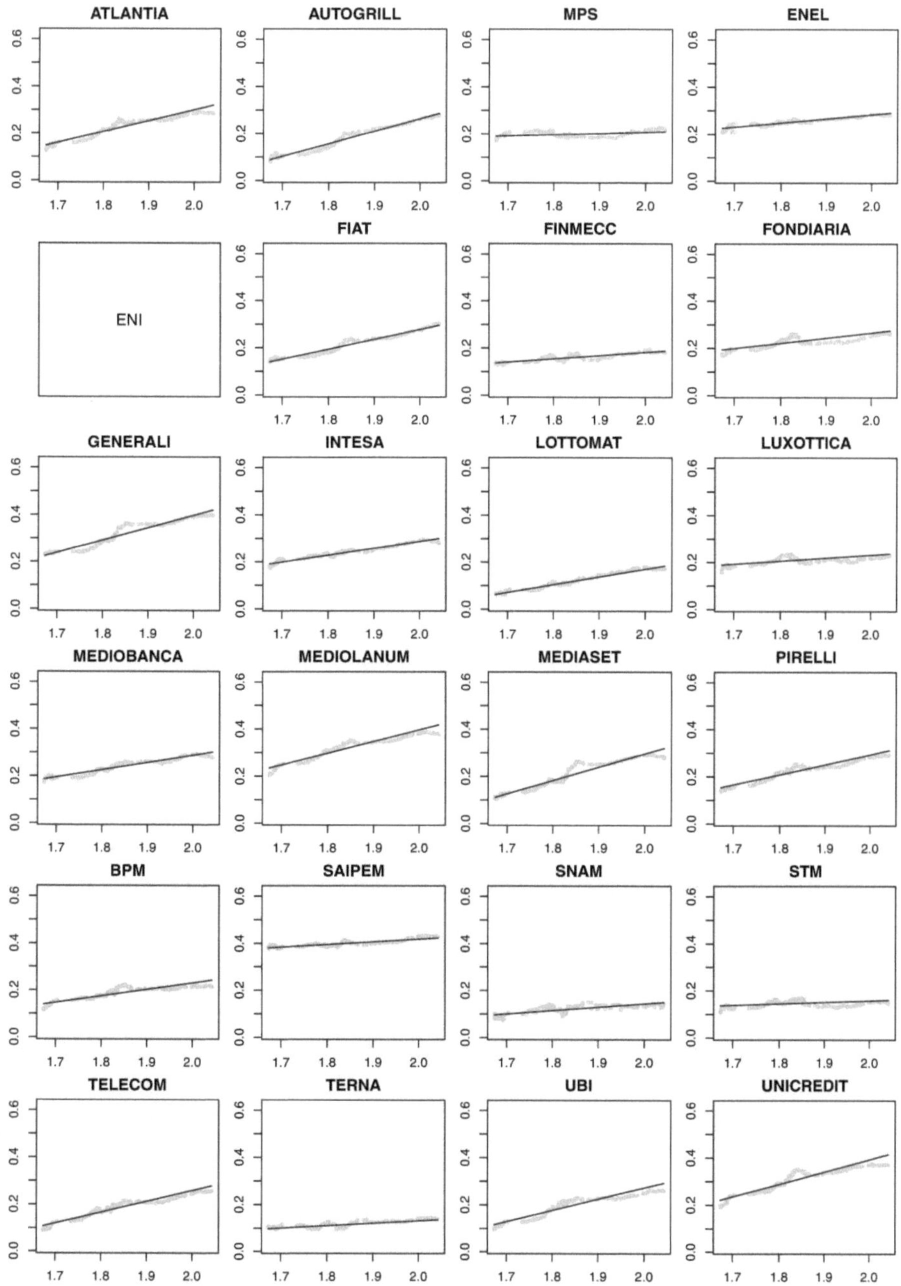

Fig. 1 Scatterplot of lower tail dependence coefficients of ENI and the other stocks, $\lambda_{Lt,ij}^{rolling}$, against volatilities $\bar{\sigma}_t$

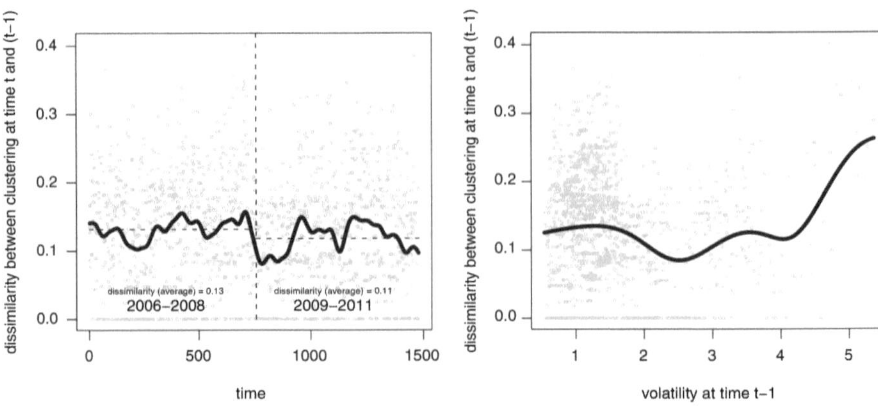

Fig. 2 Dissimilarity between clustering at time t and $t-1$: vs time (*left*) and vs volatility (*right*)

3.2 Dynamic Clustering

Relying on the empirical evidence highlighted in the preliminary analysis, we incorporate the assumption of association between volatility and lower tail dependence in the copula function formulation by defining a time-varying pattern for the parameter involved in the computation of λ_L. For each couple of stocks ij we estimate the parameters of the bivariate Joe-Clayton copula function (Joe 1997),

$$C(u_i, u_j) = 1 - \{1 - [(1-(1-u_i)^{\kappa_{ij}})^{-\theta_{t,ij}} + (1-(1-u_j)^{\kappa_{ij}})^{-\theta_{t,ij}} - 1]^{-1/\theta_{t,ij}}\}^{1/\kappa_{ij}},$$

with $\theta_{t,ij} = \exp\{\omega_{ij} + \alpha_{ij}\sigma_{t-1}\}$, where σ_t is the conditional standard deviation of the FTSE MIB index, estimated by means of a Student's t AR(1)-GARCH(1,1) model.

For the pairs with a significant coefficient α_{ij} at the 5 % level (135 out of 276) we estimate time-varying tail dependence coefficients, $\hat{\lambda}_{Lt,ij} = 2^{-1/\hat{\theta}_{t,ij}}$. Note that all the significant α_{ij} are positive, so for the 135 pairs of stocks with significant α_{ij}, the lower tail dependence coefficient tends to increase with rising volatility in the market, in accordance with the idea of contagion. For the remaining pairs, we set a constant parameter $\theta_{t,ij} = \theta_{ij}$ in the expression of the copula function and hence we obtain constant estimates $\hat{\lambda}_{Lt,ij} = \hat{\lambda}_{L,ij} = 2^{-1/\hat{\theta}_{ij}}$. On the whole, we get time-varying dissimilarity matrices $\boldsymbol{\Delta}_t$. The clustering procedure described in Sect. 2 ($\bar{s} = 0.05$) is then applied to the matrices $\boldsymbol{\Delta}_t$, thus obtaining a different clustering at each time t. The groups composition is dynamically adapted to the variations due to the changes in the volatility of the market and the discordance between the clustering at time $t-1$ and that at time t is measured with a normalized dissimilarity index d_t.

Figure 2 shows the pattern of d_t in time and with respect to the estimated volatility of the FTSE MIB index. In the period 2009–2011 (that is, after the financial crisis of 2008), d_t tends to be slightly lower in average, but less stable than

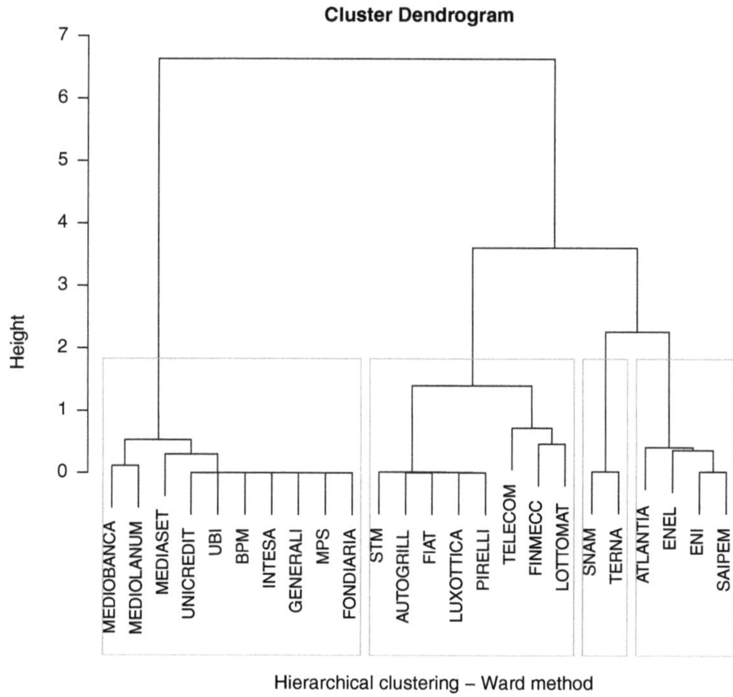

Fig. 3 Cluster dendrogram of the final clustering with distance matrix B

in the period 2006–2008. From the point of view of its relation with the volatility of the FTSE MIB index, we observe that only very high levels of volatility tend to have a strong effect on the dissimilarity of the clusterings from time $t-1$ to time t.

Finally, we can compute, for each pair of stocks ij, the index

$$b_{ij} = 1 - \frac{\sum_{t=1}^{T} I_{t,ij}}{T}$$

where $I_{t,ij}$ is the indicator function which equals 1 if stock i and stock j are assigned the same cluster at time t and 0 otherwise. The index b_{ij} denotes the fraction of the total clustering solutions with stocks i and j belonging to different clusters, a generalization of the widely used Simple Matching distance, due to Sokal and Michener (1958). So, we can perform a hierarchical clustering using the distance matrix $B = (b_{ij})_{i,j=1,\ldots,p}$, in order to summarize the T dynamic cluster solutions. Figure 3 displays the resulting cluster dendrogram, where the existence of four well-separated clusters can be easily detected.

4 Concluding Remarks

This paper deals with a clustering procedure for financial time series, based on a dissimilarity measure computed as a function of the tail dependence coefficient between the returns of two stocks. Tail dependence coefficients are estimated by means of copula functions. We have extended the procedure to a dynamic setting, using a time-varying copula function allowing for a dynamic estimation of tail dependence coefficients. The basic assumption for the proposed model is the existence of an association between tail dependence and volatility of the market, as postulated by the theories defining the concept of contagion. The proposed procedure returns a time-varying clustering structure. The main results of the presented analysis of real data concern the effective detection, through the copula function estimation, of a positive relationship between tail dependence and volatility for many couples of stocks. Thus, volatility seems to often affect the behavior of returns during extreme events and this reflects on the time-varying clustering structure. Very high levels of volatility determine more huge variations in the clusters composition. At the end, an overall distance measure between stocks has been computed using all the time-varying solutions, and this made possible to determine a final clustering with very well-separated clusters. Future research could be developed in two directions: the definition of a more complex relationship between tail dependence and volatility and the way to employ the clustering results to define investment strategies.

Acknowledgements This research was funded by a grant from the Italian Ministry od Education, University and Research to the PRIN Project entitled "Multivariate statistical models for risks evaluation" (2010RHAHPL_005).

References

Alonso, A. M., Berrendero, J. R., Hernández A., & Justel, A. (2006). Time series clustering based on forecast densities. *Computational Statistics and Data Analysis, 51*, 762—776.

Caiado, J., Crato, N., & Peña, D. (2006). A periodogram-based metric for time series classification. *Computational Statistics and Data Analysis, 50*, 2668–2684.

Cherubini, U., Luciano, E., & Vecchiato, W. (2004). *Copula methods in finance*. New York: Wiley.

Corduas, M., & Piccolo, D. (2008). Time series clustering and classification by the autoregressive metrics. *Computational Statistics and Data Analysis, 52*, 1860–1872.

De Luca, G., & Zuccolotto, P. (2011). A tail dependence-based dissimilarity measure for financial time series clustering. *Advances in Classification and Data Analysis, 5*, 323–340.

De Luca, G., & Zuccolotto, P. (2014). Time series clustering on lower tail dependence. In M. Corazza, & C. Pizzi (Eds.), *Mathematical and statistical methods for actuarial sciences and finance*. New York: Springer.

Galeano, P., & Peña, D. (2006). Multivariate analysis in vector time series. *Resenhas, 4*, 383–404.

Joe, H. (1997). *Multivariate models and dependence concepts*. New York: Chapman & Hall/CRC.

Nelsen, R. (2006). *An introduction to copulas*. New York: Springer.

Otranto, E. (2008). Clustering heteroskedastic time series by model-based procedures. *Computational Statistics and Data Analysis, 52*, 4685–4698.

Piccolo, D. (1990). A distance measure for classifying ARMA models. *Journal of Time Series Analysis, 11*, 153–164.

Sokal, R. R., & Michener, C. D. (1958). A statistical method for evaluating systematic relationships. *University of Kansas Science Bulletin, 38*, 1409–1438.

A Comparison of χ^2 Metrics for the Assessment of Relational Similarities in Affiliation Networks

Maria Rosaria D'Esposito, Domenico De Stefano, and Giancarlo Ragozini

Abstract Factorial techniques are widely used in Social Network Analysis to analyze and visualize networks. When the purpose is to represent the relational similarities, simple correspondence analysis is the most frequent used technique. However, in the case of affiliation networks, its use can be criticized because the involved χ^2 distance does not adequately reflect the actual relational patterns. In this paper we perform a simulation study to compare the metric involved in Correspondence Analysis with respect to the one in Multiple Correspondence Analysis. Analytical results and simulation outcomes show that Multiple Correspondence Analysis allows a proper graphical appraisal of the underlying two-mode relational structure.

Keywords Affiliation networks • Correspondence analysis • Social network analysis • Two-mode structural equivalence

M.R. D'Esposito
Department of Economics and Statistics, University of Salerno, Via Giovanni Paolo II, Fisciano (SA), Italy
e-mail: mdesposito@unisa.it

D. De Stefano (✉)
Department of Economics, Ca' Foscari University of Venice, Cannaregio 873, Venezia, Italy
e-mail: domenico.destefano@unive.it

G. Ragozini
Department of Political Sciences, Federico II University of Naples, Naples, Italy
e-mail: giragoz@unina.it

D. Vicari et al. (eds.), *Analysis and Modeling of Complex Data in Behavioral and Social Sciences*, Studies in Classification, Data Analysis, and Knowledge Organization,
DOI 10.1007/978-3-319-06692-9_13,

1 Introduction

Two-mode networks arise in many different contexts and, therefore are receiving more and more attention, as it is also testified by a very recent special issue devoted to this topic (Agneessens and Everett 2013). Affiliation networks, a special case of two-mode networks, are of particular interest in many economic and social sectors. They are characterized by a set of actors and a set of events in which actors are involved. These types of networks differ from the most used and known one-mode networks, the latter representing the connections that actors directly have among them. As recognized by several authors, a fundamental issue in the analysis of such networks is the direct visualization of the affiliation structure (e.g., Borgatti and Everett 1997). A widely used approach is to apply a two-mode multivariate analysis and especially factorial techniques to locate nodes. Notable attempts of using factorial methods within social network framework have been made since the early 1960. In particular, Correspondence Analysis (CA) has been frequently proposed for the analysis of both one-mode networks and two-mode networks (Roberts 2000; Faust 2005) and its use has been recognized to be suitable when the main concern is in assessing the relational similarities based on the structural equivalence principle (Roberts 2000). However, the use of simple CA for two-mode networks—i.e., CA applied considering the affiliation matrix as a contingency table—could be criticized by different points of view (Borgatti and Everett 1997). The main point, in our opinion, is that the usual metric in simple CA is not fully adequate to represent relational structures according to the structural equivalence principle.

In this paper we argue that CA applied to the affiliation matrix considered as a case-by-variable array, i.e. Multiple Correspondence Analysis (MCA), allows a better graphical appraisal of the underlying relational structure of actors or events. We aim at comparing simple CA with MCA showing how the metric adopted in MCA can be more adequate to represent relational patterns—in terms of two-mode structural equivalence—with respect to simple CA, being less affected by some network characteristics. The metric adopted in the latter, in fact, can be strongly affected by some network characteristics, such as the distribution of actor degrees or of the event sizes (i.e., the number of actor connections or the number of participants to a certain event, respectively).

The paper is organized as follows. In Sect. 2 we discuss affiliation networks as well as structural equivalence, and χ^2 metrics in both CA and MCA settings. In Sect. 3 we discuss the differences between CA and MCA distances and present the results of a simulation study to assist in the comparison. Section 4 contains some concluding remarks.

2 Affiliation Networks, Structural Equivalence, CA and MCA

Let $\mathscr{G}$ be a bipartite graph represented by a triple $\mathscr{G}(V_1, V_2, \mathscr{R})$ composed of two disjoint sets of nodes, V_1 and V_2 of cardinality n and m, and one set of edges or arcs, $\mathscr{R} \subseteq V_1 \times V_2$. By definition $V_1 \cap V_2 = \emptyset$. Two-mode networks may be modeled naturally by a bipartite graph. An affiliation network is a particular case of two-mode network in which the two disjoint sets V_1 and V_2 refer to very different entities, i.e. the set $V_1 = \{a_1, a_2, \ldots, a_n\}$ represents the actor set, whereas the other, $V_2 = \{e_1, e_2, \ldots, e_m\}$, represents the set of m events. The edge $r_{ij} = (a_i, e_j)$, $r_{ij} \in \mathscr{R}$, is an ordered couple, and indicates if an actor a_i attends an event e_j. The set $V_1 \times V_2$ can be stored in a binary affiliation matrix $\mathbf{F}(n \times m)$, with element $f_{ij} = 1$ if $(a_i, e_j) \in \mathscr{R}$, and 0 otherwise.

A fundamental concept in social network analysis is the one of equivalence among nodes. Dealing with affiliation networks this concept is mainly related to the assessment of the relational similarities among actors or among events. The common way to measure the relational similarity is based on the notion of structural equivalence. In affiliation networks, (two-mode) structural equivalence principle expresses that two actors are equivalent if they participate exactly to the same events (Pizarro 2007). If two actors a_i and $a_{i'}$ are structurally equivalent they are indistinguishable, and one equivalent actor can substitute for another because they present the same relational pattern. Analogous reasoning applies for structural equivalent events, that are structurally equivalent to the extent they are attended by the same actors.

In an affiliation network, relational similarities can be measured by the Euclidean distance, as proposed by Burt (1980):

$$\delta_E^2(a_i, a_{i'}) = \sum_{j=1}^{m} (f_{ij} - f_{i'j})^2 \tag{1}$$

In the case of events: $\delta_E^2(e_j, e_{j'}) = \sum_{i=1}^{n} (f_{ij} - f_{ij'})^2$.

If two actors a_i and $a_{i'}$ (events e_j and $e_{j'}$) are structurally equivalent their $\delta_E^2(\cdot,\cdot)$ is equal to zero.

Whenever one aims to explore and visualize actor/event relational patterns, tools able to represent structural similarities of actors or events, starting from their distances, could be employed. In this case, CA can be applied either in its simple or in its multiple version, as they deliver 2-D maps in which actors and events are projected as points. Simple CA works on the actors and event profiles $\mathbf{p}_{a_i} = [f_{i1}/f_{i\cdot}, \ldots, f_{im}/f_{i\cdot}]$, $i = 1, \ldots, n$, and $\mathbf{p}_{e_j} = \left[f_{1j}/f_{\cdot j}, \ldots, f_{nj}/f_{\cdot j}\right]$, $j = 1, \ldots, m$. Actor and event profiles are points in $\mathfrak{R}^{(m-1)}$ and $\mathfrak{R}^{(n-1)}$, respectively. A suitable distance between two profiles is not the usual Euclidean but the χ^2 distance. In particular, the distance between two actor profiles $\mathbf{p}_{a_i}$ and $\mathbf{p}_{a_{i'}}$ (and similarly for event profiles $\mathbf{p}_{e_j}$ and $\mathbf{p}_{e_{j'}}$) is computed directly from the profiles obtained by **F** and can be defined as:

$$\delta^2_{CA}(\mathbf{p}_{a_i}, \mathbf{p}_{a_{i'}}) = \sum_{j=1}^{m} \left(\frac{f_{ij}}{d_i} - \frac{f_{i'j}}{d_{i'}} \right)^2 \frac{L}{s_j} \quad (2)$$

where d_i is the degree of the actor i (i.e., number of events attended by i), s_j is the size of the event j (i.e., number of participants to j) and L is the total number of links in the whole network.

MCA works instead on a multiple indicator matrix **Z**, derived from **F** through its full disjunctive coding. The **Z** matrix is a $n \times 2m$ matrix of the form: $\mathbf{Z} \equiv \left[\mathbf{F}^+, \mathbf{F}^-\right]$, where $\mathbf{F}^+ = (e_j^+) = \mathbf{F}$, $\mathbf{F}^- = (e_j^-) = \mathbf{1} - \mathbf{F}^+ = \mathbf{1} - \mathbf{F}$, and $\mathbf{1}$ is an $n \times m$ matrix of ones. As all the columns in **F** represent dichotomous variables, **Z** turns out to be a doubled matrix (Greenacre 1984).

In **Z** row marginals $z_{i.}$ are constant and equal to the number of events m, while column marginals $z_{.j}$ are equal to the event sizes s_j, when associated to e_j^+, or to $n - s_j$, when associated to e_j^-. In MCA, the χ^2 distance between two actor profiles becomes:

$$\delta^2_{MCA}(\mathbf{p}_{a_i}, \mathbf{p}_{a_{i'}}) = \frac{n}{m} \sum_{j=1}^{m} \frac{n}{s_j(n - s_j)} (f_{ij} - f_{i'j})^2, \quad (3)$$

The distance between two event profiles becomes: $\delta^2_{MCA}(\mathbf{p}_{e_j}, \mathbf{p}_{e_{j'}}) = \frac{2n}{s_j s_{j'}} \sum_{i=1}^{n} [z_{ij} \neq z_{ij'}]$, where $[\cdot]$ is the Iverson bracket such that $[P] = 1$ if P is true and 0 otherwise.

By comparing (2) and (3), it is clear that the weighting system is somewhat different in MCA with respect to simple CA. By looking at these distances with the aim to evaluate actors' and events' relational patterns, it can be noted that δ^2_{CA} adopts a peculiar and more complex weighting system that could lead to a distorted representation of the actual relational patterns since it is strongly affected by the network structure (density, actor degree distribution, presence of clusters, and so on). In fact, the distance between actors (events) depends, not only on the pattern of participation (attendance), but also on the actor degree d_i (event size s_j). Moreover, in the δ^2_{CA}, small size events (i.e., events with small rate of attendance) are associated to larger weights, and then differences in the f_{ij} values related to such small size events have stronger impact on the overall χ^2 distance than differences related to larger size events. In the MCA case, instead, the actor degree d_i does not play any role in defining row and column profiles and in the χ^2 distance. In addition, the event size s_j has a counterpart in its complement $n - s_j$, and consequently the column weights are balanced. It turns out that δ^2_{MCA} closely resembles the Euclidean distance δ^2_E. The difference between the two distances is given by the weight $\frac{n}{s_j(n-s_j)}$ that balances the rate of attendance with respect to the rate of non attendance.

Similar arguments also apply considering the event distances. However, in the following discussion with no loss of generality, we focus on the distances among actors.

3 Relation Between χ^2 Metrics and Network Characteristics: An Experimental Comparison

As already noted, χ^2 metrics are influenced by both the actor degrees (d_i) and the event size (s_j). These two quantities enter, in rather different ways, in the computation of both CA and MCA metrics as weights. Generally speaking, d_i and s_j depend on two elements: (1) the network density θ; and (2) the presence of locally dense clusters of nodes. In the case of density, for a fixed network dimension, the larger the number of ties (i.e., dense networks), the higher the rates of membership for actors (d_i) and the larger the sizes of the events (s_j). Dealing with sparse networks, as the d_i's will generally be low in value, the corresponding weights in δ^2_{CA} will be higher in value. Moreover, small differences in two actor degrees could lead to large differences in the weights and thus in the distances. When dealing with denser networks, this effect changes since the average degree in the network increases and the relative weights become smaller in value, affecting the CA distances. On the contrary, in MCA the density does not affect the factorial solution as the actor degrees do not enter into the computation of the distance. In fact, each profile is weighted by a constant term equal to $1/m$. Furthermore, the presence of dense clusters inside the network also affects the computation of the CA metrics whereas the MCA is substantially left unchanged. This effect strongly affects CA distance, again because the "raw" d_i and s_j enter in its computation. Finally, mixing of both parameters—i.e., presence of local clusters in basically dense or sparse networks—enhances the effects of these individual elements on the distance computation.

Given the above discussed theoretical differences between δ^2_{CA} and δ^2_{MCA}, a simulation experiment has been performed to compare such metrics with δ^2_E, considering the effect of density and presence of clusters on both small and large networks. In particular, we generated networks under different configurations, considering three network features in their possible combinations: (1) dimension (small—20 actors×10 events—and large networks—200 actors×50 events) (2) density (dense *vs* sparse networks depending on the number of active links), (3) presence of structure with different amount of noise (varying number of clusters with a different amount of noise).

For each fixed dimension (small or large) we started from a full affiliation network obtaining dense and sparse networks by randomly removing a proportion l of links. The probability of a link to be removed is fixed (equal to l/L, where L is the total number of links) so that we produce a random affiliation network (dense or sparse according to the value of the parameter l). The other two parameters (h and p) allow to produce non-random networks in which nodes are grouped into structurally

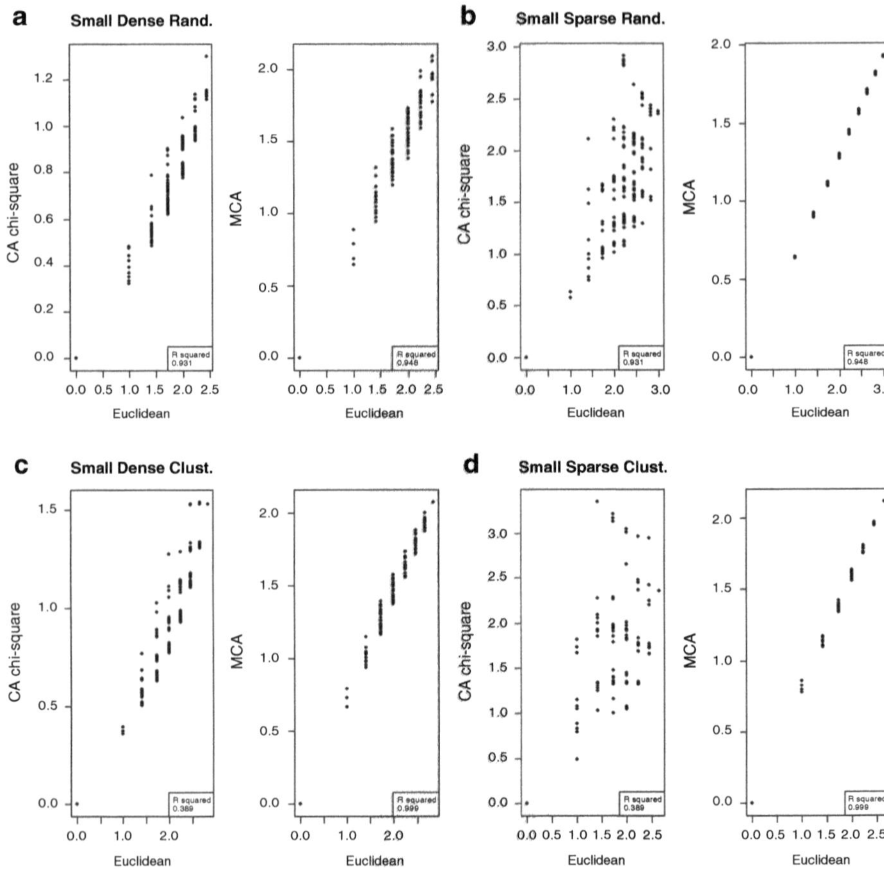

Fig. 1 Scatterplot of δ^2_{CA}, δ^2_{MCA} *vs* δ^2_E for small networks. Panel (**a**) Dense; (**b**) Sparse; (**c**) Clustered with large noise; (**d**) Clustered with small noise

equivalent clusters (or blocks). To do this we start from a random affiliation matrix of a fixed dimension and we create a partition of each matrix into a given number of clusters h. Each cluster has on the main diagonal a random size and it is created with density equal to 1. Then, an amount of noise (p) is added to each cluster, so the final inner cluster density is slightly less than 1 and there is a presence of extra clusters links.

We performed 100 runs for each design by varying l for the random networks and h and p for the clustered networks. For each simulated networks, in order to understand how and under which conditions δ^2_{CA} and δ^2_{MCA} reproduce nodes relational similarities, we plot them versus δ^2_E, representing our benchmark. As an example, Figs. 2 and 1 report the results obtained in a single run of one simulation setting for large networks and small networks, respectively. Networks in both figures are characterized by the following design parameters: random sparse ($l = 0.75$,

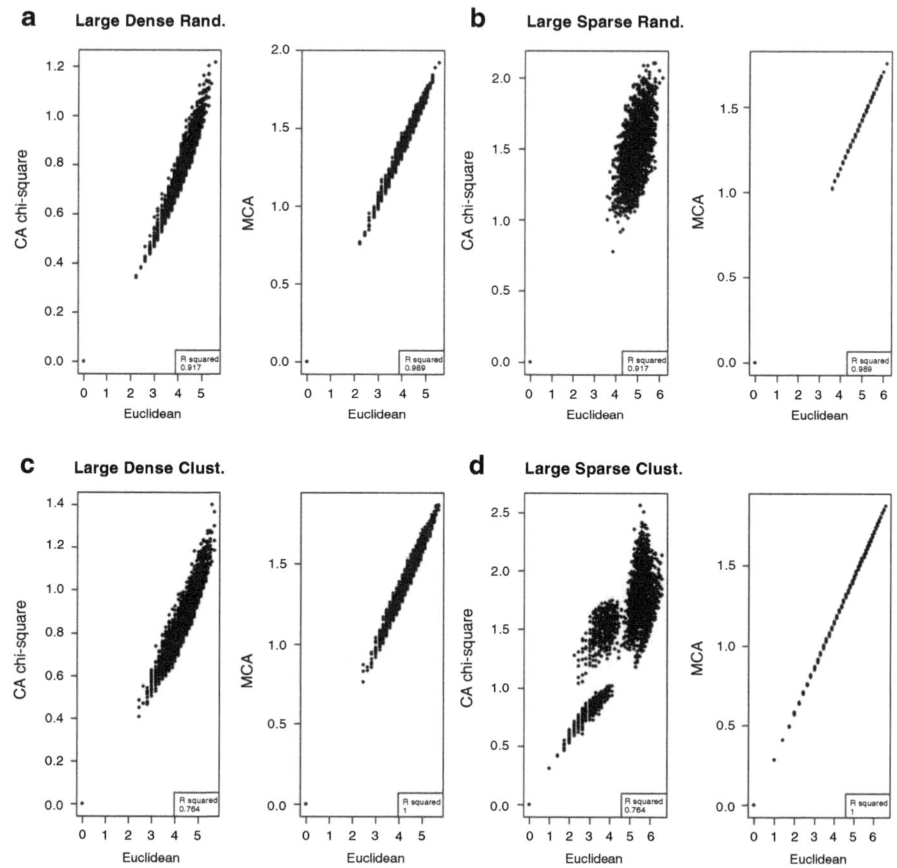

Fig. 2 Scatterplot of δ^2_{CA}, δ^2_{MCA} *vs* δ^2_E for large networks. Panel (**a**) Dense; (**b**) Sparse; (**c**) Clustered with large noise; (**d**) Clustered with small noise

$h = 1$), random dense ($l = 0.25$, $h = 1$), clustered with small noise ($l = 0.75$, $h = 3$, $p = 0.2$), and clustered with large noise ($l = 0.25$, $h = 3$, $p = 0.6$). The plots show that δ^2_{MCA} better reflects the δ^2_E metric given that a more rigorous linear pattern is observed between the δ^2_{MCA} and the Euclidean distance with respect to the δ^2_{CA} plotted against δ^2_E.

The results for 100 runs of the previous simulation setting are reported in Fig. 3 where boxplots portray the distributions of the R^2 index evaluated between the δ^2_E and δ^2_{CA} or δ^2_{MCA}. In general, we can see that δ^2_{MCA} is more robust with respect to the δ^2_{CA} in all cases, since R^2 values are closer to one with a lower variability. Considering sparse networks, this variability becomes negligible for δ^2_{MCA} with respect to δ^2_{CA}. This is due to the larger variability of actor degrees and event sizes which, in these configurations, produce a large variability in CA weights leaving MCA weights more stable.

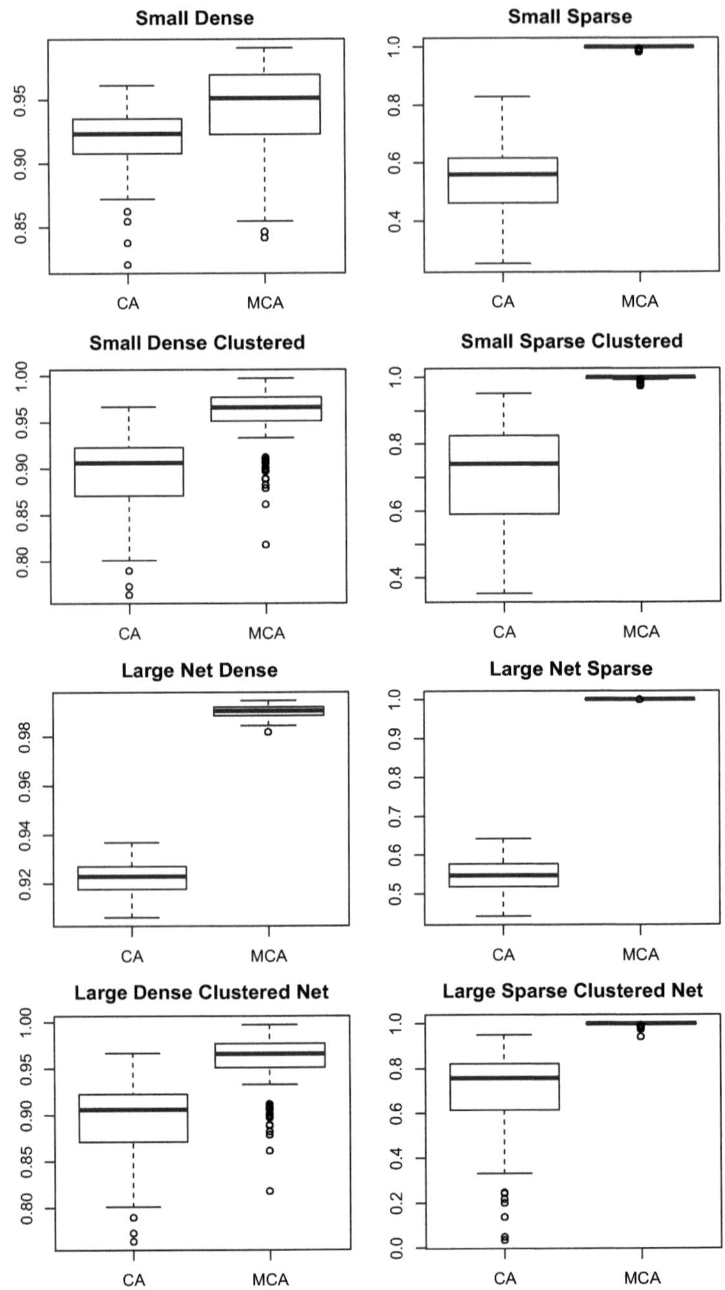

Fig. 3 Boxplots of the distributions of the R^2 index evaluated between the δ_E^2, and δ_{CA}^2 or δ_{MCA}^2 over the simulated networks

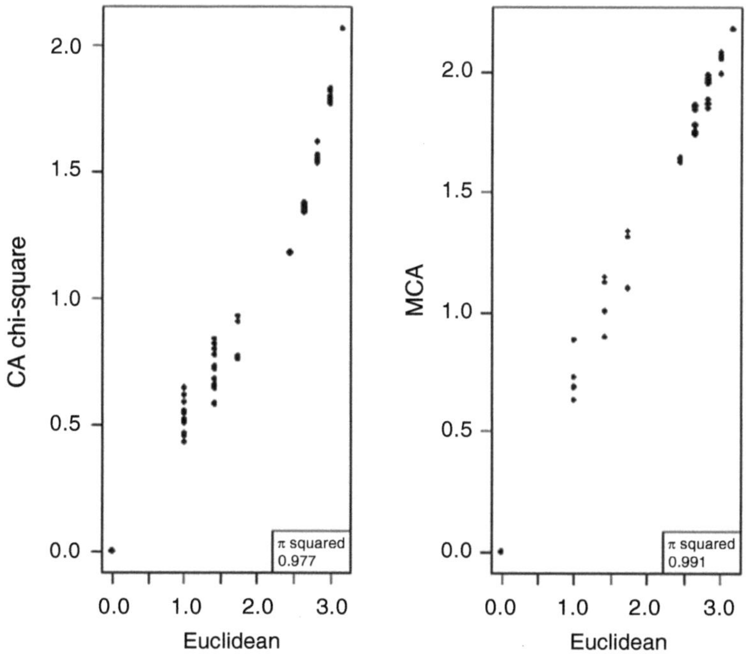

Fig. 4 Scatterplot of δ^2_{CA}, δ^2_{MCA} *vs* δ^2_E in small networks clustered in $h = 2$ clusters with $p = 0.7$

However, there are cases in which these two techniques can be interchangeably used and the advantages of the MCA are less evident. In particular this happens in random setting and when affiliation networks are (very) dense, whereas in non-random setting this occurs when the structural equivalent clusters are not well defined by the high presence of extra-clusters linkages (i.e., clustered dense design). The effect of "extreme" clusterization is even stronger as the overall density is higher and the network size is small. In other words, when the effect of clusterization is masked by large amount of noise—i.e., high presence of extra-clusters 1's—and the number of clusters h is small, both techniques capture the same "amount" of relational similarities among nodes. As an example, we report in Fig. 4 a single run over the 100 performed for a small dense network with $h = 2$ clusters and high noise ($p = 0.7$).

4 Concluding Remarks

We explored the differences in the quality of the representation of the relational similarities in the direct visualization of the affiliation networks by means of factorial techniques. We discussed how MCA adopts a more suitable weighting system—with respect to CA—able to produce more stable results in terms of relational similarities representation (even in the projection on the two-dimensional factorial space). The simulation based results confirm our theoretical discussion: δ^2_{MCA} better reproduces the actual relational pattern embedded in affiliation networks. There are few cases in which δ^2_{CA} and δ^2_{MCA} lead to comparable results, specifically in the case of (very) dense structures or in case of structural equivalent clusters masked by the high presence of extra-clusters linkages (i.e., high noise).

References

Agneessens, F., & Everett, M. G. (2013). Introduction to the special issue on advances in two-mode social networks. *Social Networks, 35*, 145–147.

Borgatti, S. P., & Everett, M. G. (1997). Network analysis of 2-mode data. *Social Networks, 19*, 243–269.

Burt, R. S. (1980). Models of network structure. *Annual Review of Sociology, 6*, 79–141.

Faust, K. (2005). Using correspondence analysis for joint displays of affiliation networks. In P. J. Carrington, J. Scott, & S. Wasserman (Eds.), *Models and methods in social network analysis* pp. 117–147. Cambridge: Cambridge University Press.

Greenacre, M. (1984). *Theory and applications of correspondence analysis*. London: Academic.

Pizarro, N. (2007). Structural identity and equivalence of individuals in social networks: Beyond duality. *International Sociology, 22*(6), 767–792.

Roberts, J. M. Jr. (2000). Correspondence analysis of two-mode network data. *Social Networks, 22*, 65–72.

Influence Diagnostics for Meta-Analysis of Individual Patient Data Using Generalized Linear Mixed Models

Marco Enea and Antonella Plaia

Abstract In meta-analysis, generalized linear mixed models (GLMMs) are usually used when heterogeneity is present and individual patient data (IPD) are available, while accepting binary, discrete as well as continuous response variables. In the present paper some measures of influence diagnostics based on log-likelihood are suggested and discussed. A known measure is approximated to get a simpler form, for which the information matrix is no more necessary. The performance of the proposed measure is assessed through a diagnostic analysis on simulated data reproducing a possible meta-analytical context of IPD with influential outliers. The proposed measure is showed to work well and to have a form similar to the gradient statistic, recently introduced.

Keywords Diagnostics • Individual patient data • Meta-analysis • Outliers

1 Introduction

Meta-analysis is a collection of techniques to combine results coming from multiple independent studies to yield an overall answer to a question of interest (Everitt 2002). The aim is to provide an overall risk-event measure of interest summarizing information coming from the studies. Often the interest is in evaluating association or interaction between the risk-event measure and covariates values, available at individual or at study level. When only Aggregate Data (AD) are available from each study, meta-regression is usually used to assess participant-level covariates (Riley and Steyerberg 2010). Meta-regression assumes that the 'across-study relationship'

M. Enea (✉) • A. Plaia
Dipartimento di Scienze Economiche, Aziendali e Statistiche, University of Palermo, Palermo, Italy
e-mail: marco.enea@unipa.it; antonella.plaia@unipa.it

D. Vicari et al. (eds.), *Analysis and Modeling of Complex Data in Behavioral and Social Sciences*, Studies in Classification, Data Analysis, and Knowledge Organization, DOI 10.1007/978-3-319-06692-9_14,

between summary estimates and mean covariate values reflects 'within-study relationship' between individual response and individual covariate values, but this may not be true in practice (ecological bias). For this reason, Meta-analysis of *Individual Patient Data (IPD)* is the gold-standard in evidence-based synthesis. Meta-analysis of IPD assumes that all the information, i.e. data, are available at both study and patient level, and also allows within-study relationships to be observed directly, separating them from across-study associations, and a suitable approach is represented by Generalized Linear Mixed Models (GLMMs). Outliers detection and influence diagnostics are a natural final step in meta-analysis. However, due to the lack of standard methods for detecting influential studies and/or individuals for non-normal mixed model, often meta-analysis practitioners neglect the diagnostics, reducing it only to a check for publication bias. The key fact, that underpin influence diagnostics at both the observation and the study level, is that it could happen that few observations (data structures) are influential for the whole study, i.e. these are the ones that let the study be influential. In such a situation, it could be sufficient to exclude few observations to adjust the model fit. In literature, the methodologies proposed to detect influential data structures for GLMMs follow essentially two approaches. The first one is the so called *deletion diagnostics* on the basis of Cook's distance (Cook 1977), the second one is the *local influence approach* (Cook 1986) for which contributions are given, for example, by Xu et al. (2006), based on a Q-displacement function, or Ouwens et al. (2001) based on log-likelihood. However, Ouwens et al.'s proposal, which requires the information matrix, could be difficult to implement due to the lack of statistical software. As far as we know this is the case of R Core Team (2012), for which only package *glmmML* returns the information matrix, but the model fit is limited to one random intercept only.

In this work, on the grounds of the measure suggested by Cook (1986) and developed by Ouwens et al., we bypass the above problem by deriving a diagnostic measure for GLMMs which does not require the information matrix, while maintaining the same large-sample behaviour. Moreover, we use the proposed measure to address the influence diagnostics for IPD meta-analysis. In fact, except (Viechtbauer and Cheung 2010), influence diagnostics for IPD has been little considered yet, and it has not still been addressed for GLMM-based meta-analysis. By using a binomial-normal random intercept model, the proposed measure will be compared with that proposed by Ouwens et al. in a hypothetical context of a meta-analysis of simulated and perturbed IPD.

2 Influence Diagnostics

Let y_{ij} be the response of subject j of the ith study, $j = 1, \dots, n_i$, $i = 1, \dots, N$, $\boldsymbol{x}_{ij}$ and $\boldsymbol{z}_{ij}$ the covariate arrays. Conditionally to the ith study, y_{ij} are assumed independent and drawn from an exponential density as defined in Ouwens et al. (2001). The GLMM is written as: $g(\mu_{ij}) = g(E[y_{ij}|\boldsymbol{x}_{ij}, \boldsymbol{z}_{ij}, \boldsymbol{b}_i]) = \eta_{ij} = \boldsymbol{x}'_{ij}\boldsymbol{\beta} + \boldsymbol{z}'_{ij}\boldsymbol{b}_i$, where

g is a link function (we use the logit for the model in Sect. 3), $\boldsymbol{\beta} = (\beta_1, \ldots, \beta_p)'$ is a vector of fixed parameters and $\boldsymbol{b}_i$ are assumed to be $N(\mathbf{0}, G)$. For such model Ouwens et al. (2001) write the quantities necessary to calculate $C_{\boldsymbol{d}}$, the measure of individual local influence with respect to the direction $\boldsymbol{d}$, proposed by Cook (1986) and based on the normal curvature. Such a measure, computable for a single study, observation or more generally for its subset M_i, is

$$C_{\boldsymbol{d}} = 2\left|\boldsymbol{d}'\Delta' H^{-1} \Delta \boldsymbol{d}\right|, \tag{1}$$

where H is the Hessian of the log-likelihood relative to the parameter vector $\boldsymbol{\zeta} = (\boldsymbol{\beta}', \boldsymbol{\delta}')'$, with $\boldsymbol{\delta}$ corresponding to the variance components, i.e. to the elements of G. Δ is the matrix whose ith column is the first-order derivative of $L_i(\boldsymbol{\zeta})$, the contribution of the ith study to the log-likelihood. Both H and Δ are calculated in $\hat{\boldsymbol{\zeta}}$, the maximum likelihood (ML) estimate of $\boldsymbol{\zeta}$. Vector $\boldsymbol{d}$ is such that $||\boldsymbol{d}|| = 1$. For our aims, $\boldsymbol{d}$ is assumed to be a vector of zeros, with one in correspondence either of study i, or of set M_i inside study i, therefore (1) reduces to the so-called *total local influence*:

$$C_i = 2\left|\boldsymbol{\Delta}_i' H^{-1} \boldsymbol{\Delta}_i\right|. \tag{2}$$

If the interest is on study i, $\boldsymbol{\Delta}_i$ will correspond to $\boldsymbol{s}_i = (\boldsymbol{s}'_{i\beta}, \boldsymbol{s}'_{i\delta})'$, the contribution to the score function for the ith study. If the interest is on set M_i, we will write

$$C_{M_i} = 2\left|\boldsymbol{\Delta}'_{M_i} H^{-1} \boldsymbol{\Delta}_{M_i}\right|, \tag{3}$$

where $\boldsymbol{\Delta}_{M_i} = \boldsymbol{s}_i - \boldsymbol{s}_{i(M_i)}$, the subvector of the difference between the contribution to the score function of study i and the score function for such study without set M_i. Now, let $\hat{\boldsymbol{\zeta}}_{(i)}$ be the estimate of $\hat{\boldsymbol{\zeta}}$ when ith study is deleted. Since $\hat{\boldsymbol{\zeta}}_{(i)} \approx \hat{\boldsymbol{\zeta}} - [H_{(i)}(\hat{\boldsymbol{\zeta}})]^{-1} \boldsymbol{\Delta}_{(i)}(\hat{\boldsymbol{\zeta}})$, by considering that $[H_{(i)}(\hat{\boldsymbol{\zeta}})]^{-1}$ can be approximate by $[H(\hat{\boldsymbol{\zeta}})]^{-1}$, as done by Zhu et al. (2001), and by omitting the dependance on $\hat{\boldsymbol{\zeta}}$, it results that

$$\hat{\boldsymbol{\zeta}} - \hat{\boldsymbol{\zeta}}_{(i)} \approx H^{-1} \boldsymbol{\Delta}_{(i)}. \tag{4}$$

By pre-multiplying both members of (4) by $\boldsymbol{\Delta}'_{(i)}$, it becomes

$$\boldsymbol{\Delta}'_{(i)} H^{-1} \boldsymbol{\Delta}_{(i)} \approx \boldsymbol{\Delta}'_{(i)} (\hat{\boldsymbol{\zeta}} - \hat{\boldsymbol{\zeta}}_{(i)}). \tag{5}$$

Notice the similarity between the second member of (5) and the gradient statistic $\boldsymbol{\Delta}'_0(\hat{\boldsymbol{\zeta}} - \boldsymbol{\zeta}_0)$, where $\boldsymbol{\Delta}_0$ is the score function calculated in $\boldsymbol{\zeta}_0$, the parameter under the null hypothesis $\boldsymbol{\zeta} = \boldsymbol{\zeta}_0$. Such a statistic, recently introduced by Terrell (2002), is asymptotically χ^2 distributed, although it is not a quadratic form and it can assume negative values for small sample sizes. Following Lesaffre and Verbeke

(1998), since $\sum_{i=1}^{N} \boldsymbol{\Delta}_i = 0$ and given that $\boldsymbol{\Delta}_{(i)} = \sum_{j \neq i} \boldsymbol{\Delta}_j = -\boldsymbol{\Delta}_i$, (5) becomes $\boldsymbol{\Delta}_i' H^{-1} \boldsymbol{\Delta}_i \approx \boldsymbol{\Delta}_i'(\hat{\boldsymbol{\zeta}}_{(i)} - \hat{\boldsymbol{\zeta}})$, which is substituted in (2) to obtain

$$C_i^a = 2 \mid \boldsymbol{\Delta}_i'(\hat{\boldsymbol{\zeta}}_{(i)} - \hat{\boldsymbol{\zeta}}) \mid . \tag{6}$$

C_i^a is a measure of influence, because of the distance $\hat{\boldsymbol{\zeta}}_{(i)} - \hat{\boldsymbol{\zeta}}$, but for its computation the information matrix is no more necessary. Notice that if the ith study is influential $\boldsymbol{\zeta}_{(i)} - \boldsymbol{\zeta}$ will be large and the accuracy of (4) will be likely to be lower. However, such an accuracy will no more be required if C_i^a is "sufficiently large to draw our attention for further consideration" (Cook and Weisberg 1982, p. 182). Similarly, the corresponding gradient-like measure for set M_i will simply be

$$C_{M_i}^a = 2 \mid \boldsymbol{\Delta}_{M_i}'(\hat{\boldsymbol{\zeta}}_{(M_i)} - \hat{\boldsymbol{\zeta}}) \mid . \tag{7}$$

The assessment of the influence on a specific parameter subset using C_{M_i} can be performed as suggested by Cook (1986) and Lesaffre and Verbeke (1998). For what concerns $C_{M_i}^a$, it will be sufficient to select the elements of $\boldsymbol{\Delta}_i$, $\boldsymbol{\zeta}$ and $\boldsymbol{\zeta}_{(i)}$ corresponding to the parameters of interest. From a computational point of view, C_{M_i} is less time consuming than $C_{M_i}^a$ since the former requires to estimate just one model, whereas the latter requires to estimate N to $\sum_i n_i$ models, depending on whether diagnostics are performed at the study or observation level, respectively. However, the point we want again to highlight is that we can carry out influence diagnostics when the information matrix is not available. Due to limitations of space, we refer to Ouwens et al. (2001) for all the computational details about Δ.

3 Simulated IPD with Artificially Created Outliers

In this section, by simulating and perturbing individual patient data, we will assess the performance of C_i^a and $C_{M_i}^a$ by comparison with the corresponding C_i and C_{M_i}. The aim of this simulated example is to show: (i) how C_i^a well approximates C_i; (ii) how to carry out the diagnostics in a meta-analysis of IPD assuming the information matrix is not available. We generate $N = 20$ simulated studies for which we want to estimate the odds ratio of the effect of a possible treatment, namely *treat*, on the eradication of a disease. Such a variable assumes the value $treat_{ij} = 0$ if the jth patient of the ith study belongs to the control (or placebo) group, $treat_{ij} = 1$ elsewhere, with $n_{i,1} = n_{i,treat=1}$ randomly chosen within the range [250; 2,500], and with a $\pm$ 10 % random variation from these sizes for $n_{i,0} = n_{i,treat=0}$. In addition, we consider a covariate *age*. To facilitate the detection of influential observations, it is preferable to work with multiple observations. Thus we proceeded to a minimum of data aggregation, in particular the variable *age* was created in years, ranging in [min($age_{i,treat}$),max($age_{i,treat}$)] with min($age_{i,treat}$) and max($age_{i,treat}$) randomly chosen in [30; 40] and [70; 80], respectively, so as to have a more realistic example. This aggregation corresponds to the very common situation of more than one patient with the same age. The response variable was created from the following random intercept binomial-normal model:

$$y_{ij}|b_i \sim \text{Bin}(n_{ij}, \pi_{ij}); \;\; \eta_{ij} = \log\left(\frac{\pi_{ij}}{1-\pi_{ij}}\right) = \beta_0 + treat_{ij} * \beta_1 + age_{ij} * \beta_2 + b_i, \tag{8}$$

where b_i is assumed to be $N(0, \sigma^2)$. The true parameter vector $(\beta_0, \beta_1, \beta_2, \sigma^2)'$ is chosen to be $(-1.7, 1.5, -0.01, 0.447)'$. Here we assume that the higher the observed response proportion, the better the clinical outcome for that patient's profile. Although models more complicated than (8) can be fitted, such a model has been used in the analysis of many real cases since it is useful when interest is in estimating the effect size for fixed individual level covariates. For convenience, we sorted such data in increasing order, according to the observed frequencies n_{ij}, and then according to *treat* and *age*, respectively. The next step concerns the perturbation of the IPD with artificially created outliers. Four studies were perturbed with two outliers in the treatment group and two in the control one, with four probability vectors: $(0.8, 0.8)'$, $(0.05, 0.8)'$, $(0.8, 0.05)'$ and $(0.05, 0.05)'$ respectively. This way of generating and perturbing data is similar to that used by Xu et al. (2006). For each study and group, those two outliers were chosen to be extreme with respect to variable *age* with the following three configurations: (1) the first two observations, that is the smallest values; (2) the last two observations, that is the highest values; (3) the first and the last observation, that is the minimum and the maximum values of *age*. Let p_{ij} be the observed proportion of events for the jth patient profile within the ith study. We perturbed the studies 5, 6, 11 and 12 because representative of four different situations: low p_{ij}, low n_i (study 5); medium p_{ij}, low n_i (study 6); medium p_{ij}, medium n_i (study 11); and low p_{ij}, medium n_i (study 12). Figure 1 displays studies 5, 6, 11 and 12 before perturbation.

Starting from the perturbed IPD, the diagnostics are carried out at both levels, by comparing C_i^a and $C_{M_i}^a$ with the corresponding C_i and C_{M_i}. To discriminate whether a study/observation is influential we can use the rule of thumb by Tukey (1977, p. 44), i.e. the cut-off 1.5*Q_3+IQR, where Q_3 is the third quartile and IQR is the inner quartile range. The study-level diagnostics are reported in Fig. 2. Observe how well C_i^a approximates C_i. The greatest difference is for study 12, which is the most influential study. The second study to be appeared is study 11, while studies 5 and 6 are not highlighted, probably due to a masking effect of the larger studies 11 and 12. Figure 3, on the left, reports the observation-level diagnostics for C_{M_i}, denoted by C_{ij}, while $C_{M_i}^a$, denoted by C_{ij}^a, is reported on the right.

The two graphics appear to be almost identical. All sixteen evidenced observations are exactly those we have perturbed and, as expected, belong to studies 12, 11, 5 and 6. In theory, in order to avoid the masking effect and highlight influential studies or observations simultaneously, we could inspect the eigenvector

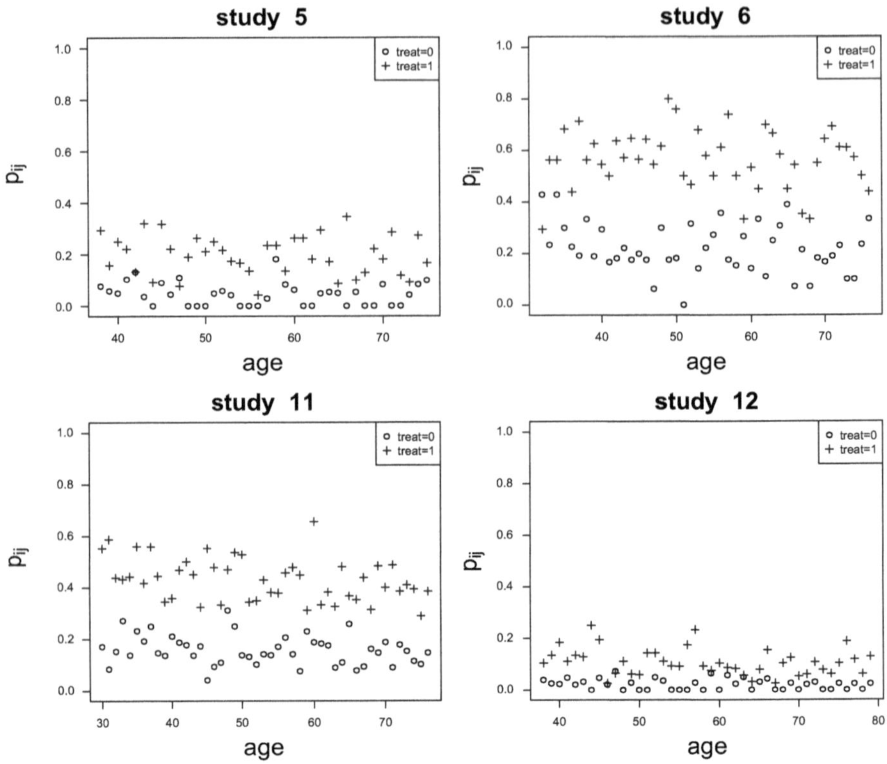

Fig. 1 Scatter plots of the studies to be perturbed

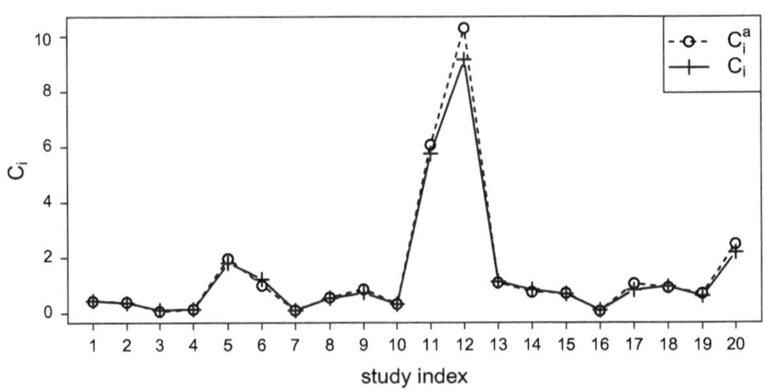

Fig. 2 Influence measures C_i and C_i^a at the study level for the perturbed IPD

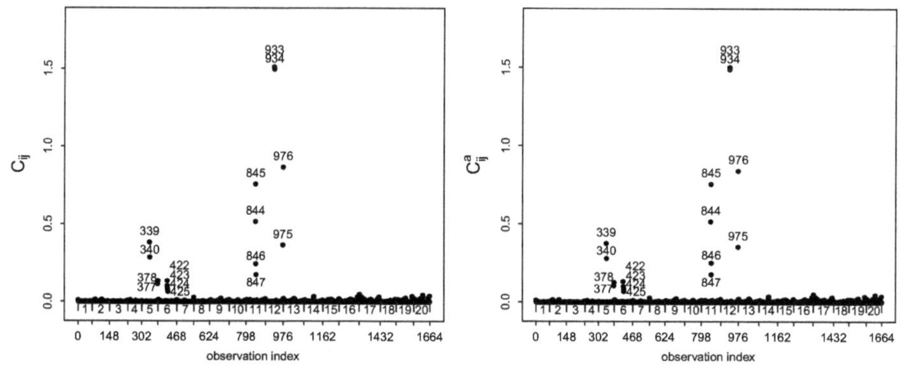

Fig. 3 Observation-oriented influence diagnostics for the IPD analysis using C_i (*left*) and C_i^a (*right*)

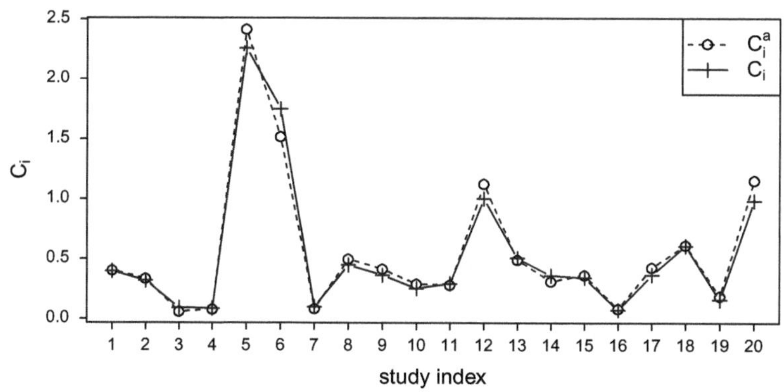

Fig. 4 Influence measures C_i and C_i^a at the study level for the perturbed IPD after removing the influential observations of studies 11 and 12

of $\Delta' H^{-1} \Delta$ corresponding to the largest eigenvalue, as suggested by Cook (1986) and Lesaffre and Verbeke (1998). However, the aim of the example is to show how to carry out the diagnostics in a meta-analysis of IPD assuming H is not available, whereas it is not possible to carry out such an analysis using C_i^a. Thus, we proceed step by step, by repeating the diagnostics at both levels, after removing the influential observations from the most influential study. This permits us to assess the conditional influence and detect possible masking or swamping effects. For space limits, we only show an intermediate step in Fig. 4, for which we have removed the observations 933–934 and 975–976, belonging to study 12, and the observations 845–847 belonging to study 11. There has been a masking effect of study 12 and 11 against studies 5 and 6. In fact, after removing the influential observations of the larger studies, the smaller perturbed studies are now evidenced. Although the scale of influence values is now reduced, such studies results to be outlying according to Tukey's rule of thumb. The approximation between the two

Table 1 Model estimates and absolute relative differences $\mid \hat{\xi}_{(M_i)} - \hat{\xi} \mid / \mid \hat{\xi} \mid$ (in brackets) for the deletion procedure on the perturbed IPD (pIPD)

Subset description	β_0	β_1	β_2	σ^2
True parameter values for the IPD	−1.7	1.5	−0.01	0.447
pIPD	−1.921	1.424	−0.005	0.423
pIPD without obs. 933–934 and 975–976	−1.843(0.04)	1.447(0.017)	−0.007(0.444)	0.479(0.132)
pIPD without obs. 933–934; 975–976 and 844–847;	−1.783(0.072)	1.472(0.034)	−0.008(0.738)	0.478(0.129)
pIPD without obs. 933–934; 975–976; 844–847; 339–340 and 377–378	−1.753(0.087)	1.485(0.043)	−0.009(0.949)	0.494(0.168)
pIPD without obs. 933–934; 975–976; 844–847; 339–340; 377–378 and 422–425	−1.722(0.104)	1.498(0.052)	−0.010(1.102)	0.494(0.167)

measures remains still good. We conclude the procedure by removing the influential observations belonging to studies 5 and 6.

We used package *lme4* to estimate the model and to calculate C_i^a, since posterior modes and their variances are available in the package output. However, the package output does not include the Hessian. Thus, in order to calculate C_i we combined the package output with that obtained from package *glmmML*, which returns the variance/covariance matrix deriving from Hessian inversion, but not the variances of the posterior modes. For the rest, both functions use the same estimation method (Laplace) and provide almost identical estimates of both fixed and random effects. Table 1 reports the parameter estimates together with their relative differences (in brackets) for model (8) fitted on the perturbed IPD (pIPD) and after the step-by-step removal of the influential observations.

As expected, after the deletion of the influential observations, the model without all the influential observations provides ML estimates which are very close to the true parameters. With respect to the model fitted on the perturbed IPD, the larger differences are observed, in relative terms, for parameter β_2, for which we can observe a 110 % estimate change. Thus, in our simulated IPD meta-analysis, the odds ratio of having a positive response for 10-year differences in older patients is $\exp(-0.01 * 10) \approx 0.9$, implying a 10 % decrease with respect to the younger ones.

4 Discussion

This paper has been motivated by the little attention given by meta-analysts in carrying out influence diagnostics for meta-analysis of IPD based on GLMMs. Influence diagnostics should be an important step to assess the influence of the studies on the estimates. Some practical difficulties could be encountered in performing such a diagnostics using the classical measure proposed by Cook (1986) since not all the statistical software programs return the necessary quantities for the calculation. As far as we know, this is the case of R Core Team (2012), used in this work for all the analyses, for which only package *glmmML* returns the information matrix but limited to a random-intercept model and, in addition, it does not provide all the quantities to calculate $s_{i\delta}$. We have proposed C_i^a, an alternative measure of influence diagnostics, here developed for the class of GLMMs, aimed at overcoming such difficulties. Although C_i^a has a computational cost higher than C_i, its calculation does not require the information matrix, while the same large-sample behaviour is maintained. A binomial-normal mixed model has been fitted to perturbed and simulated IPD in order to compare C_i^a and C_i and to address the influence diagnostics for IPD meta-analysis. As a result, C_i^a has shown a good performance in detecting the influential studies and observations while maintaining an acceptable approximation. The model chosen has been here employed also because it works with grouped data, in such a way to facilitate the detection of influential observations. To calculate $C_{M_i}^a$ and, in particular Δ_{M_i}, with other types of GLMMs, we refer to Ouwens et al. (2001). At time of writing we developed our code for binomial and Poisson distributions only. The code is available on request by sending an email to the authors.

Acknowledgements We are grateful to dr. V. Muggeo for having brought to our attention the gradient statistic.

References

Cook, R. D. (1977). Detection of influential observations in linear regression. *Technometrics, 19*, 15–18.

Cook, R. D. (1986). Assessment of local influence. *Journal of the Royal Statistical Society Series B Methodology, 4*(2), 133–169.

Cook, R. D., & Weisberg, S. (1982). *Residuals and influence in regression*. New York: Chapman and Hall.

Everitt, B. S. (2002). *The Cambridge dictionary of statistics* (2nd ed.). New York: Cambridge University Press.

Lesaffre, E., & Verbeke, G. (1998). Local influence in linear mixed models. *Biometrics, 54*(2), 570–582.

Ouwens, M. J. N. M., Tan, F. E. S., & Berger, M. P. F. (2001). Local influence to detect influential data structures for generalized linear mixed models. *Biometrics, 57*(42), 1166–1172.

R Core Team. (2012). *R: A language and environment for statistical computing*. Vienna: R Foundation for Statistical Computing. http://www.R-project.org/. ISBN 3-900051-07-0

Riley, R. D., & Steyerberg, E. W. (2010). Meta-analysis of a binary outcome using individual patient data and aggregate data. *Research Synthesis Methods, 1*, 2–19.

Terrell, G. R. (2002). The gradient statistic. *Computing Science and Statistics, 34*, 206–215.

Tukey, J. W. (1977). *Exploratory data analysis*. Reading: Addison-Wesley.

Viechtbauer, W., & Cheung, M. W.-L. (2010). Outlier and influence diagnostics for meta-analysis. *Research Synthesis Methods, 1*(2), 112–125.

Xu, L., Lee, S., & Poon, W. (2006). Deletion measures for generalized linear mixed models. *Computational Statistics & Data Analysis, 51*, 1131–1146.

Zhu, H., Lee, S., Wei, B., & Zhou, J. (2001). Case-deletion measures for models with incomplete data. *Biometrika, 88*(3), 727–737.

Social Networks as Symbolic Data

Giuseppe Giordano and Paula Brito

Abstract Starting from the main idea of Symbolic Data Analysis to extend Statistics and Data Mining methods from first-order to second-order objects, we focus on network data—as defined in the framework of Social Network Analysis—to define a graph structure and the underlying network in the context of complex data objects. A Network Symbolic description is defined according to the statistical characterization of the network topological properties. We use suitable network measures, which are represented by means of symbolic variables. Their study through multidimensional data analysis, allows for the synthetic representation of a network as a point onto a metric space. The proposed approach is discussed on the basis of a simulation study considering three classical network growth processes.

Keywords Histogram-valued data • Social network analysis • Symbolic data

1 Introduction

In recent years the network paradigm has affirmed as one of the most attractive and valuable cognitive models to describe and represent the complexity of relationships among a wide variety of actors. In a broader sense, the concept of network could be applied to any kind of actors able to establish relationships. However, the concept of network assumes special importance when the actors are individuals and relationships are related to specific state of properties attached to each pair

G. Giordano (✉)
Department of Economics and Statistics, Università di Salerno, Salerno, Italy
e-mail: ggiordan@unisa.it

P. Brito
Fac. Economia & LIAAD INESC TEC, Univ. Porto, Porto, Portugal
e-mail: mpbrito@fep.up.pt

D. Vicari et al. (eds.), *Analysis and Modeling of Complex Data in Behavioral and Social Sciences*, Studies in Classification, Data Analysis, and Knowledge Organization,
DOI 10.1007/978-3-319-06692-9_15,

of subjects (personal relations such as trustee, acquaintance, collaborations, and so on). These kinds of networks take into account human beings and the study of their birth, growth, shape and topology is the scope of Social Network Analysis. Nevertheless, the concept of network is so immediate and easy to be generalized that the underlying paradigm has been successfully applied in different fields, ranging from Communication and Transports to Economics as well as Medicine, Physics, Linguistics, Computer Science and many others. As a consequence, different disciplines contributed in different and not exclusive ways to the definition and interpretation of several network measurements. Such metrics have been found as specific features of classes of networks. The definition of network data by complex data objects is based on the different structural information that can be of interest to retrieve. The specific choice of suitable network indices expressed through statistical distributions will be addressed one by one according to the different network features that one wishes to highlight.

The idea of this work is to aggregate information attached to each node in terms of centrality and role (bridge, isolate, transmitter, etc.) in the network and express it as symbolic data by means of histogram-valued variables—see, e.g., Bock and Diday (2000), Noirhomme-Fraiture and Brito (2011)—so that the whole network can be expressed by a "vector" of high-order data (e.g. histograms). In this work, we consider the distributions of some typical network indices, measured at node-level, represented as histogram-valued variables. A symbolic data table will be defined, where each row pertains to one network and the columns hold the network indices. Symbolic data analysis of such data may be applied for the sake of comparisons among networks emerged at different occasions in time, computing similarities among networks, or representing networks as "points" on a reduced metric space, to cite just some possibilities.

The paper is organized as follows. Section 2 gives formal definitions and introduces basic statistical indices commonly used to describe networks. Section 3 recalls the general Symbolic Data framework and more specifically the concept of histogram-valued variables that will be used in the definition of the network symbolic descriptions. These descriptions and the resulting data table are defined in Sect. 4. A Simulation Study based on three network growth models (*Random Graph, Preferential Attachment and Small World*) is carried out in Sect. 5 aimed at describing the procedure and analyze its capability to discriminate among different network structures. Section 6 concludes the paper, pointing out directions for further research.

2 Statistical Description of Network Graph Characteristics

The statistical analysis of a network is basically performed with a descriptive purpose and originated in the framework of Social Network Analysis, see Wasserman and Faust (1994). Formally, network data refer to a set of actors and their relationships are commonly described and represented in the mathematical framework of *Graph Theory*. A graph data-structure is characterized by two sets: nodes and edges. Let $\mathscr{G}(N, E)$ be the graph represented by the set N of nodes (vertices) with cardinality $n = |N|$ and by the set E of edges with cardinality $m = |E|$. A fundamental concept in the description of a network is the centrality position of each node in the graph; in Social Network Analysis the definition of centrality may vary according to different criteria. The most important node-level centrality measure is the *Degree*. The Degree of a node is defined as the number of edges that connect to it. As a starting point, we consider only undirected simple finite graphs, that is, graphs where edges have no orientation, nodes have no loops, and where no more than one edge exists between any two nodes. There exist many statistical characterizations of a network according to its structural properties; among the important node-level statistics, we consider Closeness, Betweenness and Eigenvector centrality, see Freeman (1979) for definitions and interpretations.

The Degree tells about the number of connections a node has to other nodes and its distribution has been studied for real and theoretical network models, from the simplest Bernoulli random graph (Erdös and Rényi 1960) where it follows a Binomial distribution (limiting to Poisson for large n), to more complex models such as *scale-free networks* whose degree distribution follows approximately a *power law* of the form: $P(d) \sim d^{-\lambda}$, where λ is a constant. An important subclass of scale-free model is the *Preferential Attachment* generation process—also known as *Cumulative Advantage* process—first introduced to study the occurrence of power laws in scientific citation networks, see De Solla Price (1976), and then for explaining the presence of hubs in some parts of the *World Wide Web*, for which the constant λ should vary between *2* and *3*, see Barabási and Albert (1999). Network statistical measures refer either to individual nodes and edges (local measures) or to the network as a whole (global measures). They are defined for binary and weighted variables, for directed and undirected graphs (Kolaczyk 2009). Sometimes we are interested in detecting local measures in specific sub-parts of a graph. The presence of separate components in a graph defines an intermediate level of analysis. In this case the interest is in exploring not all individual nodes/edges but a small part of the graph: a connected subgraph, the giant component (i.e. the connected component with the larger number of nodes). It could also be interesting to study graph partitions induced by hierarchical clustering methods defined by similarity measures among nodes, see Batagelj (1998). Indeed, local network measures may be applied to sub-graphs too.

The definition of network data as a complex object allows considering the different structural information that can be of interest to retrieve, according to some theory-driven structural properties, depending on the field of study.

3 Symbolic Data

In classical statistics and multivariate data analysis the units under analysis are single entities described by numerical and/or categorical variables, each one taking one single value for each variable. Data are organized in a data-array, where each cell (i, j) contains the value of variable j for individual i. However, when analyzing a group rather than a single individual, the within-group variability should be explicitly considered. Consider, for instance, that we are analyzing the staff of some institutions, in terms of age, education level and category. If we just take averages or mode values within each institution, much information is lost. The same issue arises when we are interested in concepts and not in single specimen—whether it is the animal species (and not a specific animal), a model of car, etc. Symbolic Data Analysis, see, e.g., Bock and Diday (2000) and Noirhomme-Fraiture and Brito (2011), provides a framework where the variability intrinsic to a concept as a whole, or resulting from the aggregation of individual observations into groups, is considered in the data representation, and methods developed to take it into account. To describe groups of individuals or concepts, variables assume other forms of realizations; the new variable types, called "symbolic variables", may assume multiple, possibly weighted, values for each entity. Data are gathered in a matrix, now called a "symbolic data table", each cell containing "symbolic data". Each row of the table corresponds to a group, or concept, i.e., the entity of interest. A numerical variable may then be single valued, as in the classical framework, if it takes one single value of an underlying domain per entity, it is multi-valued if its values are finite subsets of the domain and it is an interval variable if its values are intervals. When an empirical distribution over a set of sub-intervals is given, the variable is called a histogram-valued variable. In this study, we shall represent information on networks, expressed by statistical distributions of node-level measures, by histogram-valued variables.

3.1 *Histogram-Valued Variables*

Let $S = \{s_1, \ldots, s_r\}$ be the set of entities under analysis. For an histogram variable Y (see Bock and Diday 2000) each element $s_i \in S$ is described by a discrete probability or frequency distribution on the set of considered sub-intervals $\{I_{i1}, \ldots, I_{ik_i}\}$ such that $Y(s_i) = (I_{i1}, p_{i1}; \ldots; I_{ik_i}, p_{ik_i})$ with $p_{i\ell}$ the probability or frequency associated to $I_{i\ell} = [\underline{I}_{i\ell}, \overline{I}_{i\ell}[$, $\ell \in \{1, \ldots, k_i\}$, and $p_{i1} + \ldots + p_{ik_i} = 1$. A *Uniform* distribution is assumed within each sub-interval $[\underline{I}_{i\ell}, \overline{I}_{i\ell}[$. For each observation s_i, $Y(s_i)$ can, alternatively, be represented by the cumulative distribution function $F_i(x)$, or by its inverse, the quantile function $q_i(t)$, both piecewise linear functions, given by

Table 1 Distribution of degree (number of friends) for two classes of students

	Degree
Class 1	([0, 4[, 0.2; [4, 8[, 0.5; [8, 12[, 0.2; [12, 16], 0.05; [16, 20], 0.05)
Class 2	([0, 4[, 0.05; [4, 8[, 0.4; [8, 12[, 0.25; [12, 16[, 0.2; [16, 20], 0.1)

$$F_i(x) = \begin{cases} 0\ , & x < \underline{I}_{i1} \\ p_{i1}\frac{x-\underline{I}_{i1}}{\underline{I}_{i2}-\underline{I}_{i1}}, & \underline{I}_{i1} \leq x < \underline{I}_{i2} \\ F(\underline{I}_{i2}) + p_{i2}\frac{x-\underline{I}_{i2}}{\underline{I}_{i3}-\underline{I}_{i2}}, & \underline{I}_{i2} \leq x < \underline{I}_{i3} \\ \vdots & \\ F(\underline{I}_{i(k_i-1)}) + p_{i(k_i)}\frac{x-\underline{I}_{ik_i}}{\overline{I}_{k_i}-\underline{I}_{ik_i}}, & \underline{I}_{ik_i} \leq x < \overline{I}_{ik_i} \\ 1, & \overline{I}_{ik_i} \leq x \end{cases}$$

$$q_i(t) = \begin{cases} \underline{I}_{i1} + \frac{t}{w_{i1}}a_{i1}, & 0 \leq t < w_{i1} \\ \underline{I}_{i2} + \frac{t-w_{i1}}{w_{i2}-w_{i1}}a_{i2}, & w_{i1} \leq t < w_{i2} \\ \vdots & \\ \underline{I}_{ik_i} + \frac{t-w_{ik_i-1}}{1-w_{ik_i-1}}a_{ik_i}, & w_{ik_i-1} \leq t \leq 1 \end{cases}$$

where $w_{ih} = \sum_{\ell=1}^{h} p_{i\ell}, h = 1, \ldots, k_i; a_{i\ell} = \overline{I}_{i\ell} - \underline{I}_{i\ell}$ for $\ell = \{1, \ldots, k_i\}$.

If this latter representation is chosen, then the observations $Y(s_i)$ should be re-written using the same weight distribution, to allow for the comparison of the corresponding quantile functions, since this procedure leads to functions with the same number of terms corresponding to the same sub-intervals of the unit interval.

Henceforth "distribution" refers to a probability or frequency distribution of a numerical variable represented by a histogram or a quantile function.

Example 1. Consider two classes of students, for which we know the friendship relation among classmates, the friendship networks are described by the respective Degree distributions, as in Table 1. In Class 1 20 % of the students have a number of friends (degree) less than 4, 50 % have degree between 4 and 7, 20 % between 8 and 11, 5 % between 12 and 15, and 5 % between 16 and 20; likewise for Class 2. The units of interest are the classes as a whole and not each individual student. Figure 1a represents the histograms of variable "Network Degree" for Class 1 and Class 2, and Fig. 1b depicts the respective quantile functions.

4 Representation of Networks by Symbolic Variables

To represent a network as symbolic data, we consider the empirical distribution of network measurements referring to each node (Degree, Closeness, etc.) and represent them by histograms. In the resulting symbolic data table, each row pertains

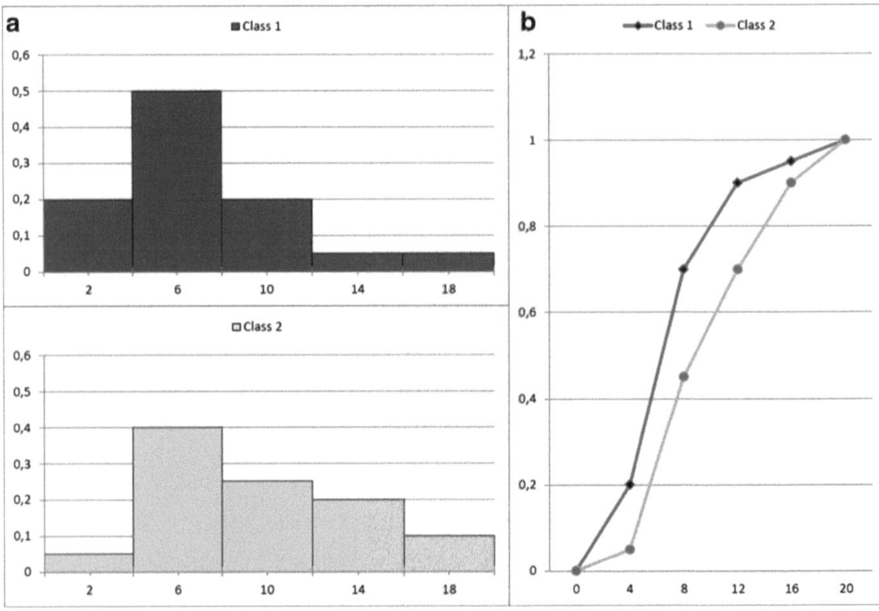

Fig. 1 Representation of the Degree of Class 1 and Class 2 by histograms (**a**) and quantile functions (**b**)

to a different network and each column to a network index, each cell recording the distribution of the corresponding index in the given network. Consider, for example, that the degree of each node of each network under study has been computed. Then, for each given network, the distribution of the degree values may be obtained, and represented in the form of a histogram (or by the corresponding quantile function). The same may be done for all network indices under analysis. Thereby, each row of the data array corresponds to a description of a network, as a "vector" of histograms. Distances between such descriptions may be used to compare several networks and represent them as "points" in a reduced metric space.

5 Simulation Study

A simulation study is carried out to generate several network data structures. The simulation scheme controls for two attributes: Graph order and Generating process. A third attribute, the parameter regulating the process is specific to each process. Each of the three factors has three levels, leading to a total of 27 different network data structures. Each type of network is replicated 100 times. The considered levels are: (1) the order of the graph: $n \in \{100; 300; 500\}$; (2) the generating process: $P \in \{\textit{Random Graph}; \textit{Preferential Attachment}; \textit{Small-World}\}$

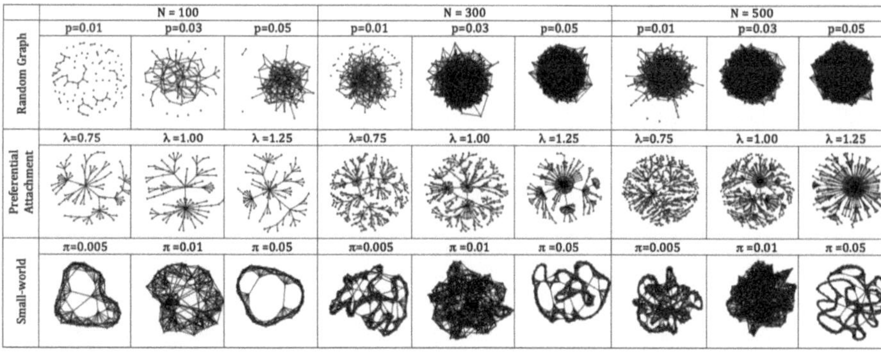

Fig. 2 Examples of the networks generated for each of the 27 configurations

and (3) the process parameter: for each generating process a specific parameter controls, respectively: the density (p) of the Random Graph: $p \in \{0.01; 0.03; 0.05\}$; the power ($\lambda$) of the Preferential Attachment: $\lambda \in \{0.75; 1.00; 1.25\}$; the rewiring probability (π) of the Small-World model: $\pi \in \{0.005; 0.01; 0.05\}$. Figure 2 shows examples of the networks generated for the different network data structures.[1]

For each of the 100×27 networks, we have computed the Betweenness Centrality, the Closeness, the Degree and the Eigenvector Centrality, and the corresponding order statistics; these were then summarized by the corresponding median values for each of the 27 network types. These data were then standardized, given the different network sizes involved. Finally, the network data are represented in a 27×4 symbolic data matrix, containing in each cell the distribution of the index in column for the network type in that row. Different multidimensional symbolic data analysis could be performed on the obtained symbolic data array. In this work, ascending hierarchical clustering has been carried out, using the Mallows' distance, which is adapted to the kind of data at hand—see also Irpino and Verde (2006). The Mallows' distance is computed using the quantile functions of the distributions in the symbolic data table. For each variable Y_j, $j = 1, \dots, 4$ (each network index in our case), we have $d_M^2(R_{i_1 j}, R_{i_2 j}) = \int_0^1 (q_{i_1 j}(t) - q_{i_2 j}(t))^2 \, dt$ where q_{ij} is the quantile function of the distribution of variable Y_j for network R_i. The global squared distance between network types is then computed additively on the variables, $D_M^2(R_{i_1}, R_{i_2}) = \sum_{j=1}^{4} d_M^2(R_{i_1 j}, R_{i_2 j})$. The use of the same quantile representation for each network makes it possible to directly compare the quantile functions. Moreover, in Irpino and Verde (2006) it is proved that if a Uniform distribution is assumed in each sub-interval of the histograms, the squared Mallows' distance may be re-written using the midpoints and half-ranges of these sub-intervals (in number

[1]Simulations and network statistics are obtained by: *R version 2.15.2 (2012-10-26). Base packages: base, datasets, graphics, grDevices, methods, stats, utils; other: igraph 0.6.5-1, sna 2.2-1.*

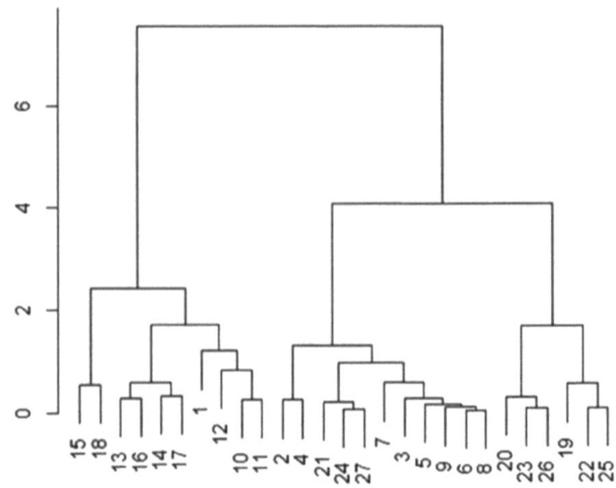

Fig. 3 Dendrogram on the 27 network types, using the Mallows' distance on standardized data, and the Ward aggregation criterion

of K_j) : $d_M^2(R_{i_1 j}, R_{i_2 j}) = \sum_{\ell=1}^{K_j} p_\ell \left[(c_{i_1 j\ell} - c_{i_2 j\ell})^2 + \frac{1}{3}(r_{i_1 j\ell} - r_{i_2 j\ell})^2 \right]$, where, for the histogram-valued variable j and network i, $c_{ij\ell} = \frac{\overline{I}_{ij\ell} + \underline{I}_{ij\ell}}{2}$ is the midpoint of the interval $I_{ij\ell}, \ell \in \{1, \ldots, K_j\}$ and $r_{ij\ell} = \frac{\overline{I}_{ij\ell} - \underline{I}_{ij\ell}}{2}$ the corresponding half-range. In our implementation, we have described each distribution by a histogram with 100 sub-intervals, defined by the distributions' percentiles, i.e., $K_j = 100, j = 1, \ldots, 4$ and $p_\ell = 0.01, \ell = 1, \ldots, K_j$. Applying the Mallows' distance $D_M(R_{i_1}, R_{i_2})$ to these data (see Irpino and Verde 2006), we obtained a 27×27 distance matrix, on which hierarchical clustering with the Ward criterion has been performed. Figure 3 represents the obtained dendrogram.

Looking at the obtained hierarchy, we may conclude that the used distance is able to discriminate the group labeled as 10–18 (the Preferential Attachment processes) and the 0–9 group (Random graphs); as regards the group 19–27 (Small World) three out of nine cases have been confused with the Random graph group, this is likely to happen when Small World processes have higher values for the parameter π, and we have the higher rewiring probability ($\pi = 0.05$) in graphs 21, 24 and 27, in this experiment.

The results are therefore promising, allowing discriminating quite well the different processes. However further study should be devoted to establish true sensitivity and robustness of the proposed approach as well as finding suitable network statistics to discriminate among different kind of networks.

6 Conclusions

A Network Symbolic Data Analysis approach has been proposed. Network symbolic descriptions have been defined that represent network indices by histogram-valued variables, leading to a symbolic data matrix allowing for multivariate data analyses. A simulation study has shown that complex information may be dealt with by this approach and a distance matrix among networks can be defined. Application to real data-sets can take advantage of the proposed approach in terms of (1) comparison among different network structures, (2) exploring sub-graphs or components of complex networks, as well as (3) in longitudinal studies where the same network is observed in different occasions. Indeed, the possibility of obtaining a distance matrix among different networks may lead to the application of classical factorial techniques. Transforming a whole network to a point in a metric space is one the major advantages of the proposed approach. Representing networks as points in a factorial subspace, for instance, may help discussing their proximity, their clustering or analysing trajectories of such *network-points* in order to explore their dynamics.

Acknowledgements This work is financed by the ERDF—European Regional Development Fund through the COMPETE Programme (operational programme for competitiveness) and by National Funds through the FCT—Fundação para a Ciência e a Tecnologia (Portuguese Foundation for Science and Technology) within project "FCOMP-01-0124-FEDER-037281".

References

Barabási, A. -L., & Albert, R. (1999). Emergence of scaling in random networks. *Science, 286*(5439), 509–512.

Batagelj, V. (1988). Generalized Ward and related clustering problems. In H.-H. Bock (Ed.), *Classification and related methods of data analysis* (pp. 67–74). Amsterdam: North-Holland.

Bock, H. -H., & Diday, E. (2000). *Analysis of symbolic data*. Berlin: Springer.

De Solla Price, D. J. (1976). A general theory of bibliometric and other cumulative advantage processes. *Journal of the American Society for Information Science, 27*(5), 292–306.

Erdös, P., & Rényi, A. (1960). *On the evolution of random graphs. Publication of the Mathematical Institute of the Hungarian Academy of Science, 5*, 17–61

Freeman, L. C. (1979). Centrality in social networks I: Conceptual clarification. *Social Networks, 1*, 215–239.

Irpino, A., & Verde, R. (2006). A new wasserstein based distance for the hierarchical clustering of histogram symbolic data. In V. Batagelj, et al. (Eds.), *Proceedings of IFCS 2006* (pp. 185–192). Heidelberg: Springer.

Kolaczyk, E. D. (2009). *Statistical analysis of network data. Methods and models*. Springer Series in Statistics. New York: Springer.

Noirhomme-Fraiture, M., & Brito, P. (2011). Far beyond the classical data models: Symbolic data analysis. *Statistical Analysis and Data Mining, 4*(2), 157–170.

Wasserman, S., & Faust, K. (1994). *Social networks analysis: Methods and applications*. New York: Cambridge University Press.

Statistical Assessment for Risk Prediction of Endoleak Formation After TEVAR Based on Linear Discriminant Analysis

Kuniyoshi Hayashi, Fumio Ishioka, Bhargav Raman, Daniel Y. Sze, Hiroshi Suito, Takuya Ueda, and Koji Kurihara

Abstract Over the past decade, therapy for thoracic aneurysms involving the use of a stent-graft has gained popularity as an alternate therapy for surgical treatment. This therapy is considered to be safe and efficient, and realizes satisfactory short-to-midterm results. However, a clinical side effect called endoleak has often been observed after alternate therapy. Based on the empirical findings of doctors, if a stent-graft is inserted into the part of the large curvature on the aortic angiography of a patient, it is believed that there is an increased risk of endoleak formation. To understand the relationship between the risk and the aortic curvature, we set a two-class discriminant problem involving no-endoleak and endoleak groups, and apply linear discriminant analysis to the two-class discriminant problem with a quantitative dataset that is associated with the curvature of aortic angiography and the insertion position of a stent-graft. Next, we propose a procedure for the

K. Hayashi (✉) • H. Suito • K. Kurihara
Graduate School of Environmental and Life Science, Okayama University, 3-1-1 Tsushima-naka, Kita-ku, Okayama City 700-8530, Japan
e-mail: k-hayashi@ems.okayama-u.ac.jp; suito@ems.okayama-u.ac.jp; kurihara@ems.okayama-u.ac.jp

F. Ishioka
School of Law, Okayama University, 3-1-1 Tsushima-naka, Kita-ku, Okayama City 700-8530, Japan
e-mail: fishioka@law.okayama-u.ac.jp

B. Raman • D.Y. Sze
Department of Radiology, Stanford University School of Medicine, 300 Pasteur Derive, Stanford CA 94305-5208, USA
e-mail: ramanb@stanford.edu; dansze@stanford.edu

T. Ueda
Department of Radiology, St. Luke's International Hospital, 9-1 Akashi-cho, Chuo-ku, Tokyo 104-8560, Japan
e-mail: takueda@luke.or.jp

D. Vicari et al. (eds.), *Analysis and Modeling of Complex Data in Behavioral and Social Sciences*, Studies in Classification, Data Analysis, and Knowledge Organization, DOI 10.1007/978-3-319-06692-9_16,

diagnostics based on the sign of the sample influence function for the average discriminant score in each class. In addition, we apply our proposed diagnostic procedure to the prediction result of the two-class linear discriminant analysis, and detect large influential individuals for the improvement of the prediction accuracy for endoleak groups. With our approach, we determine the relation between the curvature of the aorta and the risk of endoleak formation.

Keywords Average discriminant score • Quantitative analysis of aortic morphology • Sample influence function

1 Introduction

As alternate therapy, thoracic endovascular aortic repair (TEVAR) has been used in the field of surgery for thoracic aortic aneurysm therapy. In TEVAR, a specific artificial device called a stent-graft is inserted into a part of the thoracic aortic aneurysm. In comparison with therapy in existing surgeries, TEVAR is a low-risk and effective therapy with respect to short-to-midterm results. However, clinical side effects called endoleaks have been observed after TEVAR. Endoleak is a significant risk factor for postoperative aneurysmal expansion and rupture (Nakatamari et al. 2011). Thus far, many investigators have recognized the role of aortic morphology in this risk and the importance of adequate device fixation and endoleak seals (Appoo et al. 2006; Bortone et al. 2004; Serag et al. 2007). On the other hand, an adequate quantitative analysis based on the curvature of the thoracic aorta is less-advanced. Based on these facts, we have quantitatively analyzed the discrimination between the patients of no-endoleak and endoleak groups. For example, in Nakatamari et al. (2011), we performed the discrimination for no-endoleak and endoleak groups by linear discriminant analysis. In this study, we confirmed the strong relationship that exists between the curvature of the thoracic aorta of a patient and the occurrence of endoleak formation; however, we did not confirm that the risk of endoleaks is high if a stent-graft is inserted into the position of the large curvature on the aortic morphology of patients.

In this paper, to confirm the relationship between the high risk of the occurrence of endoleak and the large curvature in the region in which the stent-graft is inserted, we performed statistical diagnostics for the discrimination result of no-endoleak and endoleak groups based on linear discriminant analysis. Here, we propose a statistical diagnostic procedure that is based on the sign of the sample influence function for the average discriminant score in each class. We performed the two-class discrimination using linear discriminant analysis with a quantitative dataset in terms of the curvature of the thoracic aortic vessel and the insertion position of the stent-graft. Then, we applied our proposed procedure to the discrimination result. Finally, with our approach, we evaluated the empirical finding where the risk of an endoleak is high if a stent-graft is inserted into the position of the large curvature on large arterial vessels.

2 Average Discriminant Score in Linear Discriminant Analysis

In linear discriminant analysis, the discriminant score and its average are important statistics because these statistics represent the magnitude of the separation between classes. Therefore, in this paper, we focus on the average of the discriminant scores in each class. We denote a categorical variable and a real-valued random input variable as G and X, respectively. In addition, we assume the estimated density function of the k-th class to be $\hat{f}_k(\mathbf{x})$ in the case of the class-conditional density of X in class $G = k$ (Hastie et al. 2001). k is from 1 to K. We assume the density function in each class to be the multivariate Gaussian density function, $\hat{f}_k(\mathbf{x}) = \frac{1}{(2\pi)^{p/2}|\hat{\Sigma}_k|^{1/2}} \exp(-\frac{1}{2}(\mathbf{x} - \hat{\boldsymbol{\mu}}_k)^{\mathrm{T}}\hat{\Sigma}_k^{-1}(\mathbf{x} - \hat{\boldsymbol{\mu}}_k))$ such as Hastie et al. (2001). $\hat{\boldsymbol{\mu}}_k$ is the estimated mean vector of the population in the k-th class, and $\hat{\Sigma}_k$ is the estimated covariance matrix of the population in the k-th class. Moreover, with reference to Hastie et al. (2001), let $\hat{\pi}_g$ be the estimated prior probability of class g, with $\sum_{g=1}^{K} \hat{\pi}_g = 1$. For example, the weighted prior probability in the k-th class is calculated as $\frac{\hat{\pi}_k|\hat{\Sigma}_k|^{-1/2}}{\sum_{g=1}^{K} \hat{\pi}_g|\hat{\Sigma}_g|^{-1/2}}$. Based on the Bayes theorem, $\Pr(G = k|X = \mathbf{x}) = \frac{\hat{\pi}_k|\hat{\Sigma}_k|^{-1/2}\hat{f}_k(\mathbf{x})}{\sum_{g=1}^{K} \hat{\pi}_g|\hat{\Sigma}_g|^{-1/2}\hat{f}_g(\mathbf{x})}$. In linear discriminant analysis, we suppose that $\hat{\Sigma}_k(k = 1, \ldots, K)$ are equal to $\hat{\Sigma}$, which is the common covariance matrix of the population in all classes. Using the log-ratio between the probabilities of the k-th class and the other classes, the i-th discriminant score in the k-th class is calculated as $\hat{z}_k(\mathbf{x}_i^k) = \log \hat{\pi}_k \hat{f}_k(\mathbf{x}_i^k) - \frac{1}{K-1}\sum_{\ell=1,\ell\neq k}^{K} \log \hat{\pi}_\ell \hat{f}_\ell(\mathbf{x}_i^k)$ where i is from 1 to n_k. Therefore, the average of the discriminant scores in the k-th class is calculated as $\hat{Z}_k = \frac{1}{n_k}\sum_{i=1}^{n_k} \hat{z}_k(\mathbf{x}_i^k)$. If the value of $\hat{Z}_k$ has a large positive value, we can see that the k-th class is well separated from the other classes. Therefore, $\hat{Z}_k$ is considered as a measure of the separation between the k-th class and the other classes.

In this paper, we focus on the average of the discriminant scores in each class $\hat{Z}_k$, and evaluate the effect of the individual in each class for its statistics based on the influence function.

3 Diagnostics Based on the Sign of the Sample Influence Function

In this section, we first explain the existing single-case diagnostics, which is an evaluation of the influence of a single individual. Next, we explain the existing multiple-case diagnostics for the assessment of multiple individuals for a target statistics.

3.1 Single-Case Diagnostics

When we regard $\hat{Z}_k$ to be a target statistics, to evaluate the influence of the r-th individual in the g-th class for its statistics, the sample influence function (Hampel et al. 1986; Tanaka 1994) is

$$\mathrm{SIF}(\mathbf{x}_r^g; \hat{Z}_k) = -(n_g - 1)(\tilde{\hat{Z}}_k - \hat{Z}_k), \tag{1}$$

where $\tilde{\hat{Z}}_k$ is the average of the discriminant scores in the k-th class from the estimated prior weighted probability and the density function in each class, which is calculated by deleting $\mathbf{x}_r^g (r = 1, \ldots, n_g;\ g = 1, \ldots, K)$ where n_g is the number of individuals in class g. Therefore, in the sample influence function, at a point of $\mathbf{x}_r^g$, we evaluate the change from $\hat{Z}_k$ to $\tilde{\hat{Z}}_k$ by omitting $\mathbf{x}_r^g$.

We can regard $\hat{Z}_k$ as a functional based on the empirical cumulative distribution function in each class. Then, we can write $\hat{Z}_k$ as $\hat{Z}_k(\hat{F}^1, \ldots, \hat{F}^g, \ldots, \hat{F}^K)$, where $\hat{F}^g = \frac{1}{n_g}\sum_{i=1}^{n_g} \delta_{\mathbf{x}_i^g}$. We can also write $\hat{Z}_k$ in the case of the omission of the r-th individual in the g-th class as $\hat{Z}_k(\hat{F}^1, \ldots, \hat{F}^{g(-r)}, \ldots, \hat{F}^K)$. $\hat{F}^{g(-r)} = \frac{1}{n_g-1}(\sum_{i=1}^{n_g} \delta_{\mathbf{x}_i^g} - \delta_{\mathbf{x}_r^g})$. $\tilde{\hat{Z}}_k = \hat{Z}_k(\hat{F}^1, \ldots, \hat{F}^{g(-r)}, \ldots, \hat{F}^K)$. The superscript of $(-r)$ represents the omission of the r-th individual. Therefore, $\hat{F}^{g(-r)} = (1-\varepsilon)\hat{F}^g + \varepsilon\delta_{\mathbf{x}_r^g}$ where ε is equal to $-\frac{1}{n_g-1}$. Then, $-(n_g-1)$ in (1) corresponds to $1 / \varepsilon$. For each class, we plot $\mathrm{SIF}(\mathbf{x}_r^g; \hat{Z}_k)$ along the perturbation number r in the g-th class, and we detect the large influential individuals for the prediction accuracy. The negative value of the sample influence function shows the improvement of the prediction accuracy due to the omission of $\mathbf{x}_r^g$. On the other hand, the positive value of the sample influence function represents the degradation of the prediction accuracy realized by deleting $\mathbf{x}_r^g$.

3.2 Multiple-Case Diagnostics

In the multiple-case diagnostics, we assess the influence of the multiple individuals in the g-th class for $\hat{Z}_k$. Here, we denote the subset of the multiple individuals in the g-th class that are evaluated as A_g. In addition, in this study, we also denote the number of individuals belonging to A_g as c. Then, in the sample influence function of the multiple-version for $\hat{Z}_k$, $\hat{F}^g$ is perturbed as $\hat{F}^{g(-A_g)} = \frac{1}{n_g-c}(\sum_{i=1}^{n_g} \delta_{\mathbf{x}_i^g} - \sum_{r=1}^{c} \delta_{\mathbf{x}_r^g}) = (1-\varepsilon)\hat{F}^g + \varepsilon\frac{1}{c}\sum_{r=1}^{g} \delta_{\mathbf{x}_r^g}$ where $\varepsilon = -\frac{c}{n_g-c}$. Therefore, the sample influence function of the multiple-version (Tanaka 1994) is calculated as

$$\mathrm{SIF}(A_g; \hat{Z}_k) = -\frac{n_g - c}{c}(\tilde{\hat{Z}}_k - \hat{Z}_k), \tag{2}$$

where $\hat{\tilde{Z}}_k$ is the average of the discriminant scores in the k-th class, which is calculated by omitting the individuals belonging to A_g. As the single-case diagnostics in Sect. 3.1, for each class, with (2), we search the large influential individuals to determine the prediction accuracy in linear discriminant analysis.

3.3 *Proposed Procedure Regarding Diagnostics for Discrimination*

We searched for individuals that strongly affect the prediction accuracy in linear discriminant analysis from the sign of the sample influence function as follows.

Step 1 For each class g, detect the individuals that yield at least one negative $\mathrm{SIF}(\mathbf{x}_r^g; \hat{Z}_k)(k = 1, \ldots, K)$.

Step 2 For each class g, calculate all possible combinations A_gs of the individuals that yield negative values in Step 1.

Step 3 For each $\hat{Z}_k$, calculate $\mathrm{SIF}(A_g; \hat{Z}_k)(g = 1, \ldots, K)$ for all A_gs calculated in Step 2. On the basis of the relative separation between classes, sort $\mathrm{SIF}(A_g; \hat{Z}_k)(k = 1, \ldots, K)$ about A_g in ascending order of $\sum_{k=1}^{K} \mathrm{SIF}(A_g; \hat{Z}_k)$.

Step 4 From the sorted $\mathrm{SIF}(A_g; \hat{Z}_k)$, select the A_g corresponding to the highest-ranking $\mathrm{SIF}(A_g; \hat{Z}_k)$ $(k = 1, \ldots, K)$.

In Sect. 4, we perform the discrimination for no-endoleak and endoleak groups by linear discriminant analysis with the quantitative data in terms of the curvature of the patient and the insertion position of the stent-graft. In Sect. 5, we apply our proposed approach to the result of the discriminant analysis.

4 Discrimination of No-Endoleak and Endoleak Groups

Between April 2001 and September 2008, 121 patients were prospectively enrolled in one of six Food and Drug Administration (FDA)-sponsored clinical trials testing the thoracic EXCLUDER or TAG stent-graft devices (W. L. Gore and Associates, Flagstaff, Arizona) (Nakatamari et al. 2011). All of the patient data were handled in compliance with the Health Insurance Portability and Accountability Act (Nakatamari et al. 2011). In this study, to obtain an expert medical perspective, we included 45 patients (no-endoleak : 23 patients, endoleak : 22 patients) from the 121 patients. For each patient, we first calculated the position of the center line of the thoracic aorta from the origin. Next, we calculated the curvature index as the quantitative data in terms of the aortic morphology of a patient. The target section of the thoracic aorta was located at 30 mm before the right brachio-cephalic artery and extended to the celiac artery.

In the preprocessing, we had to normalize the length of the target part due to the different lengths of the target part of each patient. Therefore, we calculated the normalized position for each patient based on the shortest length of the target part among the 45 patients. After that, we calculated the curvature of the patient based on the normalized position of each patient. We adopted 20 variables, which are the maximum curvature values within the interval of 11 mm from the initial to the final position of the target part to be analyzed. In addition, for each patient, we added two variables, which are the initial and final positions for the stent insertion part on the normalized position. Finally, we set the 22 variables that are associated with the curvature of the aortic morphology and the position of the stent-graft insertion part.

Using the discriminant rule containing the weighted prior probability in Sect. 2, with the 22 variables, we performed the two-class discrimination between no-endoleak and endoleak groups. From the result of the discrimination in the training data, the true discriminant rates in no-endoleak and endoleak groups were found to be 91.304 and 95.455 %, respectively. On the other hand, the true discriminant rates in no-endoleak and endoleak groups based on leave-one-out cross-validation were 56.522 and 63.636 %, respectively. Therefore, with the 22 variables, we could not enhance the prediction accuracy of the classifier. Considering these results, in the next step, we performed stepwise variable selection based on leave-one-out cross-validation. As a result, 11 variables among the 22 variables were chosen in this stepwise selection. In the classifier calculated by the 11 variables after the variable selection, the true discriminant rates in the training data in no-endoleak and endoleak groups were 95.652 and 90.909 %, respectively. Moreover, based on leave-one-out cross-validation, the true discriminant rates in no-endoleak and endoleak groups were 82.609 and 86.364 %, respectively. After the variable selection, the prediction accuracy significantly improved. Of the 11 variables, 10 variables were related to the curvature of the aortic morphology, and the remaining one was the variable of the initial position of stent-graft insertion. From analysis of the linear discriminant analysis to its variable selection, we determined that some parts of the aortic morphology and the starting position of the stent-graft insertion position are important for the discrimination between no-endoleak and endoleak groups from the perspective of prediction accuracy.

5 Application of Our Proposed Procedure for Real Data Analysis

In this section, by applying our proposed procedure on the diagnostics of discriminant analysis to the linear discriminant model after the variable selection in Sect. 4, we detected the large influential individuals in no-endoleak and endoleak groups. Then, we looked into the empirical finding where the risk of endoleak formation is high if the curvature of thoracic aortic angiography is large in the part of stent-graft insertion. When the average discriminant scores in no-endoleak and endoleak groups were computed after perturbing the individuals in both groups, 17 individuals in each class yielded at least one negative value of sample influence function. Therefore, the number of subsets to be perturbed was 131,071 in no-endoleak and endoleak groups. In the results based on leave-one-out cross-validation after the variable selection, the true discrimination rates in no-endoleak and endoleak groups were 82.609 and 86.364 %, respectively. By considering the importance of the misclassification from endoleak to no-endoleak group with the condition that we can expect the true discrimination rate in no-endoleak to be equal or improved, in each class, we searched the influential subset having the effect of the largest improvement for the true discrimination rate in endoleak group among the 131,071 subsets. The subsets of the eight and two individuals were selected in no-endoleak and endoleak groups, respectively. For the no-endoleak group, we plotted the average curvatures of the detected eight individuals and the remaining individuals along the variables chosen after variable selection in the upper figure in Fig. 1. For the endoleak group, we also plotted the average curvatures of the detected two individuals and the other individuals along the variables chosen after variable selection in the lower figure in Fig. 1. In these figures, the index number 1 indicates the nearest position from the origin of the thoracic aorta after variable selection. The index number 19 represents the farthest position after variable selection. The grid interval using the broken line corresponds to the rough average position of stent-graft insertion. From these figures, we can see that there is a high risk of endoleak formation in the case of stent-graft insertion into the part of the large curvature on thoracic aortic morphology.

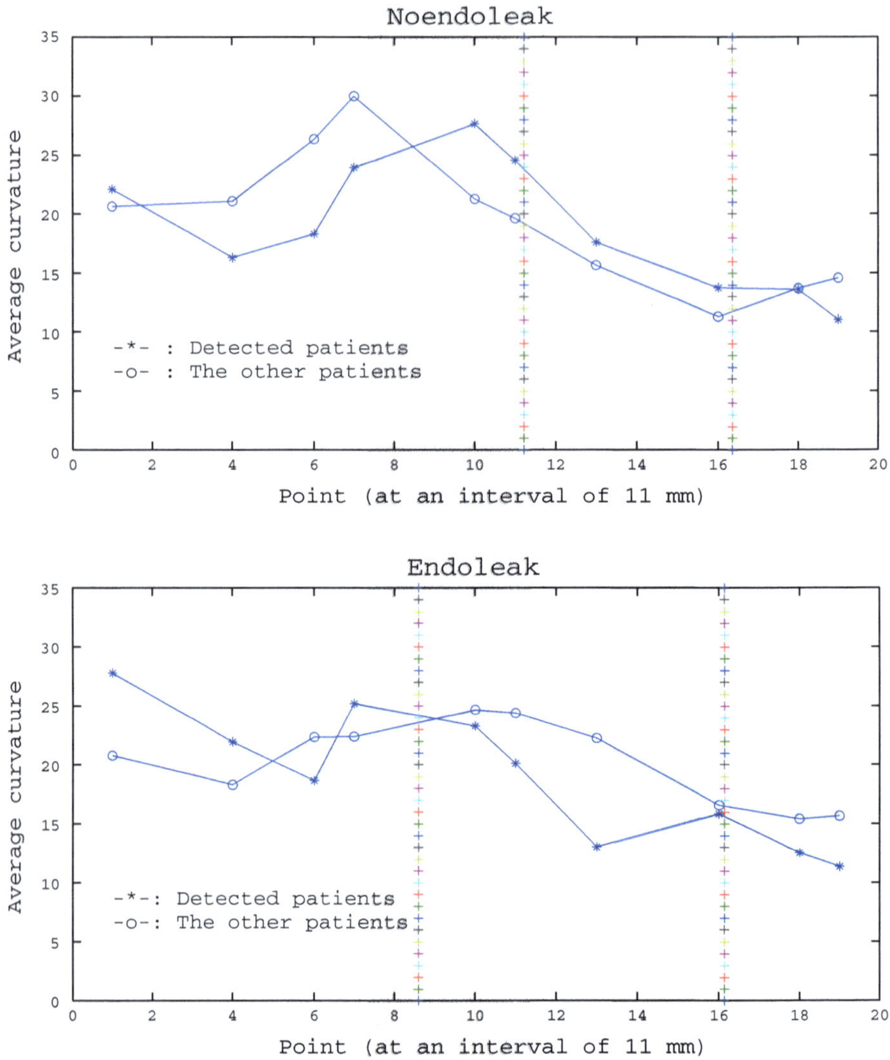

Fig. 1 Average thoracic aortic curvatures of influential individuals compared with those of other patients

6 Concluding Remarks

In Nakatamari et al. (2011), we assumed that the data in the no-endoleak and endoleak groups follow a multivariate normal distribution with a common covariance and adopted linear discriminant analysis for two-class discrimination. In this study, we confirmed that the curvature of thoracic aorta at the stent-graft insertion point is associated with endoleak formation. To demonstrate this fact, we proposed a diagnostic procedure for discriminant analysis based on the sign of the sample influence function. Through our proposed diagnostics, we detected individuals that largely affected the outcome in no-endoleak and endoleak groups. Our statistical analysis reinforced the empirical finding that endoleak risk increases if the stent-graft is inserted into a region of high curvature on the great arterial vessel. If our proposed diagnostics were applied in other discriminant methods such as quadratic discriminant analysis and logistic regression analysis, we expect that the identified influential subsets would be similar to those identified in linear discriminant analysis. However, comparing the results of diverse discriminatory diagnostics would likely reveal new findings. In addition, by selecting the part of the influential subsets that is common to all of these diagnostics, we could detect a core subset for prediction accuracy. Therefore, if a patient exhibits signals that are characteristic of essential influential patients, doctors can confidently prepare an alternative therapy for that patient in advance.

Acknowledgment This work was partly supported by the Core Research of Evolutional Science & Technology (CREST) of the Japan Science and Technology Agency (Project: Alliance between Mathematics and Radiology).

References

Appoo, J. J., Moser, W. G., Fairman, R. M., Cornelius, K. F., Pochettino, A., Woo, E. Y., et al. (2006). Thoracic aortic stent grafting: improving results with newer generation investigational devices. *The Journal of Thoracic and Cardiovascular Surgery, 131*(5), 1087–1094.

Bortone, A. S., De Cillis, E., D'Agostino, D., & de Luca Tupputi Schinosa, L. (2004). Endovascular treatment of thoracic aortic disease: four years of experience. *Circulation, 110*, II262-II267.

Hampel, F. R., Ronchetti, E. M., Rousseeuw, P. J., & Stahel, W. A. (1986). *Robust statistics: The approach based on influence functions* (pp. 92–95). New York: Wiley.

Hastie, T., Tibshirani, R., & Friedman, J. (2001). *The elements of statistical learning* (pp. 84–88). New York: Springer.

Nakatamari, H., Ueda, T., Ishioka, F., Raman, B., Kurihara, K., Rubin, G. D., et al. (2011). Discriminant analysis of native thoracic aortic curvature: Risk prediction for endoleak formation after thoracic endovascular aortic repair. *Journal of Vascular and Interventional Radiology, 22*(7), 974–979.

Serag, A. R., Bergeron, P., Mathieu, X., Piret, V., Petrosyan, A., & Gay, J. (2007). Identification of proximal landing zone limit for proper deployment of aortic arch stentgraft after supra-aortic great vessels transposition. *The Journal of Cardiovascular Surgery, 48*(6), 805–807.

Tanaka, Y. (1994). Recent advance in sensitivity analysis in multivariate statistical methods. *Journal of the Japanese Society of Computational Statistics, 7*(1), 1–25.

Fuzzy *c*-Means for Web Mining: The Italian Tourist Forum Case

Domenica Fioredistella Iezzi and Mario Mastrangelo

Abstract e-tourism is in stable growth and becoming one of the leading sectors in the *e*-commerce world. Social media and mobile technologies are holding an increasingly important role in the procurement processes of tourism, by both providing access to real-time information and promoting the exchange of experiences. Web mining allows the collection of new unstructured data and the building of users' profiles based on electronic web mouth. We apply a soft approach to solve lexical ambiguity and build a vocabulary for the tourism sector. Indeed, we propose a new version of the fuzzy *c*-means algorithm to detect the best centroid clusters, and we choose the final partition according to the validation of three indices (the partition coefficient, the classification entropy, and the Xie-Beni index). We use this method to classify 525 posts published by the Italian tourism forum from January 2010 to April 2012.

Keywords Fuzzy *c*-means • *k*-centroids • Tourist forum • Web mining

1 Introduction

In recent years, there has been a large and continuous growth of textual information, especially in online documents, e-books, journals, technical reports, and digital libraries. The Internet has changed the way people collect information or chat with other people. Nowadays, people collect and disseminate information in several ways

D.F. Iezzi (✉)
Tor Vergata University, Via Columbia 1, Roma, Italy
e-mail: stella.iezzi@uniroma2.it

M. Mastrangelo
Tor Vergata University, Via Orazio Raimondo 18, Rome, Italy
e-mail: mario.mastrangelo@uniroma2.it

D. Vicari et al. (eds.), *Analysis and Modeling of Complex Data in Behavioral and Social Sciences*, Studies in Classification, Data Analysis, and Knowledge Organization,
DOI 10.1007/978-3-319-06692-9_17,

using Internet technology (e.g., e-mail, blogs, forums, online communities). In Italy, the use of the Internet covers 51.5 % of individuals; and over half of the Internet users send messages via chat, newsgroup, or blog (52.7 %); and 41.3 % are able to post texts, games, pictures, illustrations, films, or music on social networking websites (ISTAT 2011). No existing study has been able to construct complete user profiles.

In the tourism industry, customer behavior and preferences are influenced by information from tourism practitioners and other consumers. Electronic word of mouth (eWOM) introduces, on a large scale and anonymously, both traditional and innovative ways of information transmission among tourists (Iezzi and Mastrangelo 2012).

To analyze unstructured data from the web, it is important to make a preprocessing of information to remove the sparseness of input matrix before classifying data. Fuzzy approach allows us to deal with lexical ambiguity because a single word mostly belongs to multiple semantic categories, and a hard approach does not allow us to suitably treat this issue (e.g., in the tourism sector), the expression holiday is very close to trip, but generally, holiday is synonymous with vacation or days off. The fuzzy c-means is one of the most applied algorithms in facing this issue (Chen et al. 2011).

The aim of this paper is to describe users' profiles of travel responsible tourism forums using a soft approach. We propose an improved fuzzy c-means for text mining, applying to 525 posts about sustainable tourism. The paper is organized as follows: in Sect. 2, we describe data and methods, particularly justifying the choice of a fuzzy approach; in Sect. 3, we present the main results of our method; and in Sect. 4, we expose the conclusions and the future developments.

2 Data and Methods

We analyzed 525 posts of threads concerning sustainable tourism from 10 different travel forums (il giramondo, trip advisor, voiaganto, turisti per caso, zingarate, baltazar, viaggiatori.com, vagabondo, cisonostato, Lonely Planet) from January 2010 to April 2012.

We consider each post i represented by a vector of weighted selected terms and repeated segments of the form: $d_j = w_{1j}, w_{2j}, .., w_{ij}, .., w_{pn}$, where w_{ij} represents the weight for term or repeated segment i, attached to post d_j. By joining these vectors, we get the **D** word-term-by-document-matrix (Iezzi 2012a).

We use a modified fuzzy c-means (Coppi et al. 2010; Pal et al. 1996) algorithm that is a soft version of the popular k-means clustering (Iezzi 2012b). As well known, the k-means method begins with an initial set of randomly selected exemplars and iteratively refines this set so as to decrease the sum of squared errors. k-centers clustering is moderately sensitive to the initial selection of centers, so it is usually rerun many times with different initializations in an attempt to find a good solution.

Table 1 Lexical measures of the corpus

Word token (N)	76,599
Word type (V)	11,514
(V/N)100	15.032
Percentage of hapax	53,934
N/V	6.653
V/sqr(N)	41.602

Generally, posts are characterized by a very small number of word types and by a large number of hapaxes. In this corpus, there are 76,599 tokens, 11,514 types by message, and 54 % hapaxes (see Table 1). To measure the similarity between posts, we calculate the distance matrix **C** by applying to **D** the cosine distance. We apply non metric multidimensional scaling, to reduce the high dimensional spaces of **C** matrix. The dimensions of the new matrix **M** is (525×2), and its *s-stress* is 0.005. We use c-means algorithm to classify the posts.This algorithm is based on the minimization of the following objective function:

$$J = \sum_{i=1}^{c} \sum_{j=1}^{N} u_{ij}^{m} \left\| x_j - v_i \right\|^2$$

where N is the total number of observations in the dataset and c is the number of clusters; m is a number greater than 1 giving the degree of fuzziness, u_{ij} is the degree of membership of x_j in the cluster i, x_j is the jth of d-dimensional measured data, v_i is the d-dimensional center of the cluster, and $\| \|$ is any norm expressing the similarity between any measured data and the center.

We use the partition coefficient (PC) (Bezdek 1981), the classification entropy (CE) (Bezdek 1974) and the Xie-Beni index (XB) (Xie and Beni 1991) to evaluate the quality of the obtained partition.

The PC index is defined as:

$$PC(c) = \frac{1}{N} \sum_{i=1}^{c} \sum_{j=1}^{N} u_{ij}^2$$

and measures the amount of overlap between clusters. It has a value between *1/c* and *1* and has to be maximized.

The CE index is defined as:

$$CE(c) = \frac{1}{N} \sum_{i=1}^{c} \sum_{j=1}^{N} u_{ij} \log(u_{ij})$$

and measures the fuzziness of the cluster partition. It varies between *0* and *log(c)* and has to be minimized.

While PC and CE use only the membership values, the XB index involves both the membership matrix and the dataset itself, and focuses on two properties: compactness, which measures the closeness of cluster elements, and separation, which indicates how distinct two clusters are. It is defined as:

$$XB = \frac{\sum_{i=1}^{c} \sum_{j=1}^{N} u_{ij}^2 \left\| x_j - v_i \right\|^2}{N min_{ij} \left\| x_j - v_i \right\|^2}$$

smaller XB values indicate a more compact and well-separated partition.

We propose a new method to initialize cluster centroids in the c-means algorithm in order to improve the goodness of the final partition. The steps of the procedure are as follows:

1. Construct the **D** word-term-by-document-matrix from the corpus;
2. Computes the **C** distance matrix by using the cosine distance;
3. Reduce the dimensionality of **C** matrix by applying non metric multidimensional scaling, obtaining a new matrix **M**;
4. Apply fuzzy c-means algorithm to the matrix **M**;
5. Fixed the number of clusters c, and decompose the XB, evaluating the contribution of each cluster to the overall index value;
6. Among the c centroids, identify, if they exist, the s clusters (with $s < c$) whose contributions to the XB index are much larger than others (these clusters have worst centroids, which should be replaced);
7. Identify points candidate to replace worst centroids, which are points with a high level of membership heterogeneity;
8. Apply the c-means algorithm iteratively replacing the worst centroids with the candidate points of the step 7, and leaving unchanged the remaining $c - s$ centroids;
9. Choose the final partition according to the best internal validity indexes values (PC, CE and XB indexes).

Data were processed with R software; particularly, an R program for performing steps 4–9 has been developed by the authors.

3 Results

The results show that the fuzzy c-means algorithm, applied to the **M** matrix by randomly choosing the initial centroids, produces a low quality of clustering solution (Fig. 1).

Moreover, the spanning ellipses (Fig. 1), based on the average and the covariance matrix of each cluster, are overlapped, underlying that the detected groups are not well separated. From randomly chosen centroids, we may produce poor quality

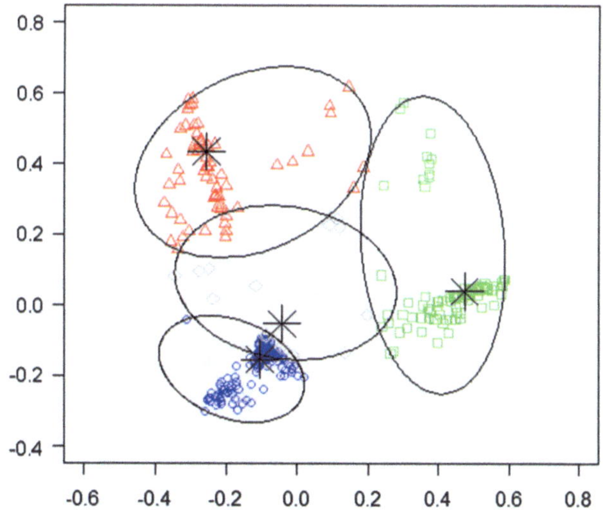

Fig. 1 Initial partition of the c-means algorithm

clusters, for this reason we should provide initial values for cluster centers, to improve the goodness of the final partition (Iezzi et al. 2013).

Considering the shape and density of the cloud of points, the centroids of the three clusters with external ellipses, which allow us to observe the degree of uncertainty of the points' position, may be considered correct, while we have to find a new fourth centroid. To improve the quality of the partitions of cluster n. 4, we select the initial centroids (Iezzi et al. 2013), that are represented by the rhombus symbol in Fig. 1. The potential candidates are points with a high level of fuzziness, near the boundary of each ellipsis, that is, points that have a threshold level of membership below 70 %. We have chosen this threshold of fuzziness level after several simulations. In Fig. 2, the 115 selected points as the potential initial centroids are represented by the filled triangle point-up symbol.

Therefore 115 simulations have been carried out, leaving the centroid unchanged for the first three clusters and choosing as fourth centroid one of potential candidates each time. PC, CE, and XB indexes have been used to evaluate the goodness of the obtained partitions. In this way, among the 115 points, we have characterized three subsets of points that provide very similar results, that is, three configurations as shown in Figs. 3, 4, 5.

The best partition obtained is shown in Fig. 5: compared to the initial partition, the PC index increases from 0.764 to 0.861, the CE index decreases from 0.437 to 0.305, and the XB index reduces from 17.351 to 7.878.

The final partition detected four profiles of users: (1) the theorists of sustainable tourism, who discuss on responsible behavior, good habits, environmental protection, education and compliance, (2) the business men, who analyze the features of green hotels, marketing, energy consumption, carbon neutrality, and organic farm; (3) the experts in sustainable tourism, who describe natural parks and reserves,

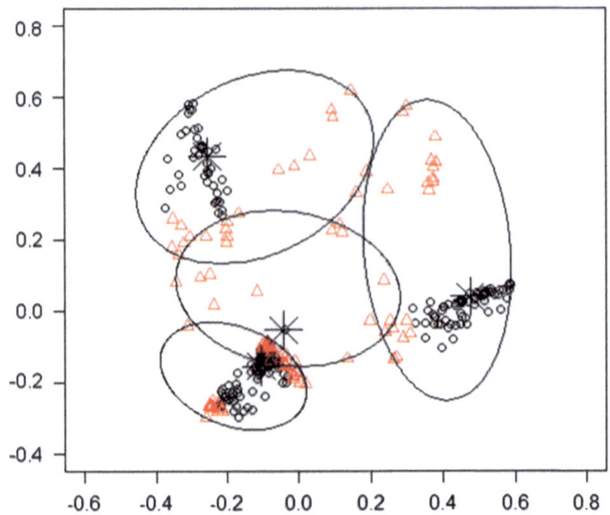

Fig. 2 New centroid candidate points (*triangle point up symbol*)

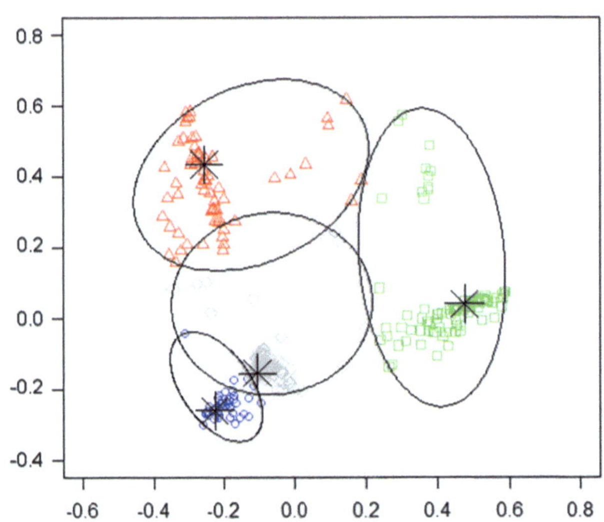

Fig. 3 Configuration 1

natural heritage and protected areas; and (4) the curious, who seek to understand how to practice sustainable tourism (tour operator, contact with nature, summer camps, excursions).

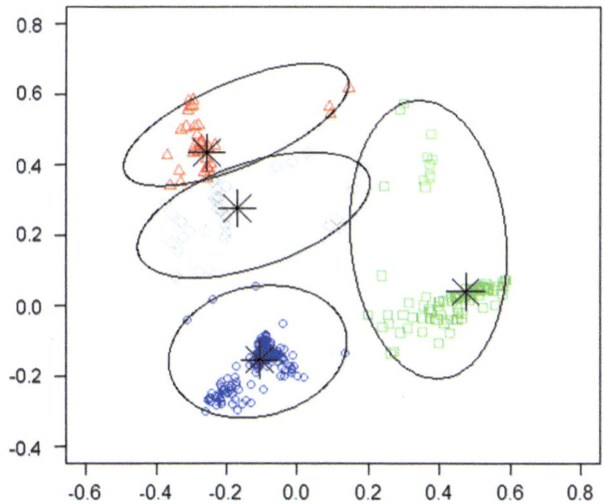

Fig. 4 Configuration 2

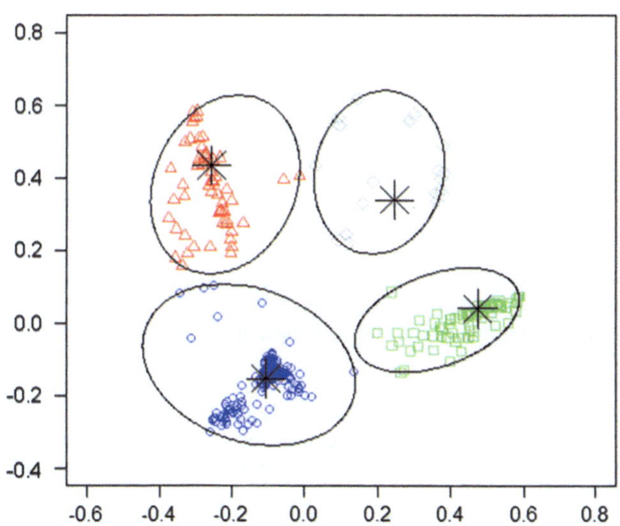

Fig. 5 Configuration 3: final partition of the modified *c*-means algorithm

4 Conclusions

There are many different ways of analyzing documents online, but all methods require a long preprocessing, which is often not automatic but linked to a specific field.

Common techniques for structuring text usually involve manual tagging with metadata or part-of-speech tagging for further text-mining-based structuring. Noises in posts such as those introduced by misspellings, abbreviations, deletions, phonetic spellings, nonstandard transliterations, etc., pose considerable issues for clustering. Such corruptions are very common also in instant messenger and short message service data. We built a vocabulary for the tourism sector, creating a list of words or repeated segment (Salem 1984) to automate the construction of the term-document matrix-that is, the input matrix for clustering.

Fuzzy approaches are suitable instruments in order to mine knowledge from words, and then disambiguate the specific lexicon. Our method allows us to obtain the best partition in order to define the users' profiles. The advantage of using fuzzy logic is that we can calculate the degree to which a given document belongs to all categories.

References

Bezdek, J. C. (1974). Cluster validity with fuzzy sets. *Journal of Cybernet*, *3*, 58–78.

Bezdek, J. C. (1981). *Pattern recognition with fuzzy objective functin algorithms*. New York: Plenum Press.

Chen, Y., Qiu, J., Gu, X., Chen, J., Ji, D., & Chen, L. (2011). Advances in research of Fuzzy *c*-means clustering algorithm. In *International Conference on Network Computing and Information Security*.

Coppi, R., D'urso, P., & Giordani, P. (2010). A Fuzzy clustering model for multivariate spatial time series. *Journal of Classification, 27*(1), 54–88.

Iezzi, D. F. (2012a). Centrality measures for text clustering. *Communications In Statistics. Theory And Methods*, *41*, 3179–3197.

Iezzi, D. F. (2012b). A new method for adapting the *k*-means algorithm to text mining. *Statistica Applicata - Italian Journal of Applied Statistics*, *22*, 69–80.

Iezzi, D. F., & Mastrangelo, M. (2012). Il passaparola digitale nei forum di viaggio: mappe esplorative per l'analisi dei contenuti. *Rivista Italiana di Economia, Demografia e Statistica, LXVI*, 143–150.

Iezzi, D. F., Mastrangelo, M., & Sarlo, S. (2013). A New Fuzzy Method to Classify Professional Profiles from Job Announcements. In P. Giudici, S. Ingrassia, & M. Vichi (Eds.), *Statistical models for data analysis* (pp. 151–159). Berlin: Springer.

ISTAT (2011). *Cittadini e nuove tecnologie*. Roma: Istat.

Pal, N. R., Bezdek, J. C., & Hathaway, R. J. (1996). Sequential competitive learning and the Fuzzy *c*-means clustering algorithms. *Neural Networks*, *5*, 787–796.

Salem, A. (1984). La typologie des segments répétés dans un corpus, fondée sur l'analyse d'un tableau croisant mots et textes. *Cahiers de l'Analyse des Données*, *IX*(4), 489–500.

Xie, X. L., & Beni, G. (1991). A validity measure for fuzzy clustering. *IEEE Transactions on Pattern Analysis and Machine Intelligence*, *13*, 841–847.

On Joint Dimension Reduction and Clustering of Categorical Data

Alfonso Iodice D'Enza, Michel Van de Velden, and Francesco Palumbo

Abstract There exist several methods for clustering high-dimensional data. One popular approach is to use a two-step procedure. In the first step, a dimension reduction technique is used to reduce the dimensionality of the data. In the second step, cluster analysis is applied to the data in the reduced space. This method may be referred to as the tandem approach. An important drawback of this method is that the dimension reduction may distort or hide the cluster structure. As an alternative, various authors have proposed joint dimension reduction and clustering approaches. In this paper we review some of these existing joint dimension reduction and clustering methods for categorical data in a unified framework that facilitates comparison.

Keywords Cluster analysis • Correspondence analysis • Homogeneity analysis

1 Introduction

A popular method for combining clustering and dimension reduction is to apply cluster analysis to the data after dimension reduction. This method may be referred to as the tandem approach (Arabie and Hubert 1994). An important drawback of

A. Iodice D'Enza, (✉)
Università di Cassino, Cassino (FR), Italy
e-mail: iodicede@unicas.it

M.V. de Velden,
Erasmus University of Rotterdam, PA Rotterdam, The Netherlands
e-mail: vandevelden@ese.eur.nl

F. Palumbo,
Università degli Studi di Napoli Federico II, Napoli, Italy
e-mail: fpalumbo@unina.it

D. Vicari et al. (eds.), *Analysis and Modeling of Complex Data in Behavioral and Social Sciences*, Studies in Classification, Data Analysis, and Knowledge Organization,
DOI 10.1007/978-3-319-06692-9_18,

this method is that the dimension reduction may distort or hide the cluster structure. Vichi and Kiers (2001) showed in a simulation study how the tandem approach may fail to retrieve the clusters in low dimensional space.

In the context of categorical data, Van Buuren and Heiser (1989) proposed a clustering and optimal scaling of variables method in which object scores are restricted using cluster memberships. Similarly, Hwang et al. (2006) proposed a joined multiple correspondence analysis (MCA) and K-means clustering (MacQueen 1967) method that uses user-specified weights for the clustering and dimension reduction parts. For binary data, Iodice D' Enza and Palumbo (2013) recently proposed a new method which they refer to as iterative factorial clustering for binary data (i-FCB). In this paper we review and re-formulate extant joint dimension reduction and clustering methods for categorical data in a unified framework that facilitates comparison. The paper is structured as follows: after introducing some notation we describe the different methods in Sects. 3–5. In Sect. 6, the methods are illustrated using an example data set. The results and most important findings of this paper are then summarized in Sect. 7 together with directions for future work.

2 Notation

Let $\mathbf{Z}_j$ denote an $n \times p_j$ indicator matrix, with n the number of observations and p_j the number of categories for the jth variable Z_j, with $j = 1, \ldots, q$ categorical variables. We introduce the following symbols for the linear algebra notation: $\mathbf{Z} = \left[\mathbf{Z}_1, \mathbf{Z}_2, \ldots, \mathbf{Z}_j, \ldots, \mathbf{Z}_q\right]$ is an $n \times Q$ block matrix, where $Q = \sum_{j=1}^{q} p_j$; $\mathbf{C}$ is an $n \times K$ indicator matrix that assigns each statistical unit to one of the K groups; $\mathbf{G}$ is a $K \times k$ matrix containing the cluster means, with k being the dimensionality of the solution, whereas $\mathbf{D}_z = diag(\mathbf{Z}^\mathsf{T}\mathbf{Z})$ and $\mathbf{D}_c = \mathbf{C}^\mathsf{T}\mathbf{C}$. Furthermore, $\mathbf{Y}$ denotes an $n \times k$ configuration matrix with k dimensional coordinates of observations, $\mathbf{B}_j$ denotes a $p_j \times k$ matrix of category quantifications. The computation of $\mathbf{Y}$ and $\mathbf{B}_j$ varies with the considered methods.

3 Constrained Homogeneity Analysis: GROUPALS

A joint dimension reduction and clustering approach is proposed by Van Buuren and Heiser (1989) in their GROUPALS method. Their method consists of a combination of optimal scaling and clustering techniques. The optimal scaling technique in question is homogeneity analysis (HOMogeneity Analysis by Alternating Least Square, Gifi 1990). The reference clustering technique is the K-means partitioning algorithm.

The loss function of GROUPALS is a modified version of the HOMALS loss function: in order to simultaneously perform a dimension reduction and clustering, a restriction is introduced in the loss function such that, in the low dimensional space, objects in the same cluster should be located at the position of the cluster centroid. The resulting loss function is

$$\begin{aligned}\min_{\mathbf{B},\mathbf{C},\mathbf{G}} \phi(\mathbf{B},\mathbf{C},\mathbf{G}) &= \frac{1}{q}\sum_{j=1}^{q} tr\left[(\mathbf{Y}-\mathbf{Z}_j\mathbf{B}_j)^{\mathsf{T}}(\mathbf{Y}-\mathbf{Z}_j\mathbf{B}_j)\right] = \\ &= \frac{1}{q}\sum_{j=1}^{q} tr\left[(\mathbf{CG}-\mathbf{Z}_j\mathbf{B}_j)^{\mathsf{T}}(\mathbf{CG}-\mathbf{Z}_j\mathbf{B}_j)\right] = \\ &= \frac{1}{q}\sum_{j=1}^{q} ||\mathbf{CG}-\mathbf{Z}_j\mathbf{B}_j||^2.\end{aligned} \tag{1}$$

The GROUPALS loss function is the HOMALS loss function, yet with the additional constraint that

$$\mathbf{Y} = \mathbf{CG}. \tag{2}$$

The minimization problem is solved via alternating least squares. The loss function can be split into two components

$$\min_{\mathbf{B},\mathbf{C},\mathbf{G}} \phi(\mathbf{B},\mathbf{C},\mathbf{G}) = \frac{1}{q}\sum_{j=1}^{q} ||\Psi-\mathbf{Z}_j\mathbf{B}_j||^2 + ||\Psi-\mathbf{CG}||^2 \tag{3}$$

with $\Psi = \frac{1}{q}\sum_{j=1}^{q}\mathbf{Z}_j\mathbf{B}_j$. Inserting the identity $\mathbf{CG} = \Psi - \Psi + \mathbf{CG}$ in Expression (1) leads to obtain the Expression in (3). For fixed **C** and **G**, the loss function is minimized with respect to $\mathbf{B}_j$ ($j = 1,\ldots,q$): an optimal scaling procedure (HOMALS) is used to achieve this task. For fixed $\mathbf{B}_j$, the first component of Expression (3) is constant, thus the second component is minimized with respect to **C** and **G** using a K-means procedure.

4 Multiple Correspondence Analysis and K-Means: MCA-K-Means

Hwang et al. (2006) proposed a joint MCA and K-means clustering approach. As Homals and MCA are equivalent, the MCA-K-means approach and GROUPALS both join MCA and k-means. However, the way in which these two methods are combined differs. The objective of this method, which we shall refer to as MCA-K-means, is

$$\min_{\mathbf{Y},\mathbf{B},\mathbf{C},\mathbf{G}} \phi\left(\mathbf{Y},\mathbf{B},\mathbf{C},\mathbf{G}\right) = \alpha_1 \sum_{j=1}^{q} \left|\left|\mathbf{Y} - \mathbf{Z}_j\mathbf{B}_j\right|\right|^2 + \alpha_2 \left|\left|\mathbf{Y} - \mathbf{CG}\right|\right|^2$$

$$s.t.\mathbf{Y}^{\mathsf{T}}\mathbf{Y} = \mathbf{I}_k.$$

This objective function is in fact a weighted sum of MCA's homogeneity criterion (Gifi 1990) and the K-means objective where the weights, α_1 and α_2 (restricted such that $\alpha_1 + \alpha_2 = 1$) are user supplied. Hwang et al. (2006) develop an alternating least-squares algorithm to solve the minimization problem. It is not difficult to show that in the optimum,

$$\mathbf{B}_j = \left(\mathbf{Z}_j^{\mathsf{T}}\mathbf{Z}_j\right)^{-1} \mathbf{Z}_j^{\mathsf{T}}\mathbf{Y} \text{ and } \mathbf{G} = \mathbf{D}_c^{-1}\mathbf{C}^{\mathsf{T}}\mathbf{Y}.$$

The configuration matrix can be obtained using the eigenequation

$$\left(\alpha_1 \sum_{j=1}^{q} \mathbf{Z}_j \left(\mathbf{Z}_j^{\mathsf{T}}\mathbf{Z}_j\right)^{-1} \mathbf{Z}_j^{\mathsf{T}} + \alpha_2 \mathbf{C}\mathbf{D}_c^{-1}\mathbf{C}^{\mathsf{T}}\right)\mathbf{Y} = \mathbf{Y}\boldsymbol{\Lambda}. \tag{4}$$

The cluster membership matrix $\mathbf{C}$ is obtained by considering distances to the cluster means in $\mathbf{G}$ and assigning the observations to the closest cluster. Starting with some initial values (e.g., random cluster memberships and $\mathbf{Y}$ the configuration obtained after applying MCA), the approximations are sequentially updated leading the objective to decrease monotonically. If the decrease is below a certain threshold, the algorithm terminates and a solution is obtained. To reduce the chance of obtaining a local minima, several random starts should be applied.

The eigenequation in (4) is of crucial importance in the proposed algorithm. However, for large n, the matrix that needs to be considered may become too large. It is therefore useful to reformulate the method in a more efficient way. This can easily be achieved by defining $\mathbf{X} = \left(\sqrt{\alpha_1}\mathbf{Z}\mathbf{D}_z^{-\frac{1}{2}} \ \sqrt{\alpha_2}\mathbf{C}\mathbf{D}_c^{-\frac{1}{2}}\right)$, and considering the singular value decomposition

$$\mathbf{X} = \mathbf{Y}\boldsymbol{\Lambda}^{\frac{1}{2}}\mathbf{V}^{\mathsf{T}},$$

where $\mathbf{Y}^{\mathsf{T}}\mathbf{Y} = \mathbf{V}^{\mathsf{T}}\mathbf{V} = \mathbf{I}$, so that, in accordance with (4),

$$\mathbf{X}\mathbf{X}^{\mathsf{T}}\mathbf{Y} = \mathbf{Y}\boldsymbol{\Lambda} \text{ and } \mathbf{X}^{\mathsf{T}}\mathbf{X}\mathbf{V} = \mathbf{V}\boldsymbol{\Lambda}.$$

Furthermore,

$$\mathbf{Y} = \mathbf{X}\mathbf{V}\boldsymbol{\Lambda}^{-\frac{1}{2}}. \tag{5}$$

The advantage of reformulating (4), is the reduction in size of the matrix for which eigenvalues need to be computed. Moreover, the alternative formulation shows that the original problem corresponds to a matrix approximation problem similar to the usual MCA problem. Observe that,

$$\mathbf{X}^{\mathsf{T}}\mathbf{X} = \begin{pmatrix} \alpha_1 \mathbf{D}_z^{-\frac{1}{2}} \mathbf{Z}^{\mathsf{T}} \mathbf{Z} \mathbf{D}_z^{-\frac{1}{2}} & \sqrt{\alpha_1\alpha_2} \mathbf{D}_z^{-\frac{1}{2}} \mathbf{Z}^{\mathsf{T}} \mathbf{C} \mathbf{D}_c^{-\frac{1}{2}} \\ \sqrt{\alpha_1\alpha_2} \mathbf{D}_c^{-\frac{1}{2}} \mathbf{C}^{\mathsf{T}} \mathbf{Z} \mathbf{D}_z^{-\frac{1}{2}} & \alpha_2 \mathbf{I}_K \end{pmatrix}. \tag{6}$$

Like in MCA and CA, if the data are not centered, a so-called trivial solution exists in MCA-K-means

Proposition 1.

$$\mathbf{X}^{\mathsf{T}}\mathbf{X}\mathbf{v} = \begin{pmatrix} \alpha_1 \mathbf{D}_z^{-\frac{1}{2}} \mathbf{Z}^{\mathsf{T}} \mathbf{Z} \mathbf{D}_z^{-\frac{1}{2}} & \sqrt{\alpha_1\alpha_2} \mathbf{D}_z^{-\frac{1}{2}} \mathbf{Z}^{\mathsf{T}} \mathbf{C} \mathbf{D}_c^{-\frac{1}{2}} \\ \sqrt{\alpha_1\alpha_2} \mathbf{D}_c^{-\frac{1}{2}} \mathbf{C}^{\mathsf{T}} \mathbf{Z} \mathbf{D}_z^{-\frac{1}{2}} & \alpha_2 \mathbf{I}_K \end{pmatrix} \begin{pmatrix} \mathbf{v}_1 \\ \mathbf{v}_2 \end{pmatrix} = \lambda \begin{pmatrix} \mathbf{v}_1 \\ \mathbf{v}_2 \end{pmatrix},$$

where

$$\lambda_1 = \alpha_1 q + \alpha_2,$$

$$\mathbf{v} = \begin{pmatrix} c_1 \mathbf{D}_z^{\frac{1}{2}} \mathbf{1}_q \\ c_2 \mathbf{D}_c^{\frac{1}{2}} \mathbf{1}_K \end{pmatrix}$$

where $\mathbf{1}_q$ and $\mathbf{1}_K$ are q-dimensional and K-dimensional vectors of ones; c_1 and c_2 are constants that satisfy

$$\frac{c_2}{c_1} = \sqrt{\frac{\alpha_2}{\alpha_1}}$$

Proof. Inserting $\mathbf{v}$ into the eigenequation above, yields

$$\begin{pmatrix} q c_1 \alpha_1 \mathbf{D}_z^{-\frac{1}{2}} \mathbf{Z}^{\mathsf{T}} \mathbf{1}_n + c_2 \sqrt{\alpha_1\alpha_2} \mathbf{D}_z^{-\frac{1}{2}} \mathbf{Z}^{\mathsf{T}} \mathbf{1}_n \\ q c_1 \sqrt{\alpha_1\alpha_2} \mathbf{D}_c^{-\frac{1}{2}} \mathbf{1}_K + c_2 \alpha_2 \mathbf{D}_c^{\frac{1}{2}} \mathbf{1}_K \end{pmatrix} = \lambda \begin{pmatrix} c_1 \mathbf{D}_z^{\frac{1}{2}} \mathbf{1}_q \\ c_2 \mathbf{D}_c^{\frac{1}{2}} \mathbf{1}_K \end{pmatrix}$$

$$\begin{pmatrix} \left(q c_1 \alpha_1 + c_2 \sqrt{\alpha_1\alpha_2}\right) \mathbf{D}_z^{\frac{1}{2}} \mathbf{1}_q \\ \left(q c_1 \sqrt{\alpha_1\alpha_2} + c_2 \alpha_2\right) \mathbf{D}_c^{\frac{1}{2}} \mathbf{1}_K \end{pmatrix} = \lambda \begin{pmatrix} c_1 \mathbf{D}_z^{\frac{1}{2}} \mathbf{1}_q \\ c_2 \mathbf{D}_c^{\frac{1}{2}} \mathbf{1}_K \end{pmatrix}$$

So that,

$$\left(q\alpha_1 + \frac{c_2}{c_1}\sqrt{\alpha_1\alpha_2}\right) = \lambda = \left(q\frac{c_1}{c_2}\sqrt{\alpha_1\alpha_2} + \alpha_2\right).$$

Now, taking $\frac{c_2}{c_1} = \sqrt{\frac{\alpha_2}{\alpha_1}}$ it follows that $\lambda = q\alpha_1 + \alpha_2$. Furthermore, as

$$c_2 = \sqrt{\frac{\alpha_2}{\alpha_1}} c_1,$$

$$\mathbf{v}^\mathsf{T}\mathbf{v} = c_1^2 \mathbf{1}_q^\mathsf{T}\mathbf{D}_z\mathbf{1}_q + c_2^2 \mathbf{1}_K^\mathsf{T}\mathbf{D}_c\mathbf{1}_K = c_1^2 nq + c_2^2 n = c_1^2 nq + \frac{\alpha_2}{\alpha_1} c_1^2 = c_1^2 \left(nq + \frac{\alpha_2}{\alpha_1}\right).$$

Choosing $\mathbf{v}^\mathsf{T}\mathbf{v} = 1$, we get

$$c_1 = \frac{1}{\sqrt{\left(nq + \frac{\alpha_2}{\alpha_1}\right)}} \text{ and } c_2 = \frac{1}{\sqrt{\left(\frac{\alpha_1}{\alpha_2} nq + n\right)}}$$

5 Iterative Factorial Clustering for Binary Data: i-FCB

Although the i-FCB method (Iodice D' Enza and Palumbo 2013) was introduced as a method that combines dimension reduction and clustering for binary data, it can also be used for categorical data with more than two categories. In i-FCB the goal is to describe the predictive power of the attributes, i.e. the categorical variables, on the clusters. This is achieved by applying K-means cluster analysis to the optimal observation scores obtained from a non symmetric correspondence analysis (NSCA, Lauro and D'Ambra 1984) of the cross-tabulation of cluster membership and the categorical variables. If there is a cluster structure underlying the data, the observations in reduced space are 'more clustered' than in the original space and the attributes have higher predictive power for the cluster membership. We reformulate the i-FCB optimization problem described in Iodice D' Enza and Palumbo (2013) as follows

$$\min_{\mathbf{B},\mathbf{C},\mathbf{G}} \phi\left(\mathbf{B},\mathbf{C},\mathbf{G}\right) = \left|\left|(\sqrt{n})^{-1/2}\mathbf{C}^\mathsf{T}\mathbf{Z}^*\mathbf{D}_z^{-1/2} - \mathbf{G}\mathbf{B}^\mathsf{T}\right|\right|^2 \quad \text{s.t.} \quad \mathbf{B}^\mathsf{T}\mathbf{D}_z\mathbf{B} = nq\mathbf{I}. \tag{7}$$

where $\mathbf{Z}^* = \left[\mathbf{Z}_1^*, \ldots, \mathbf{Z}_q^*\right]$, $\mathbf{Z}_j^* = \left(\mathbf{I}_n - \mathbf{1}_n\mathbf{1}_n^\mathsf{T}/n\right)\mathbf{Z}_j$, is the centered version of $\mathbf{Z}$. This objective function is equivalent to the objective function of a NSCA where the cluster membership $\mathbf{C}$ is used as the reference mode. To obtain optimal category quantifications ($\mathbf{B}$) as well as an optimal cluster allocation ($\mathbf{C}$), the following iterative procedure is proposed: First, for fixed $\mathbf{C}$, $\mathbf{B}$ and $\mathbf{G}$ are obtained using the singular value decomposition $n^{-1/2}\mathbf{C}^\mathsf{T}\mathbf{Z}^*\mathbf{D}_z^{-1/2} = \mathbf{U}\Sigma\mathbf{V}^\mathsf{T}$, and by defining $\mathbf{B} = \sqrt{nq}\mathbf{D}_z^{-1/2}\mathbf{V}$ and $\mathbf{G} = \mathbf{U}\Sigma$. Coordinates for the observations (i.e. the rows of $\mathbf{Z}$) can be obtained by using the so-called transition formula

$$\mathbf{Y} = n^{-1/2}\mathbf{D}_w\mathbf{Z}\mathbf{B}, \tag{8}$$

where $\mathbf{D}_w = diag(\mathbf{C}\mathbf{C}^\mathsf{T}\mathbf{C}\mathbf{1}_c)$ is used to weight the observations by the cluster sizes.

Next, we apply K-means to the observation coordinates in reduced space ($\mathbf{Y}$) to obtain updates for the cluster allocation $\mathbf{C}$ and the cluster means ($\mathbf{G}$). The whole procedure is then repeated (i.e. the improved cluster allocation, $\mathbf{C}$, is used to update $\mathbf{B}$ and $\mathbf{G}$ by applying NSCA etc.). This process is repeated until the decrease in the loss function value (7) is below a certain, small, threshold value.

Comparing (6) and (7), we see that MCA-K-means and i-FCB are closely related. Whereas MCA-K-means explicitly approximates, in a least-squares sense, all cross-tabulations between variables, i-FCB focuses on the cross-tabulation between the clusters and the variables.

6 Example

The brief descriptions of these three methods for joint dimension reduction and clustering methods exposed some similarities and differences. How these differences manifest themselves in practice remains to be seen. We apply the described methods as well as ordinary MCA, to a data set from the 1993 multinational ISSP survey on environment provided with the **R** package **ca** (Nenadic and Greenacre 2007). We consider observations of 871 respondents on four attributes. These attributes correspond to four sentences about "science": A) *We believe too often in science, and not enough in faith.*, B) *Modern science does more harm than good*, C) *Any change humans cause in nature makes things worse* and D) *Modern science will solve our environmental problems*. Attribute levels are on a five point scale, going from "strongly agree" to "strongly disagree". Figure 1 shows the solutions for the considered methods when the number of clusters is set to four.

In Fig. 1, we see that the MCA-K-means produces "tight" clusters in which individuals are positioned close to the cluster means. The positions of the attributes are also affected in this method, but no pattern can be discerned concerning these changes. The i-FCB approach, appears to yield a solution in which the clusters are less clearly separated than with MCA-K-means. The positions of the attributes, however, appear to be better separated than in the MCA solution. For these data, the positions of the attributes [in the Groupals configuration] are similar to the MCA positions. Like in MCA-K-means, tight clusters are obtained but the positioning and the size of these clusters differ notably.

7 Conclusion

In this paper, we described three joint clustering and dimension reduction techniques for categorical data in a unified framework. We applied these methods to see how the mathematical differences between the methods manifest themselves in

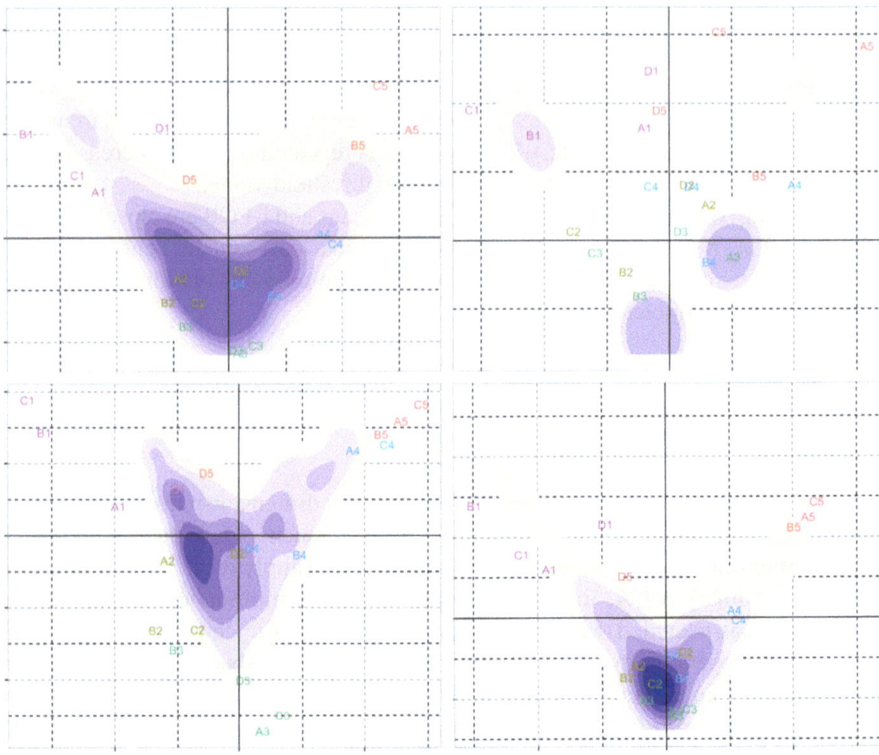

Fig. 1 Solutions for the example dataset when $K = 4$. The joint maps of individuals and attributes refer to: MCA solution (*top left*), MCA-K-means (*top right*), i-FCB (*bottom left*) and GROUPALS (*bottom right*). Attributes are labeled A through D (see text for corresponding questions) with numbers indicating the levels. Individuals are displayed by means of a density map

practice. The unified framework immediately exposed some theoretical differences and similarities between the methods. Concerning the dimension reduction part, it was found that all methods use correspondence analysis related methods. However, whereas GROUPALS and MCA-K-means apply multiple correspondence analysis, the i-FCB method is based on non symmetric correspondence analysis. For the clustering part all methods use K-means. As is clear from the different objectives and algorithms, the order and exact nature in which the dimension reduction and clustering procedures are incorporated differ considerably. These differences are also clearly reflected in our application presented in the previous section.

Through our alternative formulations of MCA-K-means and i-FCB, certain properties of these methods became apparent. In particular, the reformulation of MCA-K-means allowed us to proof an important property concerning eigenvalues and eigenvectors that is analogue to a well-known MCA result.

The application we presented show that the mathematical differences can result in different outcomes. Although results of one application do not suffice to make far reaching conclusions, our results show that more research into the performance of these methods is needed. To make such an appraisal, empirical validation on more real data sets, as well as a well designed simulation study, in which several realistic conditions are mimicked, could be considered. It should be noted, however, that such a simulation study is not a trivial task as the study should take into account the categorical nature of the data and consider various scenarios concerning the dimensionality of the data as well as underlying cluster structure. Moreover, the appraisal of results, requires the construction of objective and specific comparison measures.

References

Arabie, P., & Hubert, L. (1994). Cluster analysis in marketing research. *IEEE Transactions on Automatic Control, 19*, 716–723.

Gifi, A. (1990). *Nonlinear multivariate analysis*. (579 pp). New York: John Wiley & Sons. ISBN 0-471-92620-5.

Hwang, H., Dillon, W. R., & Takane, Y. (2006). An extension of multiple correspondence analysis for identifying heterogenous subgroups of respondents. *Psychometrika, 71*, 161–171.

Iodice D' Enza, A., & Palumbo, F. (2013). Iterative factor clustering of binary data. *Computational Statistics, 28*(2), 789–807.

Lauro C. N., & D'Ambra, L. (1984). L'analyse non symétrique des correspondances. *Data Analysis and Informatics, III*, 433–446.

MacQueen, J. (1967). Some methods for classification and analysis of multivariate observations. In L. M. L. Cam & J. Neyman (Eds.), *Proceedings of the Fifth Berkeley Symposium on Mathematical Statistics and Probability* (Vol. 1, pp. 281–297).

Nenadic, O., & Greenacre, M. (2007). Correspondence analysis in R, with two- and three-dimensional graphics: the ca package, *Journal of Statistical Software, 20*(3).

Van Buuren, S., & Heiser, W. J. (1989). Clustering *n* objects in *k* groups under optimal scaling of variables. *Psychometrika, 54*, 699–706.

Vichi, M., & Kiers, H. (2001). Factorial k-means analysis for two way data. *Computational Statistics & Data Analysis, 37*, 49–64.

A SVM Applied Text Categorization of Academia-Industry Collaborative Research and Development Documents on the Web

Kei Kurakawa, Yuan Sun, Nagayoshi Yamashita, and Yasumasa Baba

Abstract A method of automatically extracting Japanese documents describing University-Industry (U-I) relations from the Web is proposed. The proposed method consists of Japanese text processing and support vector machine (SVM) classification. The SVM feature selections were customized for U-I relations documents. The strongest experimental result was 79.95 of accuracy and 81.17 of f-measure.

Keywords U-I relations • Web documents • text categorization • machine learning

1 Introduction

When making new policies related to science and technology research and development, it is important to investigate the relevant university-industry-government (U-I-G) relations (Leydesdorff and Meyer 2003). Web documents are a key research target used to clarify the state of such relationships. In the clarification process, extracting resources related to U-I-G relations is the first requirement. Our objective in this study is to automatically extract U-I relation resources from the Web. We set "press release articles" of organizations as a target and constructed a framework

K. Kurakawa (✉) • Y. Sun
National Institute of Informatics, Tokyo, 101-8430, Japan
e-mail: kurakawa@nii.ac.jp; yuan@nii.ac.jp

N. Yamashita
GMO Research, Tokyo, 150-8512, Japan
e-mail: nagayoshi3@gmail.com

Y. Baba
The Institute of Statistical Mathematics, Tokyo, 190-8562, Japan
e-mail: baba@ism.ac.jp

D. Vicari et al. (eds.), *Analysis and Modeling of Complex Data in Behavioral and Social Sciences*, Studies in Classification, Data Analysis, and Knowledge Organization,
DOI 10.1007/978-3-319-06692-9_19,

to automatically crawl these articles and determine which were relevant to U-I relations.

U-I relations is a part of academia-industry relations that means relationships between academic institutions and industries. The framework for automatic crawling and U-I relations detection can be extended to the case of academic-industry relations.

2 Automatic Extraction Framework for Academia-Industry Collaborative R&D Documents on the Web

2.1 Crawling Web Documents and Extracting Relevant Texts

When we look for documents related to U-I relations, especially in terms of collaborative research and development, there can be several potential sources, including press release news sites of the organization, faculty introductory sites, laboratory home pages, researchers' own sites, and general scientific news sites of commercial sectors. Generally speaking, these sources vary in terms of their value as evidence of university and industry collaboration, but it seems reasonable to conclude that press release sites are a promising source for automatic extraction and coverage. This is because both university and industry news sites contain articles on research and development results that are both formally presented and available to the public.

The flow of the proposed method is shown in Fig. 1. In the first step, we set seed URLs for popular crawling programs such as "wget" to crawl all press releases. Next, we extract relevant texts from each article to classify whether or not it is related to U-I relations. Texts extracted from Web documents are typically very noisy, which hampers content analysis, and even though we scrape the text from HTML tagged documents, irrelevant text—e.g., menu label text, page headers or footers, ads—remain. Such irrelevant text does not typically form full sentences, and the exact evidence of U-I relations tends to be spread out over several chunks of multiple sentences. For example, in "the MIT researchers and scientists from MicroCHIPS Inc. reported that...", our Japanese target of "東京大学とオムロン株式会社は、共同研究により、重なりや隠れに強く..." can be isolated. In that sense, the only text we need for detection is text that includes punctuation marks, i.e., text that forms complete sentences. This makes extracted text considerably less noisy.

2.2 Classifying the Documents

We used a support vector machine (SVM) (Vapnik 1995) to discriminate documents of U-I relations. SVM is a kernel-based algorithm that have sparse solutions, so

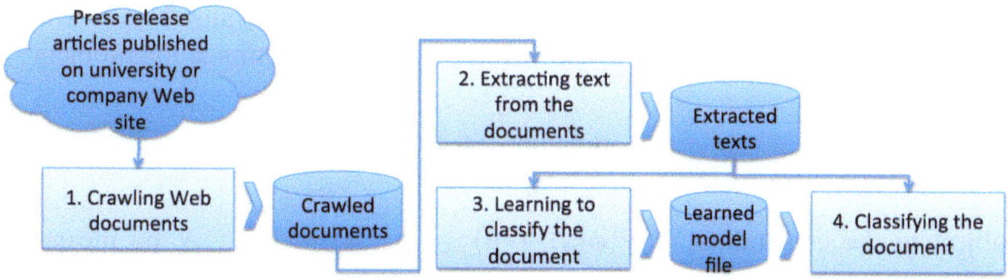

Fig. 1 Automatic extraction framework for U-I relations documents on the Web

that predictions for new inputs depend only on the kernel function evaluated at a subset of the training data points. The determination of the model parameters corresponds to a convex optimization problem, and so any local solution is also a global optimum.

Our approach is not aiming at theoretical extension of the SVM model, but experimental feasibility study to build a new features for document classification of U-I relations. We apply existing tf-idf scheme of information retrieval for feature vectors of SVM, which will be here based on bag-of-words and rule-base of document features especially to detect the U-I relations.

2.2.1 Support Vector Machine

A support vector machine is the two-class classifier simply using linear models of the form

$$y(\mathbf{x}) = \mathbf{w}^{\mathrm{T}}\boldsymbol{\phi}(\mathbf{x}) + b \tag{1}$$

where $\boldsymbol{\phi}(\mathbf{x})$ denotes a fixed feature-space transformation, and b is the bias parameter. The training data set comprises N input vectors $\mathbf{x}_1, \ldots, \mathbf{x}_N$, with corresponding target values $t_1, \ldots, t_N$ where $t_n \in \{-1, 1\}$ as binary class labels, and new data points $\mathbf{x}$ are classified according to the sign of $y(\mathbf{x})$. Weight vector $\mathbf{w}$ and input vector $\mathbf{x}$ are M dimensional.

Assuming that the training data is linearly separable in feature space, so that by definition there exists at least one choice of the parameters $\mathbf{w}$ and b that satisfies $y(\mathbf{x}_n) > 0$ for points having $t_n = +1$ and $y(\mathbf{x}_n) < 0$ for points having $t_n = -1$, so that $t_n y(\mathbf{x}_n) > 0$ for all points.

To try to find the one solution of $\mathbf{w}$ and b, SVM gives a concept of the *margin*, which is defined to be the smallest distance between the decision boundary and any of the samples. The decision boundary is chosen to be the one for which the margin is maximized. The location of this boundary is determined by a subset of the data points (support vectors), which are on the hyperplanes $y(\mathbf{x}) = 1$ and $y(\mathbf{x}) = -1$.

In this case, the optimization problem simply requires to maximize $\|\mathbf{w}\|^{-1}$, which is equivalent to minimize $\|\mathbf{w}\|^2$. Then we have to solve the optimization problem

$$\arg\min_{\mathbf{w},b} \frac{1}{2}\|\mathbf{w}\|^2 \tag{2}$$

subject to the constraints, $t_n(\mathbf{w}^{\mathrm{T}}\boldsymbol{\phi}(\mathbf{x}) + b) \geq 1, \ n = 1, \ldots, N$ by means of Lagurangian method.

In order to classify new data points using the trained data, we evaluate the sign of $y(\mathbf{x})$. This can be expressed by

$$y(\mathbf{x}) = \sum_{n=1}^{N} a_n t_n k(\mathbf{x}, \mathbf{x}_n) + b \tag{3}$$

where kernel function is defined by $k(\mathbf{x}, \mathbf{x}') = \boldsymbol{\phi}(\mathbf{x})^{\mathrm{T}}\boldsymbol{\phi}(\mathbf{x}')$, and $a_n > 0$ is Lagurange multipliers.

The solution of the problem can be applied efficiently to feature spaces whose dimensionality exceeds the number of data points, including infinite feature spaces because the original optimization problem for M dimensional weight vector is reformulated to dual problems using kernels, which has N variables.

2.2.2 Feature Selection

Feature Vector. Our problem is to classify web documents of U-I relations by means of SVM, so that the text document should be represented in a vector $\mathbf{x}_n$. Mapping from text document to a vector is known for feature selection, and several methods are proposed, i.e term selection based on document frequency, information gain, mutual information, a χ^2-test, and term strength (Yang and Pedersen 1997). In our approach, we adopt tf-idf (term frequency—inverse document frequency), which is one of the most commonly used term weighting schemes in today's information retrieval systems (Aizawa 2003).

tf-idf is formally defined as follows.

$$\text{tf-idf}(t, d, D) = \text{tf}(t, d) \times \text{idf}(t, D) \tag{4}$$

$$\text{tf}(t, d) = n_{t,d}, \quad \text{idf}(t, D) = \log \frac{|D|}{|\{d \in D : t \in d\}|}. \tag{5}$$

t, d, D respectively denotes term, a document, and a total of documents. $\text{tf}(t, d)$ is term frequency. $n_{t,d}$ is a number of a term t occurred in a document d. $\text{idf}(t, D)$ is inverse document frequency, which is calculated by log of a total document size $|D|$ divided by a number of document including the term t.

The feature is expressed by

$$\mathbf{x}_d = (x_{t_1,d}, x_{t_2,d}, \cdots, x_{t_M,d}), \quad x_{t,d} = \text{tf-idf}(t, d, D) \times b_{t,d}. \tag{6}$$

$b_{t,d}$ means boolean existence of the terms in d. The term can be a term in a document, type of POS (part-of-speech) of morpheme, or analytical output of external tools in our experiment.

According to the classical reference of information retrieval (Salton and Buckley 1988), the above tf-idf scheme corresponds to the classical tf-idf without document length normalization. In that sense, tf-idf scheme is just a classical reference to represent document features.

Vector Elements. We sought several features for SVM input for both learning and classifying. In this experiment, tf-idf is a base element of input vectors $\mathbf{x}_n$, so the variations here depend on term selections. We used Mecab, a Japanese morphological analyzer,[1] to divide sequential Japanese text into morphemes (word units).

Table 1 describes the features. In the first group of features, we examined three types of bag of words (BoW) features. In the second group of features, we prepared keywords related to U-I relations and calculated tf-idf for all documents. All other words were ignored. In the third group of features, we determined the existence of a corporation or a university.

These group of features are nominated because they are candidate features to detect U-I relations in our previous studies. Bag of words is the most widely used features to detect subjects and categories of documents in many literatures. Keywords related to U-I relations are a result of heuristic studies over surveying several press release articles of U-I relations on the Web. Existence of corporation and university in the text is another heuristic results of looking at many press release articles of U-I relations. Almost all the cases in the press release articles of U-I relations formally express the name of university and corporate body.

In our observation, there exists a problem-specific feature which makes it difficult to discriminate documents of U-I relations by only means of bag of words features on the tf-idf scheme. Documents of U-I relations are often characterized by only a few sentences that describe the U-I relations. They used to be provided in the introduction part of the document, and in the rest of the document, results of the activities such as research and development achievement are described. Most part of it doesn't give any evidence of U-I relations.

[1]http://mecab.googlecode.com/svn/trunk/mecab/doc/index.html.

Table 1 Features for vector elements

	Features	Description
(1)	BoW	Bag of words. A Japanese morphological analyzer Mecab is used to extract words from a sentence
(2)	BoW(N)	Bag of words, where a noun was chosen
(3)	BoW(N-3)	Bag of words, which were restricted to proper-noun, general-noun, and sahen-noun (verb formed by adding "する" ([suru], do) to the noun)
(4)	K(14)	Fourteen keywords were detected. They are "研究" ([kenkyu], research), "開発" ([kaihatsu], development), "実験" ([jikken], experiment), "成功" ([seikou], success), "発見" ([hakken], discover), "開始" ([kaisi], start), "受賞" ([jushou], award), "表彰" ([hyoushou], honor), "共同" ([kyoudou], collaboration), "協同" ([kyoudou], cooperation), "協力" ([kyouryoku], join forces), "産学" ([sangaku], UI relationship), "産官学" ([sankangaku], UIG (University-Industry-Government) relations), and "連携" ([renkei], coordination)
(5)	K(18)	Eighteen keywords were detected. They are K(14) plus the following keywords, i.e., "受託" ([jutaku], entrusted with), "委託" ([itaku], consignment), "締結" ([teiketsu], conclusion), and "研究員" ([kenkyuin], researcher)
(6)	K(18)+NM	The keywords and part of speech (POS) of the next morpheme in a sequential text were checked to ensure that grammatical connections from the keywords were restricted to verb, auxiliary verb, and sahen-noun
(7)	ORG	The existence of an organization by means of Cabocha's entity extraction function (http://code.google.com/p/cabocha/) was checked
(8)	Corp.	Cooperation marks were checked. They are "株式会社" ([kabushikigaisha], incorporated)," （株） " (a Unicode character such as U+3231), and its character variations
(9)	Univ.	University names were checked, i.e., "大学" ([daigaku], university), or "大" ([dai], a shortened representation of university)
(10)	C.+U.	Both cooperation marks and a university name occurred in the same sentence were checked

3 Experiment

3.1 Data Set

We prepared real web documents for experimental data set. Press release articles in several kinds of organizations are crawled and manually tested whether it is U-I relations or not. Then we picked up the same amount of articles in both positives and negatives. The numbers are shown in the Table 2. In order to consider quality and quantity of press release articles, we tried to choose famous and world-ranked universities and the companies likely to have liaison with those universities. The universities and the company listed in Table 2 are a part of available cases in Japan. To keep equally positive document size and negative document size for optimal learning, we reduce negative document size by selecting some negative documents at random from all.

Table 2 Data set for experiment

	Crawled articles		Articles for experiment	
Organization	Positive article	Negative article	Positive article	Negative article
Tohoku Univ.	44	499	44	44
The Univ. of Tokyo	106	848	106	106
Kyoto Univ.	40	329	40	40
Tokyo Inst. of Tech.	37	343	37	37
Hitachi Corp.	103	450	103	103
Total	330	2,469	330	330

3.2 *Test Cases and Results*

We constructed test cases to examine features described in Table 1 and kernel functions. The left side of Table 3 shows the features and SVM kernel selections for each classification test. Since support vector machines are driven by a kernel function, we here examined the linear kernel, polynomial kernel, and radial basis function (RBF) kernel pre-defined by SVM^{light}.[2] We tuned a parameter γ of RBF kernel.

For each combination of feature elements in Table 3, we then conducted 10 runs of tenfold cross validation for the above data set. Classification test results are shown in the right side of Table 3 from the viewpoint of accuracy, precision, recall, and f-measure.

Tenfold cross validation is a widely used validation method for machine learning algorithms when the amount of learning and test data is a few. But, according to an error analysis of tenfold cross validation (Bouckaert 2003), it is not enough for replicability of test measures. A way of separating data into each fold affects accuracy of classifications and leads to a variety of the error of the first kind, so that it suffers from low replicability. The reference recommends to use 10 runs of tenfold cross validation, in which a set of learning and test data is differently separated. For error analysis of measures, we need to conduct a t-test for a null hypothesis that means of cross validation results of two different algorithms are equal. The degree of freedom in t statistic influences on the test result. In increasing of the degree of freedom under the same sample variances, the standard error is decreasing. A tenfold cross validation has 9 $(10-1)$ degree of freedom while 10 runs of tenfold cross validation has 99 $(100-1)$ degree of freedom. We, therefore, examine 10 runs of tenfold cross validation.

In the test ID 1-1, 1-2, 1-3, feature elements consist of BoW which count over 15,800, 13,000, and 12,000 respectively. The f-measures are worse than the other features with the same linear kernel function. They seem to be out of learning.

---

[2]http://svmlight.joachims.org/.

Table 3 Feature selections and classification results

Test ID	TF-IDF feature element (1) BoW	(2) BoW (N)	(3) BoW (N-3)	(4) K (14)	(5) K (18)	(6) K(18) +NM	(7) ORG	(8) Corp.	(9) Univ.	(10) C. +U.	Size	Kernel	Results Acc.	Prec.	Rec.	F.
1-1	✓										ave. 15795 (sd. 267)	L	60.85	69.83	39.15	50.07
1-2		✓									ave. 13271 (sd. 222)	L	60.73	70.63	38.54	49.78
1-3			✓								ave. 12011 (sd. 212)	L	60.96	71.38	38.39	49.88
2-1				✓							14	L	67.76	72.24	62.30	66.90
2-2				✓							14	P	57.76	74.06	23.21	35.32
2-3				✓							14	R	67.11	62.76	87.18	72.99
3-1					✓						18	L	67.73	71.72	62.76	66.94
3-2					✓						18	P	57.76	72.93	23.36	35.37
3-3					✓						18	R	66.41	61.95	87.97	72.70
4-1						✓					41	L	–	–	–	–
4-2						✓					41	P	–	–	–	–

4-3	✓					41	R	70.65	65.28	90.18	75.74
5-1	✓	✓				42	L	70.61	74.61	63.64	67.31
5-2	✓	✓				42	P	–	–	–	–
5-3	✓	✓				42	R	70.76	65.49	90.30	75.66
6-1	✓	✓	✓	✓		44	L	–	–	–	–
6-2	✓	✓	✓	✓		44	P	–	–	–	–
6-3	✓	✓	✓	✓		44	R	70.15	64.64	93.64	76.09
7-1	✓	✓	✓	✓	✓	45	L	78.94	85.03	71.82	77.16
7-2	✓	✓	✓	✓	✓	45	P	–	–	–	–
7-3	✓	✓	✓	✓	✓	45	R	71.82	65.73	**94.85**	77.35
7-4	✓	✓	✓	✓	✓	45	R(γ)	79.85	78.51	83.94	80.86
8-1	✓		✓	✓	✓	44	L	78.95	**85.15**	71.37	77.65
8-2	✓		✓	✓	✓	44	P	–	–	–	–
8-3	✓		✓	✓	✓	44	R	72.80	66.35	**94.85**	78.08
8-4	✓		✓	✓	✓	44	R(γ)	**79.95**	78.34	84.21	**81.17**

Results are average points in 10 runs of tenfold cross validation
Kernel = {L:Linear, P: Polynomial, R: Radial basis function, R(γ): Radial basis function (γ parameter tuned)}
– Not calculated because of precision zero or learning optimization fault
Bold values are top scores in the accuracy, precision, recall, and F-measure, respectively

In the test ID from 2-1 to 8-4, feature element size is about 14–45. They seem to be effectively caused by learning, except for some tests resulting in some learning optimization fault. Accuracy and f-measure are gradually inclined while feature elements are additionally complex. Top f-measure is in the test ID 8-4, and its accuracy is high too.

3.3 Discussion

In comparing with BoW features (test ID 1-*) and keyword features (test ID 2-*, 3-*), BoW features was out of learning while keyword features produced good results in a sense. Even though BoW is known for useful general elements for text categorization, this time BoW features failed in learning. The reason why they are failed in learning can be that training data size is too much smaller than enough to learn. If we have enough size of training data, it become larger than the feature vector size. This means training data size surpass the number of basis function of SVM, so that learning could be done without over-fitting. To reduce feature elements into order 10^1, we have to choose most sensitive term for feature elements. In our result, 18 keywords of U-I relations that we choose got effectiveness for learning.

For evaluating effect of co-occurence of keywords and POS of the next morpheme in a sequential text, we check the results of test ID 3-* and 4-*. F-measure of test ID 4-* is 3 points better than that of test ID 3-*. POS was restricted to verb, auxiliary verb, and sahen-noun, so as to exclude noun phrases partially constructed with the keywords. The keywords with the next morpheme of the POS tends to be a part of statement that expresses U-I relations. Otherwise, the keywords tend to be a part of name or title of research groups and organizations.

Test ID 6-* and 7-* is related to an occurrence of university and company symbols. In comparing them to ID 5-*, the f-measure of them become higher so that we can recognize the occurrence of the two symbols in a sentence is sensitive to U-I relations. However, in comparing test ID 4-* to test ID 5-*, or test ID 7-* to 8-*, the results showed there is nothing to change. This means organization tag count as a result of named entity extraction is meaningless. Press release articles may intend to be written with organization name in various genre.

If we compare test ID *-1, *-2, *-3, and *-4 of the same major number, we can recognize how much kernel functions and parameters effect on scores. The results showed that they should be optimized to get highest scores. Loss function weight parameter influenced on balance between precision and recall rate. γ of Radial Basis Function was optimized to get highest f-measure.

4 Conclusion

To automatically extract U-I relation documents from the Web, we proposed the framework that consists of crawling Web documents, extracting text from the documents, learning to classify the documents, and classifying new documents. We conducted experiments under the framework for actual Web documents. We performed experiments using several combinations of SVM feature vector elements and kernel function types. Results showed that the most effective elements are U-I relation keywords with consideration of grammatical connection in a sequential text, and university and company symbols used in a sentence. Selecting a kernel function and tuning parameters are important to get highest score. The highest f-measure we obtained was 81.17. Simultaneously, the accuracy was 79.95.

References

Aizawa, A. (2003). An information-theoretic perspective of tf-idf measures. *Information Processing and Management, 39*(1), 45–65. doi:10.1016/S0306-4573(02)00021-3.

Bouckaert, R. (2003). Choosing between two learning algorithms based on calibrated tests. In *Proceedings of the 20th International Conference on Machine Learning (ICML-2003)* (pp. 51–58), Washington, DC.

Leydesdorff, L., & Meyer, M. (2003). The triple helix of university-industry-government relations. *Scientometrics, 58*(2), 191–203.

Salton, G., & Buckley, C. (1988). Term-weighting approaches in automatic text retrieval. *Information Processing & Management, 24*(5), 513–523.

Vapnik, V. N. (1995). *The nature of statistical learning theory*. New York: Springer.

Yang, Y., & Pedersen, J. O. (1997). A comparative study on feature selection in text categorization. In D. H. Fisher (Ed.), *Proceedings of ICML-97, 14th International Conference on Machine Learning* (pp. 412–420). San Francisco: Morgan Kaufmann.

Dynamic Customer Satisfaction and Measure of Trajectories: A Banking Case

Caterina Liberati and Paolo Mariani

Abstract The most important company asset seems to be Customer Satisfaction (CS), which banks, in the recent years, have frequently analyzed. For reaching such target, a dynamic Factor Analysis offers an effective way of merging information about clients and their preferences evolution. In our work we performed a dynamic Customer Satisfaction study, by means of a three-way factorial analysis, and we also introduced a new index of shift and shape (SSI), to synthesize information about every customer, cluster or typology. We considered a national bank case, with spread network, evaluating results provided by a questionnaire framed according to the SERVQUAL model. The information employed was obtained via a Customer Satisfaction survey repeated three times (waves). We performed the dynamic factorial model and we illustrated the usage of SSI as a new measure of trajectories' dissimilarity. Finally, we showed our results which highlight promising performances of our index.

Keywords Customer satisfaction • Dynamic patterns • Multiway factor analysis • Trajectories analysis

1 Customer Satisfaction: A Scenario

Since the 1990s, the increasing level of information and competence of the average retail customer has lead to a more complex needs architecture and a demand of diverse financial and tailored services (Carú and Cugini 2000; Cosma 2003; Mottura 1982). Thus, banks have to now act according to customer preference and innovate to retain customers (ABI 2009). Customer concerns and wishes change

C. Liberati (✉) • Poalo Mariani
DEMS, Università Milano-Bicocca, Milano, Italy
e-mail: caterina.liberati@unimib.it; paolo.mariani@unimib.it

D. Vicari et al. (eds.), *Analysis and Modeling of Complex Data in Behavioral and Social Sciences*, Studies in Classification, Data Analysis, and Knowledge Organization,
DOI 10.1007/978-3-319-06692-9_20,

continuously: this requires a non-stop improvement, quality enhancement, and a monitoring of Customer Satisfaction effectiveness (Munari 2000; Oliver 1977). Thus, the diversifying of product attributes has been substituted by customer requests, that leads to homologated offers which render competition difficult. The bank's ability to focus on 'how to serve the client' and not on 'what the client receives' clearly creates customer satisfaction-dissatisfaction and consequently, his loyalty (Mihelis et al. 2001). For these reasons, the following analysis explores the main aspects of the banking sector in terms of customer expectation and satisfaction, the identification of improvement areas, and appropriate recovery actions (Munari 2002). Through different survey waves, management can monitor evolution and measure the impact of specific actions for opinion improvement.

2 Dynamic Factor Analysis and Shift and Shape Index

The main purpose of the Multi-way analysis is to draw a dynamic path. It summarizes the variability of a complex phenomenon by highlighting both similarities/dissimilarities among the "occasions" (the waves) considered and the main components of the average behavior in the time interval chosen (Bolasco 1999; Kroonenberg 2007). In our work we employed tools based on multivariate dynamic analysis, three way arrays, to analyze the data "volumes". Dynamic multivariate techniques allow the management and the analysis of complex data structures, to study a given instance phenomenon in both a structural and a dynamic way.

We analyze three-way data matrices individuals $\times$ variables $\times$ occasions, therefore, X_{ijk} type, with $i = 1, \ldots, N$ individuals, $j = 1, \ldots, M$ variables, $k = 1, \ldots, K$ occasions.

A common factorial space was built: the Compromise Space, in which the elements are represented according to their average configuration relative to data volume as a whole. The Compromise Space, obtained by means of a Principal Component Analysis of the K tall matrices (subjects $\times$ variables), can be reviewed as the weighted mean of the similarity/distance-matrices among individuals. [Lacangellera et al. (2011); Liberati and Mariani (2012)].[1]

The union of the partial factors, referring to the same unit related to different moments in time, allows the tracking of the temporal trajectories which describe the dynamic behavior of the phenomenon (Coppi and D'Urso 2001; D'Urso 2000). Following such an approach, we were able to monitor not only clients positions onto the Compromise Space, but also their trajectories over the time. Therefore, the most

[1]As Kroonenberg (2007) highlighted, a three-way array can be seen as composed of two-mode sub-matrices called slices, and of one-mode sub-matrices (or vectors), called fibers. These two-way sub-matrices will be referred to as frontal slices, horizontal slices, and lateral slices. In our case we composed the X (tall matrix) using the horizontal juxtaposition of the slices. We analyzed such X via a simple PCA which is a no weighted version of the STATIS method (Escoufier 1980).

innovative aspect of our work consists in employing the information obtained via PCA for feeding an index which is able to summarize trajectories characteristics. The Shift and Shape Index (SSI) is a measure of synthesis of shape and length of a path, and its algebraic formulation is shown as follows:

$$SSI = \sum_{k=1}^{K-2} \frac{2A_k}{d_{\min}} + d_{tot} \tag{1}$$

where

- A_k = Area of the triangle obtained by joining successive triplets onto the Compromise Space for each observation.
- $d_{\min}$ = minimum distance covered by the i-th observation obtained by connecting the first wave point and the last wave point of the scatter plot (for K=3 $d_{13}(i) = \sum_j \|x_{ij1} - x_{ij3}\|^2$)
- d_{tot} = total distance covered by the i-th observation obtained summing up all the successive partial segments (for K=3 $d_{12}(i) = \sum_j \|x_{ij1} - x_{ij2}\|^2 + d_{23}(i) = \sum_j \|x_{ij2} - x_{ij3}\|^2$).

Moreover, $\frac{2A_k}{d_{\min}}$ is equivalent to the height of the triangle obtained by linking three points obtained onto the compromise plan, for each observation. Such height indicates how far the subject has moved away from the optimal path ($d_{\min}$), then such value is summed to the real distance d_{tot}.

SSI does not have any upper or lower bounds and it can be computed having three wave observations at least for each subject. The higher the index value becomes, the greater the individual displacement on the factorial plane and the degree of reactivity of the analyzed phenomenon. In order include in SSI, also, indication about the direction of the shifts covered by subjects across the different waves, we propose the usage of the sign of Pearson correlation coefficient ρ which measures the symmetric linear relationship between two vectors X and Y. X and Y represent respectively the shifts of each subject:

$$X(i) = \begin{bmatrix} \sum_j \|x_{ij2} - x_{ij1}\| \\ \sum_j \|x_{ij3} - x_{ij2}\| \end{bmatrix} Y(i) = \begin{bmatrix} \sum_j \|y_{ij2} - y_{ij1}\| \\ \sum_j \|y_{ij3} - y_{ij2}\| \end{bmatrix} \tag{2}$$

Therefore an alternative version of SSI could be

$$SSI_s = sign(\rho) \sum_{k=1}^{K-2} \frac{2A_k}{d_{\min}} + d_{tot} \tag{3}$$

The limit of usage of ρ coefficient is evident when both shifts are negative because the $sign(\rho)$ provides a positive value which is difficult to justify. Further studies will be necessary to address such incoherences.

Table 1 Questionnaire's items

Tangible aspects	Responsiveness
Bank 'XX' has facilities with advanced technology?	The bank 'XX' guarantees short waiting time in the queue?
The buildings of the bank 'XX' are attractive?	The employees of the bank 'XX' provide service to customers quickly and efficiently?
The bank 'XX' has offices with standardized aspect?	
The employees of the bank 'XX' are always clean-looking?	

3 An Italian Bank Case

In this section, we illustrate the effectiveness of our Shift and Shape Index. We show the results on professional clusters obtained via a post-stratification of the data collected, according to the bank's job categories; we also plot clusters trajectories onto the compromise as average values of the shifts computed for each factors scores.

After a progressive loss of some customer segments, managers of an Italian bank decided to conduct a survey by choosing a sample of retail customers.

Therefore, the questionnaire was framed according to the SERVQUAL model (Berry et al. 1985, 1988, 1991, 1993) with five dimensions to analyze the perceived quality and expectation of the banking service. The choice of the dimensions has been realized according to the bank's need. All the items are measure via a Likert scale from 1 to 10.[2]

The same questionnaire was submitted to the sample for three waves, therefore we could perform a dynamic model to quantify variable changes across three different occasions. We used a sub sample of 2,000 customers and 2 macro dimensions: tangibles aspects (measured by 4 items) and responsiveness (measured by 2 items) (Table 1). We analyzed the sample across nine different professional clusters: Entrepreneurs, Managers, Employees, Workers, Farmers, Pensioners, Housewives, Students, Others, obtaining the average score of the SERVQUAL model grouped in nine clusters.

The first step of the study consisted in estimating the compromise space to represent the average position of the professional clusters according to the selected variables. The PCA was performed on differences between expected and perceived items (disconfirm analysis) (Oliver 1977; Seth et al. 2005), results were sturdy: the first two components (of the six obtained) explained about 80 % of the total variance,

[2]Currently, in literature, two different ways to model a Likert scale are present: some refer to it as quantitative scale some as ordinal one. Based on Zani and Berzieri (2008) work we considered our data cardinal because it does not produce a substantial bias, therefore the 10-categories could be approximated to a 10 score

Table 2 Factor loading matrix

Variables	f_1	f_2
Technology	0.793	0.333
Physical structures	0.230	0.913
Offices standardized	0.190	0.858
Employees clean-looking	0.362	0.736
Waiting time	0.898	0.193
Speed of service	0.903	0.256

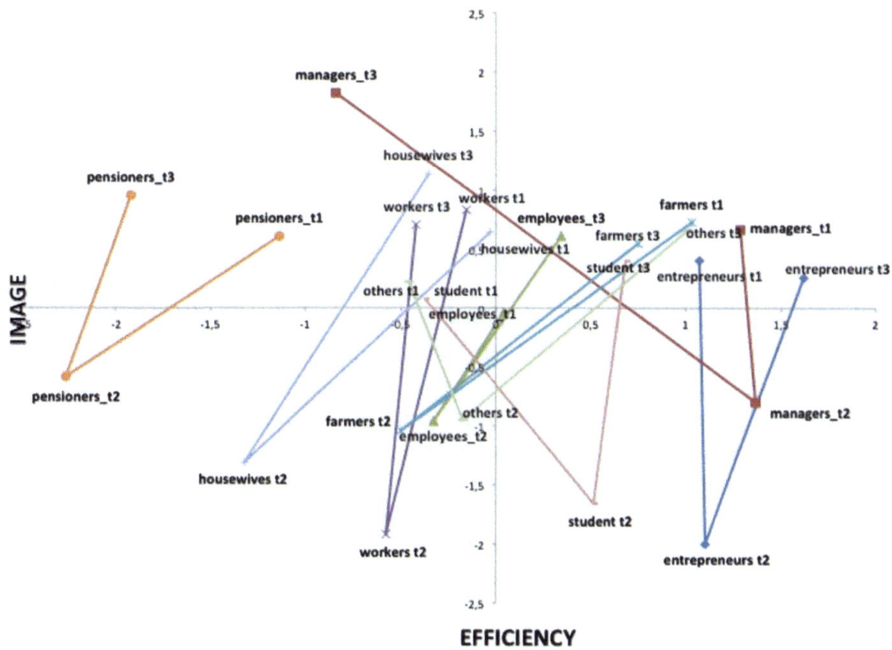

Fig. 1 Professional clusters trajectories

in particular, the first explains 61.8 % alone. Also the KMO index (0.664) and the Bartlett test (109.1; sig. 0.000) showed the quality of the created factorial model. By representing the circle of correlations between the original items and the first two components of the compromise plan, a clear polarization of the variables, respect to the two axes, appeared (Table 2).

Based on such evidences we named the first component as the "Efficiency" of the service because it includes variables: "Technology", "Waiting time" and "Speed of Service". The second component, instead, seems to capture those aspects that contribute to delineate the "Image" of the service provider. In fact, it includes: "Physical Structures", "Offices Standardized" and "Employees Clean-looking". Finally, we represented as scatter plot the partial factors for each observation onto the Compromise Space and we composed them in paths. Figure 1 shows the trajectories drawn by each professional clusters.

Table 3 Elementary quantities for Shift and Shape Index (SSI)

Professional clusters	d_{12}	d_{23}	d_{min}	d_{tot}	Area	$\frac{2A}{d_{min}}$	SSI
Entrepreneurs	2.4047	2.3106	0.5679	4.7154	0.6554	2.3040	7.0235
Managers	1.4578	3.4233	2.4259	4.8811	1.5056	1.2413	6.1223
Employees	0.9923	1.7054	0.714	2.6977	0.023	0.0644	2.7621
Workers	2.7749	2.6226	0.2884	5.3975	0.3299	2.2878	7.6858
Farmers	2.3314	2.0153	0.3324	4.3468	0.1115	0.6709	5.0174
Pensioners	1.6427	1.5748	0.8591	3.2175	0.6639	1.5456	4.7630
Housewives	2.3473	2.6368	0.5900	4.9841	0.6361	2.1563	7.1404
Students	1.9452	2.0652	1.1136	4.0104	1.0664	1.9152	5.9256
Others	1.1996	1.9993	1.5372	3.1990	0.9207	1.1979	4.3969

The trajectories, seem to have very similar shapes: in general, with reference to the first factor, we notice a clear improvement when it is passed from lag $k = 1$ at time $k = 2$, and a worsening between $k = 2$ and $k = 3$. The routes are drawn in a space derived with a factorial disconfirm analysis, so every shift toward the negative side of the first axis produces a reduction of the difference between expected and perceived items, which induces an improvement.

This trend is perfectly coherent with the actions that the bank has implemented between the first and second administration of the questionnaire, which has been focused on the reduction of the waiting time at the counters and on the rationalization of retail space. The loss of satisfaction, found among the second and third survey, is instead due to the bank's decision to cease monitoring of banking services. The second factor seems to have a much more variable trend. Satisfaction related to the Image hit differently the professional categories examined (Fig. 1), in fact, for some of them we noticed an improvement in the perceived image of the bank (Entrepreneurs, Managers, Housewives, Pensioners and Other), for the others, however, a worsening (Farmers, Students and Workers).

To analyze the opinions expressed by bank customers surveyed regarding the dimensions considered, we computed the Shift and Shape Index, in order to provide a synthesis measure of Customer Satisfaction detected. Table 3 collects values of the main quantities that compose SSI for each job category:

As we highlighted above, SSI is a suitable tool to describe and estimate the evolutions of the clusters. According to it, we interpreted high values as a proxi of the customer sensibility to the bank marketing stimuli; therefore, to the higher values correspond more strong reactions of the clusters. In our case of study, Entrepreneurs, Workers and Housewives show a high sensibility to marketing stimuli: their SSI values are close each other, moreover, if we decompose those values into $\frac{2A}{d_{min}}$ and d_{tot} we have to register a similar percentage of both quantities over the total index value. Such tendency reveals a similar elasticity to changes: therefore, the suggestion to the management, highlighted by means of SSI, is to monitor those professional clusters more carefully.

Table 4 collects values of an alternative version of Shift and Shape Index that we call for simplicity Shift and Shape Index with sign (SSI_s). It provides the same

Table 4 SSI_s values for different professional clusters

Professional clusters	ρ	SSI	SSI_s
Entrepreneurs	0.6345	7.0235	7.0235
Managers	−0.8903	6.1223	−6.1223
Employees	0.9995	2.7621	2.7621
Workers	0.9233	7.6858	7.6858
Farmers	0.9989	5.0174	5.0174
Pensioners	0.8159	4.7630	4.7630
Housewives	0.9680	7.1404	7.1404
Students	−0.4853	5.9256	−5.9256
Others	0.6518	4.3969	4.3969

evaluations as SSI, but it also produces information about the direction of the shift covered by every clusters. We have to highlight the negative values of the index in correspondence of Managers and Students groups: the signs are coherent with the negative correlation of the X and Y shifts.

4 Final Remarks

While the graphical analysis of the trajectory shows how the individual moved over time, SSI index provides a measure of the degree of reactivity of the individual according to the phenomenon under examination. The trend of trajectories and the index calculation show the same conclusions. The values assumed by the index are very close together while the categories had a similar degree of reactivity compared to the size of the SERVQUAL model considered. In particular, "Workers" as a category (whose index is equal to 7.686) proved more sensitive to interventions on the quality of services implemented by the bank. The former were followed respectively by "Housewives" (7.1404) and by "Housewives" (7.1404), while "Managers" (−6.122) and "Students" (−5.9256) proved to be less sensitive. The bank obviously needs to intervene by taking actions to try to improve the satisfaction of consumers belonging to those categories that had a negative reaction to existing policies.

References

ABI (2009). Dimensione Cliente 2009. Roma: Bancaria Editrice.

Berry, L. L., Parasuraman, A., & Zeithaml V. A (1985). A conceptual model of service quality and its implications for future research. *The Journal of Marketing, 49*, pp. 44–50.

Berry, L. L., Parasuraman, A., & Zeithaml, V. A. (1988). SERVQUAL: a multiple-item scale for measuring consumer perceptions of service quality. *Journal of Retailing, 64*(1), 12–37.

Berry, L. L, Parasuraman, A., & Zeithaml V. A. (1991). Refinement and reassessment of the SERVQUAL scale. *Journal of Retailing, 67*, 420–450.

Berry, L. L., Parasuraman, A., & Zeithaml V. A. (1993). Research note: more on improving service quality measurement. *Journal of Retailing, 69*, 140–147.

Bolasco, S. (1999). *Analisi multidimensionale dei dati. Metodi, strategie e criteri d'interpretazione*. Roma: Carocci.

Carú, A., & Cugini, A. (2000). *Valore per il cliente e controllo dei costi: una sfida possibile*. Milano: EGEA.

Coppi, R., & D'Urso, P. (2001). The geometric approach to the comparison of multivariate time trajectories. In S. Borra, R. Rocci, M. Vichi, & M. Schader (Eds.), *Advances in data science and classification*. Heidelberg: Springer.

Cosma, S. (2003). *Il CRM: un modello di relazione tra banca e cliente*. Roma: Bancaria Editrice.

D'Urso, P. (2000). Dissimilarity measures for time trajectories. *Journal of the Italian Statistical Society, 9*(1–3), 53–83.

Escoufier, Y. (1980). L'analyse conjointe de plusieurs matrices. In Jolivet et al. (Eds.), *Biomiétrie et Temps*. Société Francaise de Biométrie. Paris.

Kroonenberg, P. M. (2007). *Applied multiway data analysis*. New York: Wiley.

Lacangellera, M., Liberati, C., & Mariani, P. (2011). Banking services evaluation: a dynamic analysis. *Journal of Applied Quantitative Methods, 6*(4), 3–13.

Liberati, C., & Mariani, P. (2012). Banking customer satisfaction evaluation: a three-way factor perspective. *Advances in Data Analysis and Classification, 6*(4), 323–336.

Mihelis, G., Grigoroudis, E., Siskos, Y., Politis, Y., & Malandrakis, Y. (2001). Customer satisfaction measurement in the private bank sector. *European Journal of Operational Research, 130*, 347–360.

Mottura, P. (1982). *La gestione del marketing nella banca. inStruttura organizzativa, controllo di gestione e marketing nella banca*. Giuffrè.

Munari, L. (2000). *Customer satisfaction e redditività nelle banche*. Banche e Banchieri, n. 3.

Munari, L. (2002). CRM e redditività di cliente: opportunità per una revisione degli orientamenti gestionali nel retail banking. APB News, n. 1.

Oliver, R. L. (1977). Effect of expectation and disconfirmation on post-exposure product evaluations: an alternative interpretation. *Journal of Applied Psychology, 4*, 480–486.

Seth, N., Deshmukh, S. G., & Vrat, P. (2005). Service quality models: a review. *International Journal of Quality & Reliability Management, 22*(9), 913–949.

Zani, S., & Berzieri, L. (2008). Measuring customer satisfaction using ordinal variables: an application in a survey on a contact center. *Statistica Applicata, 20*, 331–351.

The Analysis of Partnership Networks in Social Planning Processes

Rosaria Lumino and Concetta Scolorato

Abstract This article focuses on using social network analysis (SNA) to evaluate social planning. A case study on a regional policy, the Territorial Youth Plans (PTGs), is presented. PTGs were introduced by the Campania Region (Southern Italy) in 2009 as an effort to reform the regional youth policies to foster increased participation in decision-making processes. In our case study, we use a combination of SNA tools and multivariate data analysis techniques to analyze *if* and *how* the structure of interactions between actors at the local level shapes the quality of planning in terms of coherence and innovation.

The relational data have been gathered through the analysis of official documents. Attention has been directed to the following aspects: (1) describing the networks characteristics, (2) detecting networks with a typical and homogenous structure configuration, and (3) determining the relationship between several network structure configurations and different forms of social planning, assuming that relational structures of networks shape the coherence of social planning activity and local innovation capacity.

Keywords Partnerships networks • Social planning • Youth policies • Evaluation • Clustering

1 Social Network Analysis for Social Policies Evaluation

In line with the growing consensus centred around the important analytic role that "network" plays in many fields of social activities, there has been an increasing interest in the application of social network tools within evaluation practice, both

R. Lumino (✉) • C. Scolorato
University of Naples Federico II, Naples, Italy
e-mail: rosarialumino@gmail.com; concetta.scolorato@unina.it

D. Vicari et al. (eds.), *Analysis and Modeling of Complex Data in Behavioral and Social Sciences*, Studies in Classification, Data Analysis, and Knowledge Organization,
DOI 10.1007/978-3-319-06692-9_21,

for national and international scenarios (Penuel et al. 2006; Durland and Fredericks 2005; Cross et al. 2009; Di Nicola et al. 2010).

Our own research explores how we can use social network analysis (SNA) to evaluate initiatives based on cooperation among several actors with a focus on decision-making processes. This study is part of a larger research project[1] focusing on the *ex ante* evaluation of a new policy instrument, the Territorial Youth Plans (*Piani territoriali per le politiche giovanili*; PTGs), which was introduced by the Campania Region (Southern Italy) in 2009 to reform the regional youth policies to foster increased participation in decision-making processes.

A central motivation for our research is that today, the capacity for decision-making, program formulation and implementation is widely distributed among mutually interdependent public and private actors (Kenis and Schneider 1991). For the past several decades, a long season of reforms in most European countries has resulted in a reorganization of public policy making in an ever-increasing number of policy domains, with a shift towards a sharing of tasks and responsibilities (Peterson 2003) and a transfer from systems of local *government* to systems of local *governance* (Rhodes 2007).

These trajectories of change have involved the whole domain of social policy, implying the involvement of new actors both in the initial implementation of social services and their management, and in the following planning and codefinition of goals (Kazepov 2008). An actual involvement of these actors would help increase policy effectiveness and efficiency.

This redefinition of the public sphere has occurred within the emergence of new policy instruments based on a high degree of negotiation among the actors involved. However, this negotiation is not simple or spontaneous, and it requires continuous adjustments in the network management. The effectiveness of these processes of negotiation can be readily assessed using SNA, especially when it is possible to assume that the structure of the interactions between policy actors explains policy outcomes (Brandes and Erlebach 2005).

SNA allows evaluators to identify what the network is, how it operates (Honeycutt 2009) and how governance outcomes are conditioned by both the structural characteristics of networks and the way in which governance arrangements are managed (Fawcett and Daugbjerg 2012).

In this study, we seek to keep the network at the core of the evaluation exercise, focusing on the partnership networks that have emerged as an outcome of the PTG initiative, combining the analysis of a single network with the analysis of the whole set of networks that has been funded by the same policy programme and then analysing the relationship between the structural properties of the networks and the quality of planning.

[1]The research project has been supported by the Campania Region Grant for the "Permanent Observatory on Youth Policies and Youth Condition" and by the project REPOS, Reti di Politiche Pubbliche e Sviluppo.

The article is divided into three sections. The first lays out our case study, the second presents our analytic strategy and the third describes our findings and concluding remarks.

2 The Case Study

PTGs are one of the most recent examples of innovation in social planning. They were introduced by the Campania Region (Italy) in 2009 as a way to increase youth policies' effectiveness and efficiency through the enlargement of the policy-making process. They initiate a process of local policy learning, balancing the needs to overcome the overall fragmentation of planning and to value the properties of local contexts.[2]

PTGs represent an instrument for the participated planning of youth policies within a decentralized form of governance that encourages the municipalities of a territorial area, namely a school district,[3] together with local communities, to take part in the planning of a shared territorial system of youth services (e.g., not-for-profit and for-profit organizations as well as citizens).

The PTGs scheme assigns a key role to one municipality in each school district (e.g., a 'leading municipality' or *comune capofila*), whose responsibilities include mobilizing and coordinating local resources and partnerships. Specifically, in 2009, the Campania Region issued the call for PTG planning by modifying a regional application form. Each leading municipality began a process of participatory planning with the other municipalities in the same district and other local actors with the end goal of developing a common project. After this preliminary phase, the Campania Region revised the projects, initially financing 48 PTGs (2010) across the 62 territorial districts.

Due to the recent introduction of PTGs in the youth policy domain, they offer a very specific setting for evaluation, highlighting various relevant issues, such as whether the introduction of PTGs actually creates a new pathway for youth policies, whether they have led to new forms of partnership, whether the structure of the networks can be described and how the membership structures of the partnerships have shaped the social planning activity at local level. These are the most relevant issues that have motivated our research. To address them, we have used a multi-method approach, combining SNA tools with multivariate data analysis techniques.

[2]PTGs trace the model of similar instruments, such as Piani di Zona (Area Plans), adopted since 2000 in the social policies domain.

[3]Campania Region is divided in 62 school districts, each of them involves a groups of municipalities in close geographical proximity.

3 PTGs Network Definition and Data Analysis Strategy

The relational data were gathered through the analysis of official documents (PTGs, protocols, contracts, etc.). We constructed 36 complete undirected networks[4] (one for each PTG, using $\mathcal{N}_j, j = 1, \ldots, 36$ to encode the partnerships by looking at the list of actors involved in the plan (they were listed in a specific section of the regional application form), the formalized agreement protocols attached to the form and the descriptions of the planned activities within the PTGs, assuming that the common actors' involvement in a module of the project was a proxy for the existence of a relationship between them. The documents reported both formal and informal partnerships and coparticipation or cofinancing of the activities. For the sake of simplicity, we chose to not distinguish between these types of ties.

More specifically, for the jth PTG, we constructed an adjacency matrix X_j with n_j actors a_i^j, $i = 1, \ldots, n_j, j = 1, \ldots, 36$, and with $x^j{}_{ih} = 1$ if we found a connection between the actors a_i^j and a_h^j in the documents attached to jth PTG, and $x^j{}_{ih} = 0$ if otherwise. For convention, we set the leading municipality as the first actor; that is, a_i^j indicated the leading municipality of the jth PTG. Beyond the relational data collected through the documents analysis, we evaluated the project plans using an analytic template which was comprised of the following dimensions: the content of the planning, the priorities and strategic goals, the financial resources put in place and their specific destination, the coherence of the documents and the innovation of the planned activities. The last two dimensions were measured using a four-point scale, with one indicating the lower level of coherence (or innovation) and four indicating the higher level.

The coherence of the documents referred to the relationship between both the identified needs in each local context and the planned actions and the declared specific goals and planned actions. The innovation was linked to both the presence of strategies based on active involvement of several local actors and to the integration between different sectors of intervention.

The analysis of official documents was enriched through interviews with some youth policy experts who were involved in the PTG planning at the local level to audit our results.[5]

Due to the limited length of this article, we concentrate on some specific aspects concerning the PTGs and check them against measures provided by SNA.

In this article, we (1) explore and describe the partnership networks' characteristics, we (2) highlight the presence of typical and homogenous partnership/network structures and we (3) analyse the relationship between the network structures and the plans' coherence and innovation.

---

[4]The number of analyzed plans, on a total of 48 districts involved in the experimentation of the PTGs, depends on the completeness of the available information in accordance with our research interests.

[5]A full research report is cfr. Bisceglia et al. (2013).

Regarding the first objective, each network $\mathcal{N}_j$ is described using indices at both the network level and the leading municipality level (Wasserman and Faust 1994). We looked at the centrality of the leading municipality to determine its role and position in the partnership network and to highlight the power of the leading municipalities that arose from their relationships with other partners. Specifically, we considered the actor degree centrality $C_D\left(a_1^j\right) = \sum_{h=1}^{n_j} x_{1h}^j / \left(n_j - 1\right)$ to analyse the leading municipality's capability of activating partnerships; the actor betweenness centrality $C_B\left(a_1^j\right) = \sum_{h<k} g^j{}_{hk}\left(a_1^j\right) / g^j{}_{hk}$, with $g^j{}_{hk}\left(a_1^j\right)$ being the geodesic path between a_h^j and a_k^j passing through a_1^j and $g^j{}_{hk}$ all the geodesic path linking a_h^j and a_k^j in the network, $\mathcal{N}_j$, to highlight the leading municipality's brokerage role; and the actor closeness centrality $C_C\left(a_1^j\right) = \left(n_j - 1\right)\left[\sum_{h=1}^{n_j} d\left(a_1^j, a_h^j\right)\right]^{-1}$, with $d\left(a_1^j, a_h^j\right)$ being the length of geodesic path between the two actors to measure their capability of easily communicating with each other. For the network level, we looked at the centralization to analyze the leadership structure in the partnership beyond the role of the leading municipality by considering the network degree centralization $C_D = \sum_{h=1}^{n_j}\left[C_D\left(a_*^j\right) - C_D\left(a_k^j\right)\right] / \left[\left(n_j - 1\right)\left(n_j - 2\right)\right]$, with $C_D\left(a_*^j\right)$ being the largest observed degree of centrality in the network, $\mathcal{N}_j$, to measure how the leadership was spread among the partners; and the network betweenness centralization $C_B = 2\sum_{h=1}^{n_j}\left[C_B\left(a_*^j\right) - C_B\left(a_h^j\right)\right] / \left[\left(n_j - 1\right)^2\left(n_j - 2\right)\right]$, with $C_B\left(a_*^j\right)$ being the largest observed betweenness centrality in the network, $\mathcal{N}_j$, to highlight the presence of other partners having a mediating role.

We also considered the reachability of network actors to analyse the partnership cohesion and the capability of sharing information. In particular, we looked at the component ratio $CR(N_j) = (\gamma^j - 1)/(n_j - 1)$, with γ^j being the number of components (connected subgraphs) in the network, $\mathcal{N}_j$; the fragmentation $FR\left(N_j\right) = 1 - \left[2\sum_{i>h} r_{ih}^j / n_j\left(n_j - 1\right)\right]$ with $r_{ih}^j = 1$ if the actor a_i^j can reach the actor a_h^j by a path of any length; and the average geodesic distance $AD\left(N_j\right) = 2\sum_{i<h} d\left(a_i^j, a_h^j\right) / n_j\left(n_j - 1\right)$.

In general, a low level of cohesion measured through the presence of isolate nodes, a high level of fragmentation and large distances among partners indicates that the actors were only involved in planning in a formal way and that they did not actually cooperate.

All the indices were computed using UCINET 6.0 (Borgatti et al. 2002).

We assumed that network structural characteristics shape the sharing of information and expertises that are regarded as useful for joint decisions or joint actions (Reagans and McEvily 2003). Specifically, highly centralized networks might increase the spread of information among actors and improve their performance, but they are more exposed to the risk of breakdown when the central actors fail or exit from the networks (Capuano et al. 2013). Cohesive networks might promote cooperation and motivate actors to share information and expertise, but they also obstruct innovation that results from stepping across highly closed subgroup boundaries (Cross and Cummings 2004).

In order to highlight the presence of typical and homogenous partnership/network structures, we constructed a matrix in which the PTG networks were on rows and the above-listed indices were on columns; i.e., each network was represented by a set of variables describing its structure. We performed a cluster analysis of the 36 networks, mapping each one into a multidimensional space in which the dimensions were the parameters above mentioned. We applied a usual agglomerative hierarchical cluster algorithm (Lebart et al. 1984), obtaining four groups of homogenous partnership networks.

With respect to the objective related to the analysis of the relationship between the network structures and the plans' coherence and innovation, the groups identified through clustering were linked to some positive aspects of the PTGs. We evaluated the average innovation score and the average coherence score within each group to discover clusters characterised by the highest scores. This allowed us to explain these better scores in terms of the leading municipality's role, network centralization and network cohesion.

4 Main Results and Concluding Remarks

The analysis of partnership networks showed their various compositions in terms of number and type of actors involved. The networks' sizes varied between a minimum of 6 to a maximum of 71 organizations, with a median value of 28 organizations for each district. They were primarily public institutions (about 55 % of the total on the median value) and not-for-profit organizations active in the youth policy domain (about 33 % of total cases).

The network indices showed the presence of a high degree of network centralization in almost all networks and a relatively low degree of cohesion resulting from the weak attitudes towards cooperation at the local level in the youth policy domain. Furthermore, we found that the negotiated relationships based on shared goals and preferences require long settling time that is not necessarily consistent with the 'bureaucratic' start-up of a new political instrument such as the PTGs.

Our analysis of actor-level indices, based on centrality measures, highlighted the power of the leading municipalities in terms of being in contact to many other actors and maintaining connections between them. This result is in accordance with the role assigned to them by the regional administrative design of the PTGs. With some exceptions, the leading municipalities did not have a very central location with respect to closeness, which hindered the communication of information to the other actors, thereby hindering joint decisions.

Then, we compared the partnerships networks to each other using the networks' indices and leading municipalities' indices, based on centralization, as cluster variables. According to the dendrogram behaviour and the variations in the ratio of the total variance explained by the clusters, four subgroups represent the optimal solution.

We observed two extreme typical configurations; In the third cluster, the leading municipality occupied a very peripheral location and the networks were highly disconnected, and the fourth cluster had opposite network characteristics. The first and second clusters were in the middle—they were dissimilar because the second cluster displayed a higher degree of both centralization and cohesion than the first cluster. In order to visualize their configurations, we looked at their central objects (Fig. 1).

Finally, we analysed the cluster characteristics along with the mean values of the plans' coherence and innovation within each cluster (Table 1). At first glance, we noted that the planning performance was related to the position of the leading municipalities within networks and to the structure of the linkages among actors and their reachability.

In general, the higher centrality scores were associated with higher scores on the plans' coherence and the lower ones on innovation; conversely, it occurred when we considered the cohesion measures. Specifically, the best planning performance was achieved by the second cluster, where in the leading municipalities played a central, but *not* dominant, role in mobilizing actors to cooperate and the networks had a relatively high degree of structural cohesion. In addition, the lowest score for coherence was observed in the third cluster, in which the leading municipality had the lowest value of centralization, along with the highest level of fragmentation. This can be interpreted as a leading municipality that was not able to coordinate and drive the planning process.

Regarding innovation, we observed that the lowest value occurred in the fourth cluster, which was characterised by the highest degree of centralization both at leading municipality and network level. In this cluster, we observed small star networks in which the leading municipalities dominated the partnership without mobilizing other territorial actors.

To conclude, in accordance with the current literature's assumptions regarding the link between partnerships and group performances, we have also observed that a centralized structural configuration is linked to a more effective coplanning of tasks, but there is the risk of weakening the stability of the networks over time, as well as producing inequalities in the decision-making processes. As noted by Meuleman (2008), the key players within networks have a strategic advantage

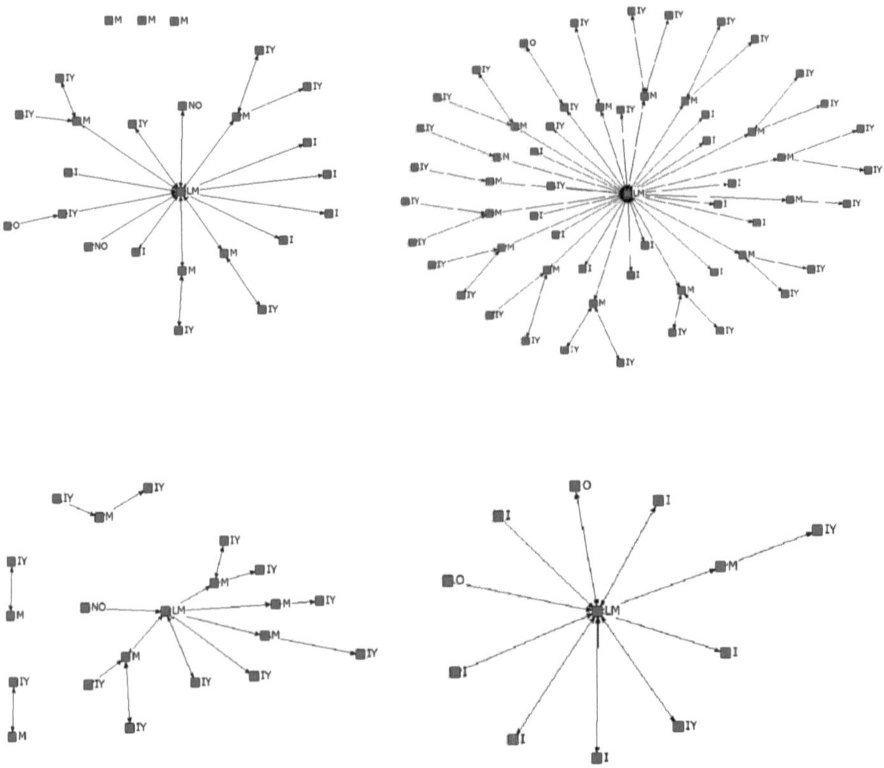

Fig. 1 Partnership graphs of the central objects of four identified clusters: *left* to *right*, *top* to *bottom*, cluster I–IV. Legend: *LM* leading municipality, *M* municipality, *I* institution, *OI* other institution, *NP* No Profit organization, *O* other actor

over other players. This creates a problem for the gathering of participants within network governance processes, as it may result in a decrease of citizen participation compared to the classical representative forms of democratic decision-making. The leading municipalities must adopt a hierarchical model of coordination that would the start-up of partnerships networks in a fragmented context in a shorter timeframe; however, this approach could possibly decrease the active involvement of local communities over time. This is an example of the typical tension between hierarchical and network governance systems.

Policymaking appears to be in transition towards more open instruments, but it remains an open-ended experiment that requires more significant effort towards a long-term transition that involves both the private sector and the state in establishing new relationships.

Table 1 Networks structural characteristics and mean scores for each Plan's coherence and innovation

	Leading municipality level			Network level					Plans' coherence		Plans' innovation	
Clusters	Degree	Betweenness	Closeness	Centrality degree	Centrality betweenness	Component ratio	Fragmentation	Avg distance	Mean	Std. Dev.	Mean	Std. Dev.
I	0.590	0.722	0.229	0.561	0.703	0.098	0.219	2.467	2.07	0.80	2.80	0.41
II	0.665	0.880	0.677	0.636	0.854	0.012	0.015	2.639	2.56	0.53	2.56	0.73
III	0.367	0.255	0.208	0.304	0.244	0.145	0.425	2.409	2.00	0.89	2.17	0.98
IV	0.849	0.847	0.876	0.767	0.813	0.015	0.014	1.937	2.17	0.98	2.00	1.26

References

Bisceglia, A., Lumino, R., & Ragozini, G. (2013). *Il nuovo corso delle politiche giovanili in Campania: l'esperienza dei Piani Territoriali Giovani*. Milano: Franco Angeli.

Brandes, U., & Erlebach, T. (2005). *Network analysis. Methodological foundations*. Berlin: Springer.

Borgatti, S. P., Everett, M. G., & Freeman, L. C. (2002). *Ucinet for windows: Software for social network analysis*. Harvard: Analytic Technologies.

Capuano C., De Stefano D., Del Monte A., D' Esposito M.R., Vitale M.P. (2013). The analysis of network additionality in the context of territorial innovation policy: The case of Italian technological districts, in P. Giudici, S. Ingrassia, M. Vichi (eds), Statistical models for data analysis, Springer, pp. 81–88.

Cross, R., & Cummings, J. N. (2004). Tie and network correlates of individual performance in knowledge intensive work. *Academy of Management Journal, 47*(6), 928–937.

Cross, J. E., Ellyn, D., Rebecca, N., & Fagan, J. M. (2009). Using mixed-method design and network analysis to measure development of interagency collaboration. *American Journal of Evaluation, 30*, 310–329.

Di Nicola, P., Stanzini, S., & Tronca, L. (2010). *Forme e contenuti delle reti di sostegno. Il capitale sociale a Verona*. Milano: Franco Angeli.

Durland, & Fredericks, K.A. (Eds.). (2005). Special Issue: Social network analysis in program evaluation. *New Directions for Evaluation, 107*

Fawcett, P., & Daugbjerg, C. (2012). Explaining governance outcomes: Epistemology, network governance and policy network analysis. *Political Studies Review, 10*, 1985–2007.

Honeycutt, T. (2009). Making connections: Using social network analysis for program evaluation. *Mathematica Policy Research, 1*

Kazepov, Y. (2008). The subsidiarization of social polices: Actors, processes and impacts. *European Societies, 10*(2), 247–273.

Kenis, P., & Schneider, V. (1991). Policy networks and policy analysis: Scrutinizing a new analytical toolbox. In B. Marin & R. Mayntz (Eds.), *Policy networks: Empirical evidence and theoretical considerations* (pp. 25–59). Boulder: Westview Press.

Lebart L., Morineu A., Warwick K.M. (1984). Multivariate descriptive statistical analysis: Correspondence analysis and related techniques for large matrices, Wiley: New York

Meuleman L. (2008). Public management and the metagovernance of hierarchies, networks and markets: The feasibility of designing and managing governance style combinations, Phisyca-Verlag: Heidelberg

Penuel, W. R., Sussex, W., Korbak, C., & Hoadley, C. (2006). Investigating the potential of using social network analysis in educational evaluation. *American Journal of Evaluation, 27*(4), 437–451.

Peterson, J. (2003). *Policy networks*. Working papers of IHS Political Science Series, 90. Vienna: Institute for advanced Studies.

Reagans R., Mc Ivily B. (2003). Network structure and knowledge transfer: The effects of cohesion and range, *Administrative science quarterly, 48*(2), 240–267.

Rhodes, R. A. W. (2007). Understanding governance: Ten years on. *Organization Studies, 28*(8), 1243–1264.

Wasserman, S., & Faust, K. (1994). *Social network analysis: Methods and applications*. Cambridge: Cambridge University Press.

Evaluating the Effect of New Brand by Asymmetric Multidimensional Scaling

Akinori Okada and Hiroyuki Tsurumi

Abstract Brand switching data among potato snacks are analyzed to evaluate a new brand. The brand switching matrix among existing brands is analyzed by asymmetric multidimensional scaling. The analysis shows the outward and the inward tendencies of existing brands, which tell the strength of switching from the corresponding brand to the other brands and the strength of switching to the corresponding brand from the other brands respectively. The inward tendency of a new brand is estimated by analyzing the brand switching data from the existing brands to the new brand obtained soon after its introduction based on the outward tendency of existing brands. The estimated inward tendency of the new brand is compared with those derived by analyzing the brand switching matrix including the new brand obtained 2 months after the introduction of the new brand. The comparison shows the estimated inward tendency is similar to the actually derived one.

Keywords Asymmetric multidimensional scaling • Brand switching • New brand introduction

A. Okada (✉)
Graduate School of Management and Information Sciences, Tama University, 4-1-1 Hijirigaoka, Tama-shi, Tokyo, 206-0022, Japan
e-mail: okada@rikkyo.ac.jp; okada@tama.ac.jp

H. Tsurumi
College of Business Administration, Yokohama National University, 79-4 Tokiwadai, Hodogaya-ku, Yokohama-shi, 240-8501, Japan
e-mail: tsurumi@ynu.ac.jp

D. Vicari et al. (eds.), *Analysis and Modeling of Complex Data in Behavioral and Social Sciences*, Studies in Classification, Data Analysis, and Knowledge Organization, DOI 10.1007/978-3-319-06692-9_22,

1 Introduction

In marketing, introducing a new brand is one of the most important issues. After the seminal work of Parfitt and Collins (1968), there are many researchers that analyze and evaluate the effect of newly introduced brands into a market (e.g. Bass 1969; Hahn et al. 1994). Most of them focus their attention on the sales quantity or the market share of new brands, while these are effects resulted from the brand switching between new brands and the existing ones. It is desirable to disclose what caused the resultant sales quantity or the market share.

The present study analyzes a brand switching matrix soon after the introduction of new brands. The brand switching to the new brands from the existing brands occur even if soon after the introduction of new brands, but neither the brand switching from the new brands to the existing brands nor brand switching among new brands do not occur or are negligible. How long the period of '*soon after the introduction of the new brand*' varies depending on the category of goods. In this period the brand switching matrix is represented by a rectangular matrix but not by a square matrix, because the brand switching from existing brands to new brands and that among existing brands occur, but the brand switching from new brands and that among new brands do not occur or are negligible. The purpose of the present study is to evaluate a new brand by estimating the strength of the new brand in the brand switching based on the analysis of brand switching matrix among existing brands as well as those from the existing brands to the new brand.

2 Data

The brand switching data among nine potato snack brands (A, B, ..., G, O, and N; N is a new brand, and O represents brands other than A, ..., G, and N) were collected in April through August of 2009 at three stores of a supermarket chain in Tokyo metropolitan area. New brand N was introduced at the beginning of period 2 (period 1: April 1–June 1; period 2: June 2–July 31; period 3: August 1–August 31 of 2009). The characteristics (tastes) and the market share based on the amount of money of sales in period 2 (the new brand has not been introduced in period 1) of each brand are shown in Table 1.

Of 47,633 customers who are members of the frequent shoppers program of the supermarket chain, 8,431 customers purchased potato snacks in April through August of 2009 at any of the three stores. Of 8,431 customers, 1,395 customers purchased potato snacks at both period 1 and 2, and 822 customers purchased potato snacks at both period 2 and 3. Table 2 shows the brand switching matrix among eight existing brands and those from eight existing brands to the new brand (the rightmost column). The brand switching matrix consists of the frequency of the switch from the row brand in period 1 to the column brand in period 2 measured by the following procedure. The frequency of the brand switching is the number of consumers who

Table 1 Characteristics of the brands

Brand	Taste	Market share (%)
A	Salt	26.4
B	Salad	20.5
C	Butter	10.8
D	Cheese	9.5
E	Savory salt	12.0
F	Basil and tomato	2.9
G	Salt	1.8
O	–	5.1
N	Butter and soybean sauce	11.0

Table 2 Brand switching matrix from period 1 to 2

	Period 2								
Period 1	A	B	C	D	E	F	G	O	N
Brand A	220	46	23	15	48	10	1	16	27
Brand B	20	208	38	19	32	11	8	13	26
Brand C	2	24	45	8	7	3	3	8	131
Brand D	5	10	10	35	10	4	2	7	8
Brand E	6	15	9	5	38	1	1	4	12
Brand F	13	39	17	13	19	19	3	22	15
Brand G	7	24	9	7	19	0	13	6	11
Brand O	4	8	5	6	9	1	3	8	19

changed the largest purchase brand from period 1 to 2. While we are dealing with nine brands, the brand switching matrix is 8×9 rectangular. This is because the ninth brand (N) is a new brand, and therefore the brand switching from a new brand N to the existing eight brands cannot occur soon after the introduction of the new brand.

3 Analysis

Since the brand switching matrix shown in Table 2 is asymmetric, i.e. its (j, k) element is not always equal to the (k, j) element, it can be analyzed by asymmetric multidimensional scaling (e.g., Borg and Groenen 2005, Chap. 23). The matrix is analyzed by asymmetric multidimensional scaling (Okada and Tsurumi 2012) using singular value decomposition (Eckart and Young 1936). The analysis gives the outward tendency which shows the strength of switching from a brand to the other brands and the inward tendency which shows the strength of switching to the brand from the other brands along each dimension. The brand switching from existing brands to a new brand and the outward tendency of the existing brands are used to estimate the inward tendency of the new brand.

Let **A** be the 8×8 matrix of asymmetric brand switching matrix among eight existing brands, and **B** be the 8×1 brand switching matrix from eight existing brands to the new brand. The (j, k) element of **A** represents the frequency of the brand switching from brand j to k, and the j-th element of **B** represents that from brand j to a new brand. By using the singular value decomposition, **A** is approximated by

$$\mathbf{A} \simeq \mathbf{XDY}', \tag{1}$$

where **D** is the $r \times r(r < 8)$ diagonal matrix of the r largest singular values $(d_1, \ldots, d_r)$ in descending order as its diagonal elements, **X** is the $8 \times r$ matrix of the orthonormalized left singular vectors, and **Y** is the $8 \times r$ matrix of the orthonormalized right singular vectors. The j-th element of the i-th column of **X**, x_{ji}, represents the outward tendency of brand j along Dimension i, and the k-th element of the i-th column of **Y**, y_{ki}, represents the inward tendency of brand k along Dimension i. When $r = 3$, a_{jk}, the (j, k) element of **A**, can be approximated by

$$a_{jk} \simeq d_1 x_{j1} y_{k1} + d_2 x_{j2} y_{k2} + d_3 x_{j3} y_{k3}. \tag{2}$$

We assume that the brand switching matrix **B** from existing brands to the new brand is approximated by

$$\mathbf{B} \simeq \mathbf{XDZ}', \tag{3}$$

where **Z** represents the inward tendency of the new brand, which is estimated by

$$\mathbf{Z} = \mathbf{B}'\mathbf{XD}^{-1} \tag{4}$$

(Okada and Tsurumi 2011). The i-th element of **Z**, z_i, is the estimated inward tendency of the new brand along Dimension i.

4 Results

The five largest singular values are 269.5, 184.6, 52.5, 34.1, and 28.8. The three-dimensional result was chosen as the solution. The reason for choosing the three-dimensional result as the solution is; the first through third dimensions account for brand switching among brands which have rather large market share, and the fourth dimension accounts for brand switching among brands which only have rather small market share. Figures 1, 2 and 3 represent the outward and inward tendencies along Dimensions 1, 2, and 3 respectively.

Outward and inward tendencies along Dimension 1 shown in Fig. 1 tells that brands B, C, D, E, and O are above the 45° line emitting from the origin. This means that the inward tendency of these brands is larger than the outward tendency

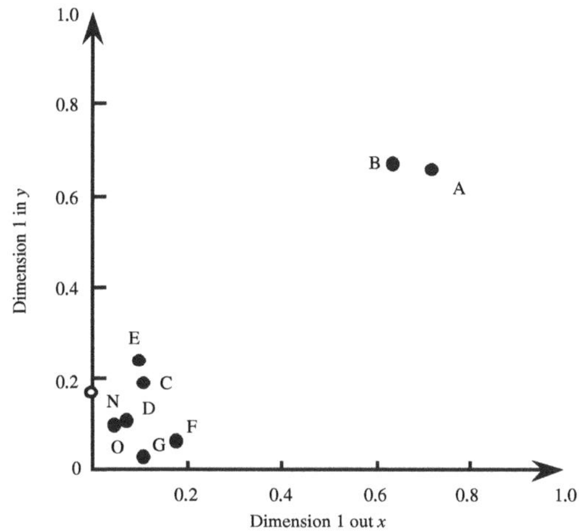

Fig. 1 Outward and inward tendencies along Dimension 1. The existing brand (A, B, ..., G, and O) is represented by a *solid circle*. The estimated inward tendency of the new brand (N) is represented by an *open circle* on the vertical axis

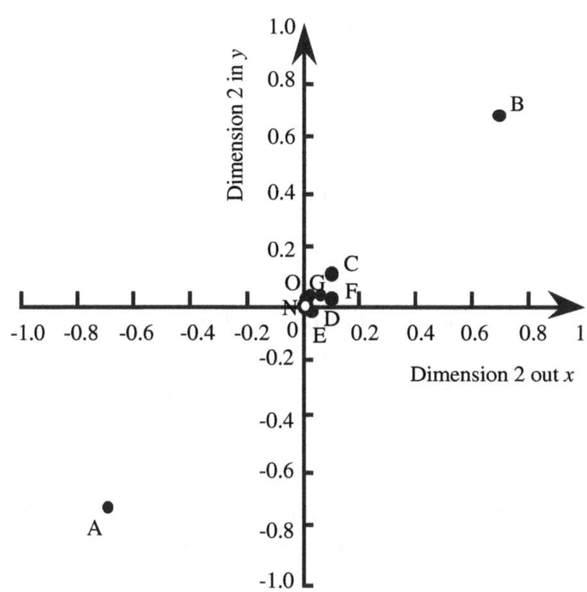

Fig. 2 Outward and inward tendencies along Dimension 2. The existing brand (A, B, ..., G, and O) is represented by a *solid circle*. The estimated inward tendency of the new brand (N) is represented by an *open circle* on the vertical axis

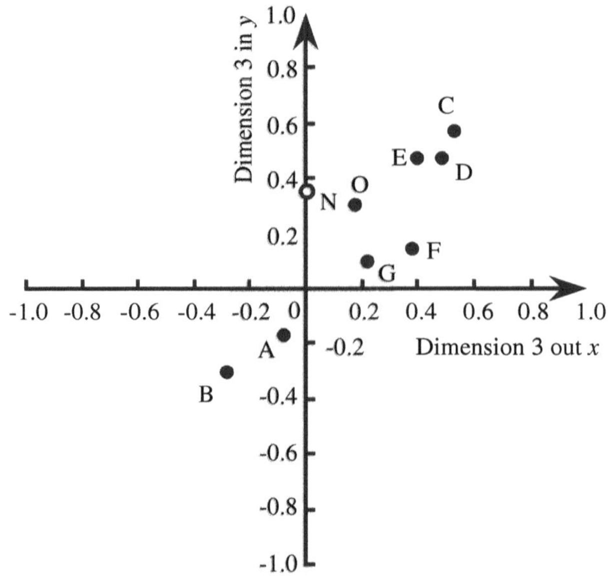

Fig. 3 Outward and inward tendencies along Dimension 3. The existing brand (A, B, ..., G, and O) is represented by a *solid circle*. The estimated inward tendency of the new brand (N) is represented by an *open circle* on the vertical axis

of them, suggesting these brands are dominant over brands below the line. The large outward and inward tendencies of brands A and B show that they have large market shares (Table 1), and that they are switched to/from the other brand a lot. Outward and inward tendencies along Dimensions 2 and 3 shown in Figs. 2 and 3 tell that eight brands are classified almost into two groups; one consists of brands in the first quadrant, and the other consists of those in the third quadrant. The outward and inward tendencies of the brand in the first quadrant are positive, and those in the third quadrant are negative. For two brands, one is in the first quadrant and the other is in the third quadrant, the product of the outward tendency of a brand in the first/third quadrant and the inward tendency of a brand in the third/first quadrant is negative, because the outward and inward tendencies have opposite signs. This suggests that any two brands between two groups is distant in the brand switching. The brand switching from the brand in the first quadrant to the brand in the third quadrant along Dimension 2 or 3 is represented by the product of the outward tendency of the former (positive) and the inward tendency of the latter (negative) multiplied by the second or the third largest singular value respectively which are positive. As a consequence these products, which represent the second and the third terms of Eq. (2) respectively, are negative. The brand switching from the brand in the third quadrant to the brand in the first quadrant along Dimension 2 or 3 is negative as well. Any two brands in the same group or in the same quadrant are closer than two brands, one is in the first quadrant and the other is in the third quadrant, in the brand switching along Dimension 2 or 3, because outward and inward tendencies are positive in the first quadrant and are negative in the third quadrant, and the product of outward and inward tendencies between any two brands both in the first or the third quadrant along Dimension 2 or 3 is positive (Okada and Tsurumi 2011).

Along Dimension 2 brands A and B belong to different groups, A is in the third quadrant and B is in the first quadrant, but along Dimension 3 they belong to the same group. This suggests that, along Dimension 2, brand B and the brands in the first quadrant are close in the brand switching, while brand A and those in the first quadrant are distant in the brand switching. Along Dimension 3 brands C, E, and O are dominant in the brand switching among brands in the first quadrant. Brands A and B are in the third quadrant, while the other brands are in the first quadrant, suggesting the two brands and the other brands are distant in the brand switching along Dimension 3.

The estimated inward tendencies of the new brand along Dimensions 1, 2 and 3 (z_1, z_2, and z_3 respectively) are represented on the vertical axis by an open circle designated as N. The estimated inward tendency of the new brand along Dimension 1 is almost equal to that of brand C. That along Dimension 2 is almost 0, suggesting the brand switching from the other brands to the new brand is almost zero. That along Dimension 3 is positive, suggesting the new brand will be either in the first or the second quadrant.

5 Discussion

Dimension 1 outward seems to correspond to the market share and the average price (the correlation coefficient of the coordinate of eight brands (brands A, B, $\cdots$ G, O; excluding the new brand N) of Dimension 1 outward with the market share is 0.87, and that with the average price is 0.64). Dimension 1 inward seems to represent the market share and the sales quantity (the correlation coefficient of the coordinate of the eight brands of Dimension1 inward with the market share is 0.96, and with the sales quantity is 0.78). In the configuration along Dimension 2, all brands are almost on the diagonal line running from lower left to upper right directions passing through the origin. This indicates that Dimension 2 outward and inward have the same meaning. It seems both Dimension 2 outward and inward differentiate the taste of salad (upper right), salt (lower left), and the other tastes (center), even if there is an anomaly (brand G (salt) is in the center). Dimension 3 outward seems to differentiate the complex taste (right) and the simple taste (left): Dimension 3 inward differentiates the rich taste (upper) and the plain taste (lower).

The 9×9 brand switching matrix among nine brands (A, B, $\ldots$, G, O, and N) from period 2 to 3 is analyzed by the asymmetric multidimensional scaling, which gives the inward and outward tendencies of the new brand as well. The new brand is almost at the same location of brand C in the configuration along Dimension 1, and almost at the same location of G in the configuration along Dimension 2, and between E and D in the configuration along Dimension 3. The derived inward tendencies of the new brand along Dimensions 1, 2, and 3 from the brand switching among nine brands from period 2 to 3 are 0.127, 0.074, and 0.387 (the derived figures were multiplied by $\sqrt{9/8}$ to adjust the length of the singular vector), while those estimated are 0.163, 0.025, and 0.344 (Figs. 1, 2, and 3). These figures show the estimation of the inward tendency of the new brand was successful.

The new brand N along Dimension 1 (Fig. 1) has the estimated inward tendency which is much smaller than those of brands A and B, and is similar to that of brand C. This means that the new brand will not have the large market share like brands A and B have (as shown in Table 1, they have large market shares of 26.4 % and 20.5 % respectively), or the new brand will not directly contend with brands A and B which are market leaders (Kotler and Keller 2011). This is compatible with that the estimated inward tendency of the new brand which is positive along Dimension 3 (Fig. 3). The new brand will contend directly with brands in the first quadrant of the configuration along Dimension 3 including C and E, while the new brand will not contend with brands A and B in the third quadrant of the configuration along Dimension 3. This implies that the new brand will be a niche brand (Kotler and Keller 2011), and will not be a market leader like brands A and B. As stated earlier, the estimated inward tendency of the new brand along Dimension 3 is positive, suggesting the new brand will be in the first or the second quadrant. If the new brand is in the second quadrant, it is always dominant over brands in the first quadrant in the brand switching.

The estimated inward tendency of the new brand N along Dimension 2 (Fig. 2) is almost zero. This tells that the brand switching from the other existing brands to the new brand are almost zero. The taste of the new brand is butter and soybean sauce, and it is natural that the new brand is located near to the origin. The estimated inward tendency of the new brand N along Dimension 3 (Fig. 3) is positive; it is larger than that of brand F and is smaller than those of brands E and D. The taste of the new brand (butter and soybean sauce) seems to be richer than the taste of brand F (basil and tomato). The tastes of brands E and D are savory salt and cheese respectively, it seems the new brand with its butter and soybean sauce taste has not less rich taste than brands E and D have. When consumers recognize the taste of the new brand as rich as those of brand E and D, the inward tendency of the new brand might increase, suggesting the new brand becomes more dominant in the brand switching among brands in the first quadrant.

Acknowledgements The authors would like to express their appreciation to two anonymous referees who gave them useful and thoughtful reviews on the earlier version of the present paper, which helped us to improve the earlier version. They also would like to express their gratitude to The Distribution Economics Institute of Japan for permitting them to use the present data.

References

Bass, E. M. (1969). A new product growth for model consumer durables. *Management Science, 15*, 215–227.

Borg, I., & Groenen, P. J. K. (2005). *Modern multidimensional scaling: theory and applications* (2nd ed.). New York: Springer.

Eckart, C., & Young, G. (1936). The approximation of one matrix by another of lower rank. *Psychometrika, 1*, 211–218.

Hahn, M., Park, S., Krishnamurthi, L., & Zoltners, A. A. (1994). Analysis of new product diffusion using a four-segment trial-repeat model. *Marketing Science, 13*, 224–247.

Kotler, P., & Keller, K. L. (2011). *Marketing management* (14th ed.). Upper Saddle River: Prentice Hall.

Okada, A., & Tsurumi, H. (2011). *External analysis of asymmetric multidimensional scaling based on singular value decomposition.* Book of Abstracts of the 8th Scientific Meeting of the Classification and Data Analysis Group of the Italian Statistical Society (p. 42).

Okada, A., & Tsurumi, H. (2012). Asymmetric multidimensional scaling of brand switching among margarine brands. *Behaviormetrika, 39*, 111–126.

Parfitt, J. H., & Collins, B. J. K. (1968). The use of consumer panels for brand share prediction. *Journal of Marketing Research, 5*, 131–145.

Statistical Characterization of the Virtual Water Trade Network

Alessandra Petrucci and Emilia Rocco

Abstract The water that is used in the production process of a product (a supply, commodity or service) is called the "virtual water" contained in the product. If one country (or region, company, individual, etc.) exports a water intensive product to another country, it exports water in virtual form. Virtual water trade as both a policy instrument and practical means to balance the local, national and global water budget has received much attention in recent years. Several studies have been conducted by researchers from various disciplines including engineers, economists and demographers. The aim of this paper is to improve the statistical characterization of the virtual water flow networks by suggesting a statistical modeling approach for examining their stochastic properties.

Keywords Exponential random graph models • Mixed effects models • Valued links

1 Introduction

Producing goods and services generally requires water. The water used in the production process of a product is known as the 'virtual water' contained in the product. If goods and services are exchanged then water in virtual form is traded from one country (or region, company, individual, etc.) to another. In recent years, the concept of virtual water trade has gained weight in both the scientific and political debate. It is a fact that water is a renewable but finite resource and that both water availability and quality vary enormously in time and space. It is also a fact

A. Petrucci (✉) • E. Rocco
Department of Statistics, Informatics, Applications, University of Firenze, Viale Morgagni, 59-50134 Firenze, Italy
e-mail: alessandra.petrucci@unifi.it; rocco@disia.unifi.it

D. Vicari et al. (eds.), *Analysis and Modeling of Complex Data in Behavioral and Social Sciences*, Studies in Classification, Data Analysis, and Knowledge Organization,
DOI 10.1007/978-3-319-06692-9_23,

that growing populations coupled with continuous socioeconomic development put pressure on the globe's scarce water resources and that in many parts of the world there are signs that water consumption and pollution exceed a sustainable level and indicate a growing water scarcity. These considerations have led researchers from various disciplines, including engineers, economists and demographers, to conduct several studies on virtual water flows. The aim of this paper is to improve the statistical characterization of the networks of virtual water flows by suggesting a statistical modeling approach for examining their stochastic properties. In particular we intend to extend and adapt some statistical models inherent in the social networks analysis in order to study the virtual water flow networks. In the last decade, researchers from various disciplines have been increasingly more involved in the collection and statistical analysis of relational data, or complex systems where it is meaningful to analyze the relationships between units or elements, even heterogeneous, to detect any underlying structural model, and to draw implications from these reports. This work includes both descriptive analyses of the structure of networks and inferential studies focusing on the construction of network models. According to our knowledge, only the former type of network analysis has been applied for studying virtual water trade. Konar et al. (2011) and Carr et al. (2012), among others, use the analytical tools of complex network analysis to characterize respectively the global structure and the spatio-temporal patterns of the virtual water trade associated with the international food trade. The descriptive analysis, while interesting in its own right, is by necessity the first step towards developing and validating modelling approaches. In particular, in this paper, we suggest a first simple "network-data model" for the analysis of the average annual virtual water flows related to international crop trade among major world regions in the period from 1995 to 1999.

2 Virtual Water Trade Networks

Network data typically consist of a finite set of n actors (nodes) and a collection of m relational ties (links) that specify how these actors are relationally tied together. Only of interest to us here is the case of $m = 1$. Let $y_{i,j}$ denote the relational tie measured on each ordered pair of actors i, j. In the simplest case $y_{i,j}$ is a dichotomous variable indicating the presence or absence of the relation of interest. More generally relations can either be binary or valued and both can be directed (i's tie to j may differ from j's tie to i) or non-directed (at the most, there is one non-directed tie connecting i and j). Normally, actors are not assumed to have relations with themselves, so for each i $y_{i,i} = 0$. Network data are usually represented by an $n \times n$ matrix. In this matrix, denoted by $\mathbf{Y}$, and called adjacency matrix, each row and the corresponding column represent an actor in the network, each element $y_{i,j}$ is the tie variable from actor i to actor j and the pair of elements (tie variables) $(y_{i,j}, y_{j,i})$ is called a dyad.

In the network of virtual water trade each country participating in trade is represented by a node. Links between nodes are the volume of virtual water embodied in the traded commodities and are directed on the basis of direction of trade flow. The virtual water flows are usually calculated by more steps. For each product, k, trade is expressed by constructing a trade matrix, $\mathbf{T_k}$, whereby the export of that product from country i, to country j, is stored in the (i, j) element of the matrix. If no export occurs, that element is set to zero. Then, for each product k, the conversion of the product trade matrix, $\mathbf{T_k}$, to virtual water trade matrix, $\mathbf{Y_k}$, is accomplished by multiplying each element of $\mathbf{T_k}$ by the virtual water content, W_{ck}, of the product in the country of export. The virtual water content of a product can change from one place to another depending on climatic condition and choices regarding the production process. Hoekstra and Chapagain (2008) have defined the virtual-water content of a product as "the volume of freshwater used to produce the product, measured at the place where the product was actually produced". It refers to the sum of the water used in the various steps of the production chain. Virtual water content of a product is usually estimated by hydrological models (Hanasaki et al. 2008) or using the country specific average water footprint (Mekonnem and Hoekstra 2010). Thus, the virtual water trade matrix is calculated as the sum $\mathbf{Y} = \sum_k W_{ck} \times \mathbf{T_k}$. Because virtual water moves through a weighted directed network, this matrix is non-symmetrical and two directed links can connect two nodes in two opposite directions.

3 Network Models

Complete network data can be analyzed by many statistical and non-statistical methods and models. For an overview, see Chaps. 13 and 15 of Wassermann and Faust (1994). The goal of statistical models for network data is to examine the stochastic properties of relations between the actors of a particular network. They allow us to capture both the regularities giving rise to network ties and the presence of variability that we are unlikely to model in detail. Different types of models for social network structures have been suggested in literature. They may be organized along several major axes. First all they may be classified into static and dynamic models. Static network models concentrate on explaining the observed set of links based on a single snapshot of the network, whereas dynamic network models, considering situations in which relational ties between actors are observed repeatedly over time (and are therefore represented though more matrices $\mathbf{Y_t}$, one for each $t = 1, \ldots, T$), are often concerned with the mechanisms that govern changes in the network over time. Here we only consider static models, the most commonly used of which are the Exponential Random Graph Models (ERGMs) also called p* models (originating in the work of Frank and Strauss 1986), as well as the random effects, p_2, models (van Duijn et al. 2004; Hoff 2005).

The fundamental conceptual difference between the ERGM and p_2 modeling perspectives is the use of fixed effects in the former and random effects in the latter.

This is evident from the explicit notations of the two models shown here for the simplest case of a binary relation represented as a directed graph (digraph). In this case the probability distribution for the ERGM can be defined by

$$P(\mathbf{Y} = \mathbf{y}) = \frac{exp(\sum_k \theta_k s_k(\mathbf{y}))}{\kappa(\theta)},$$

where $\mathbf{Y}$ is the adjacency matrix; $s_k(\mathbf{y})$ are statistics of the digraph (the most frequently used $s(\mathbf{y})$ are the density of ties, the number of mutual dyads, or structures representing transitivity); θ is a vector of statistical parameters and $\kappa(\theta)$ is a normalizing factor ensuring that the sum of probabilities equals 1. In the same case the probability distribution of each dyad $(Y_{i,j}, Y_{j,i})$ under the p_2 model is

$$P(Y_{i,j} = y_{i,j}, Y_{j,i} = y_{j,i}) = k_{ij}^{-1} exp[y_{i,j}(\mu + \mathbf{x}'_{\mathbf{i,j}}\boldsymbol{\beta} + s_i + r_j) + \\ + y_{j,i}(\mu + \mathbf{x}'_{\mathbf{j,i}}\boldsymbol{\beta} + s_j + r_i) + y_{i,j} y_{j,i} \rho_{i,j}],$$

where $y_{i,j}$ and $y_{j,i}$ are 0 or 1; μ is a density parameter, constituting an overall mean; $\mathbf{x}'_{\mathbf{i,j}}$ contains explanatory variables which could depend on i only or on j only but also on i and j simultaneously; $\boldsymbol{\beta}$ is a vector of parameters, s_i and r_i are random variables characterizing i as a sender and receiver respectively, they are assumed independent for different i and normally distributed, however s_i and r_i whose variables refer to the same actor can be correlated; $\rho_{i,j} = \rho_{j,i}$ is a parameter indicating the force of reciprocation; k_{ij} is a normalizing factor ensuring that the sum of the four probabilities (for outcomes $(0,0), (0,1), (1,0), (1,1)$) equals 1.

We can observe how p_2 models for the network structure are focused on the dyads in the network. However, alongside dyads, higher level substructures of the network may often be interesting. p*/ERGMs models build on this idea, and are at present the most promising way to model network structure via a series of substructures, for cross-sectional data. They have, in principle, an unlimited possibility for network effects (and p_2 does not); the practical limitation resides in the estimability of these effects. On the other hand, p_2 models have random effects for representing between-actor differences (and p* does not). From these considerations it is evident that: p_2 models focus on testing effects of covariates, which can be actor-bound or dyad-bound, and which can interact with reciprocity, while controlling for differential actor tendencies to send and receive ties, and for reciprocity; p*/ERGMs models focus on structurally modeling networks (which may include covariate effects).

For valued (nonbinary) dyadic datasets, a perceived lack of statistical tools has sometimes led to ad hoc reductions of valued responses to binary data. Conversely, random effects models have been a widely successful tool in capturing statistical dependencies for a variety of data types. In a series of papers (Hoff 2003, 2005; Westveld and Hoff 2011) the social relations models were expanded in this direction. In particular, the use of generalized linear mixed models (McCulloch and Searle 2001) was suggested to allow for binary, ordinal and continuous relational data. Here, in order to study patterns of virtual water trade we propose a linear mixed effects model which assumes that each element of $\mathbf{Y}$ matrix (the volume of virtual water traded), $y_{i,j}$, can be written as a function of a linear predictor $\mathbf{x}'_{\mathbf{i,j}}\boldsymbol{\beta}$ and be modelled as conditionally independent, given appropriate random effects terms. The linear predictor (the fixed part of the model) estimates the linear relationship between the volume of virtual water embodied in the trade commodities $y_{i,j}$ and a possibly vector-valued set of variables $\mathbf{x_{i,j}}$ which could include characteristics of the exporting country, characteristics of the importing country and characteristics specific to the pair. The random effect terms take into account that in the context of virtual water trade, exports from one country to another may be determined by factors beyond a country's control, such as environmental effects, socio-demographic evolution, and economical and political scenarios. Moreover, they are essential for considering possible dependence structures among the $y_{i,j}$. The model is:

$$y_{i,j} = \mathbf{x}'_{\mathbf{i,j}}\boldsymbol{\beta} + \gamma_{i,j} \qquad \text{where} \qquad \gamma_{i,j} = s_i + r_j + \varepsilon_{i,j}. \tag{1}$$

In this model $\mathbf{x}'_{\mathbf{i,j}}\boldsymbol{\beta}$ is a fixed effect expressing the mean for $y_{i,j}$, while the error term $\gamma_{i,j}$ is decomposed into a set of mean zero Gaussian random effects. This linear decomposition consists of a sending effect s_i, a receiving effect r_j and a residual error term $\varepsilon_{i,j}$ and allows for including some structures of dependencies in the model, such as within-node dependence and reciprocity. The network dependencies can be characterized by specifying covariance structures for the random effects in (1). In particular, we assume the following covariance structure:

$$\begin{array}{ll} E[\gamma_{i,j}^2] = \sigma_s^2 + \sigma_r^2 + \sigma_\epsilon^2 & E[\gamma_{i,j}\gamma_{k,i}] = \sigma_{sr} \\ E[\gamma_{i,j}\gamma_{i,k}] = \sigma_s^2 & E[\gamma_{i,j}\gamma_{j,i}] = \rho\sigma_\epsilon^2 + 2\sigma_{sr} \\ E[\gamma_{i,j}\gamma_{k,j}] = \sigma_r^2 & E[\gamma_{i,j}\gamma_{k,l}] = 0 \end{array}$$

4 Analysis of the Virtual Water Flows Related to International Crop Trade

In order to investigate the potentiality of the suggested model approach, model (1) was applied to the data available in Hoekstra and Hung (2005) on average annual virtual water flows related to international crop trade between major world regions in the period from 1995 to 1999. In particular, these data refer to a classification of the world into 13 regions: North America, Central America, South America, Eastern Europe,Western Europe, Central and South Asia, the Middle East, South-east Asia, North Africa, Central Africa, Southern Africa, the Former Soviet Union (FSU) and Oceania. The virtual water flows between these regions have been calculated by summing up the country flows obtained by multiplying the international crop trade flows by their associated virtual water content and data on crop water requirements have been calculated with FAO CropWat model (www.fao.org). For the calculated crop water requirements for different crops in different countries of the regions that

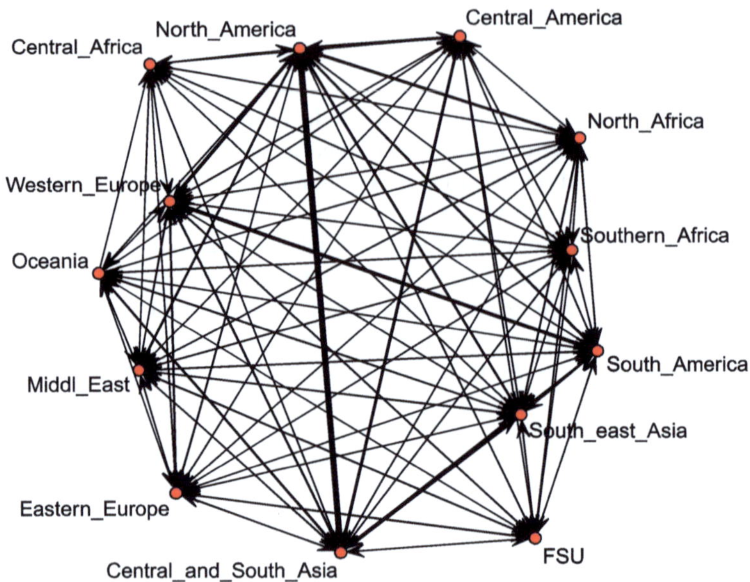

Fig. 1 Graphical representation of average annual virtual water flows

are used in this paper, the reader should refer to Hoekstra and Hung (2005). The flows data are shown in Fig. 1 from which it is evident that North America is by far the largest virtual water exporter in the world, while Central and South Asia is by far the largest virtual water importer. In the specification of the model we consider as explicative variables the population of the export world region, the population of the import world region and the difference between the percentages (on total land) of agricultural land in the two regions. Other possible explanatory variables are available, but the aggregation of data at the level of macro-regions of the world has led us to choose only the most relevant.

We take a Bayesian approach to parameter estimation and use Markov chain Monte Carlo (MCMC) simulation to approximate the posterior parameter distributions. The results suggest a modest, albeit significant effect for all three explicative variables in that their posterior distributions are centered around a mean of 0.47, 0.42 and 0.41 respectively, and the corresponding 95 % quantile-based intervals are $(0.29, 0.66)$, $(0.17, 0.79)$ and $(0.05, 0.76)$. We now examine the posterior distributions of the country-specific sender and receiver random effects. These effects describe the average deviation of a region's export and import levels from those that would have been predicted by the regression model alone. Figure 2 presents 200 random samples from the bivariate posterior distribution of the sender and receiver effects for each region (region labels are located at the posterior mean). The spread of the posteriors of the sender-receiver estimates gives evidence of

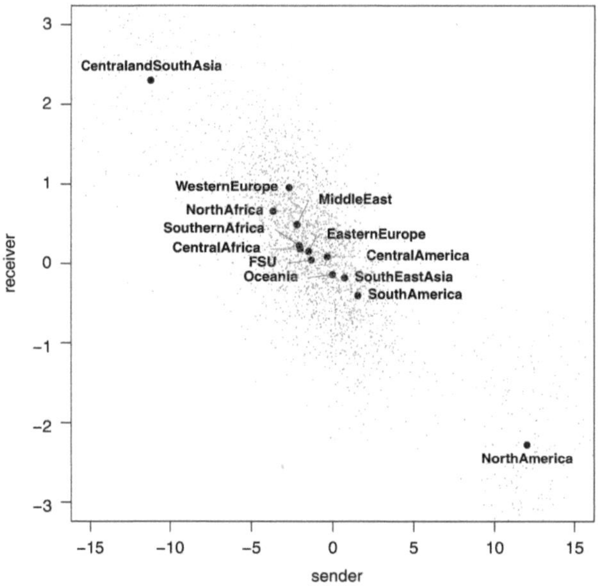

Fig. 2 Posterior distribution of the sender-receiver effects

Table 1 Posterior means and 95 % credibility intervals of variance terms of the sender-receiver effects

Parameter	Posterior mean	95 % credibility interval
σ_s^2	26.80	(10.20, 38.8)
σ_r^2	1.38	(0.61, 1.96)
σ_{sr}	−4.86	(−7.17, 1.89)

receiver-specific variability as well as a more relevant sender-specific variability and shows that a strong negative relationship exists between exporting (sending) and importing (receiving) virtual water. This last result reflects and confirms a peculiarity of the virtual water trade compared to other trades: the virtual water flows are not reciprocal since water-scarce countries might wish to import products that require a lot of water in their production whereas water-rich countries could profit from their abundance of water resources by producing water-intensive products for export. Figure 2 also shows the ability of model (1) to capture the characteristics of the data and, according to the descriptive analysis performed by Hoekstra and Hung (2005), indicates which regions have net virtual water import and those which have net virtual water export, highlighting in particular how North America is by far the largest virtual water exporter in the world while Central and South Asia is by far the largest virtual water importer. The spread of the posteriors of sender-receiver estimates for the regions is confirmed by the posterior means of σ_s^2, σ_r^2 and σ_{sr} showed in Table 1.

5 Final Remarks and Ongoing Questions

In this paper we suggest a random effects model to capture dependence in a virtual water network. The results of its application to the average annual virtual water flows related to international crop trade between major world regions are encouraging and support the potentiality of this approach to model the peer dependencies in the networks of virtual water flows as well as investigating the effect of several covariates. Such covariates may characterize the nodes or be determined by factors/circunstances beyond a node's control such as environmental effects, socio-demographic evolution, economic and political scenarios, etc. The estimation of the impact of these factors on the virtual water flows may be useful for evaluating how to optimize the use of water resources.

We are aware that the proposed model is simple and unable to capture any substructures of the network data of higher level than the dyad one, and this is just the starting point of our study. The next step could be to define a random effects model for network data able to capture more complex dependence structures. Another extension of the work could be the use of a modeling approach for the study and management of domestic virtual water trades. For countries which are relatively dry in some parts and relatively wet in others, the domestic virtual water trade may be a relevant issue.

References

Carr, J. A., D'Odorico, P., Laio, F., & Ridolfi, L. (2012). On the temporal variability of the virtual water network. *Geophysical Research Letters, 39*. doi:10.1029/2012GL051247.

Frank, O., & Strauss, D. (1986). Markov graphs. *Journal of the American Statistical Association, 81*, 832–342.

Hanasaki, N., Kanae, S., Oki, T., Masuda, K., Shirakawa, N., Shen, Y., et al. (2008). An integrated model for the assessment of global water resources - Part 1: Model description and input meteorological forcing. *Hydrology and Earth System Sciences, 12*, 1007–1025.

Hoekstra, A. Y., & Chapagain, A. K. (2008). Globalization of water: sharing the planet's freshwater resouurces. Malden: Blakwell.

Hoekstra, A. Y., & Hung, P. Q. (2005). Globalisation of water resources: International virtual water flows in relation to crop trade. *Global Environmental Change, 15*(1), 45–56.

Hoff, P. D. (2003). Random effects models for network data. In R. Breiger, K. Carley, P. Pattison (Eds.), *Dynamic Social Network Modeling and Analysis: Workshop Summary and Papers* (pp. 303–312). Washington, DC: National Accademies Press.

Hoff, P. D. (2005). Bilinear mixed-effects models for dyadic data. *Journal of the American Statistical Association, 100*, 286–295.

Konar, M., Dalin, C., Suweis, S., Hanasaki, N., Rinaldo, A., & Rodriguez-Iturbe, I. (2011). Water for food: The global virtual water trade network. *Water Resources Research, 47*. doi:10.1029/2010wr010307.

McCulloch, C. E., & Searle, S. R. (2001). Generalized, linear, and mixed models. New York: Wiley.

Mekonnem, M. M., & Hoekstra A. Y. (2010). The green, blue and grey water footprint of crops and derived crop products. Value of Water Research Report Series no. 47. Unesco-Ihe Institute for water research, Delt, The Netherlands.

van Duijn, M. A. J., Snijders, T. A. B., & Zijlstra, B. J. H. (2004). p2: a random effects model with covariates for directed graphs. *Statistica Neerlandica, 58*, 234–254.

Wassermann, S., & Faust, K. (1994). *Social network analysis: Methods and applications*. Cambridge: Cambridge university Press.

Westveld, A. H., & Hoff, P. D. (2011). A mixed effects model for longitudinal relational and network data, with application to international trade and conflict. *The Annals of Applied Statistics, 5*, 843–872.

A Pre-specified Blockmodeling to Analyze Structural Dynamics in Innovation Networks

Laura Prota and Maria Prosperina Vitale

Abstract In recent decades economic theory has highlighted the benefits produced by networks of organizations in fostering innovation. A number of public policies were put in place to favor these innovation networks throughout Europe. The top-down institution of a number of specialized technological districts in Italy has been one of the main outcomes of this new wave of policies, in mid-2000. The aim of this paper is to explore what impact the institution of technological districts had on collaborative patterns over time. Using a pre-specified blockmodeling, observed network configurations obtained by the co-participation to R&D projects undertaken by organizations involved in a technological district are compared with a theoretical core-periphery structure in a 8-years time interval. The analyses of networks over time show that collaborative patterns have evolved from a core-periphery structure towards a complete network in which each research group is connected with the others.

Keywords Behavioral additionality • Blockmodeling • Social network analysis • Technological districts

1 Introduction

In 1936 Keynes invoked an innate "animal spirit" to explain how entrepreneurs decide to take risks for new ideas (Keynes 1936). Since the 1990s however, this individualistic view of innovation has been challenged by recognizing the importance of collaboration and networking in fostering innovation in the era of

L. Prota (✉) • M.P. Vitale
Department of Economics and Statistics, University of Salerno, Via Giovanni Paolo II, 132, IT-84084, Fisciano (SA), Italy
e-mail: laprota@unisa.it; mvitale@unisa.it

D. Vicari et al. (eds.), *Analysis and Modeling of Complex Data in Behavioral and Social Sciences*, Studies in Classification, Data Analysis, and Knowledge Organization,
DOI 10.1007/978-3-319-06692-9_24,

the knowledge economy. This shift of paradigm ushered in, after 2000, a new wave of public policies aiming at strengthening innovation networks throughout Europe. Innovation networks can be defined as groups of interconnected organizations—including universities, research institutions, firms and government agencies—that share scientific research and technological development. The top-down institution of a number of specialized technological districts (TDs) in Italy has been one of the main outcomes of this new wave of policies. The *rational* inspiring the policy is that by changing the patterns of collaboration, long term innovative behaviours will be fostered.

The impact that the institution of TDs has on collaborative behaviours is, however, theoretically and methodologically difficult to capture. Therefore, the concept of behavioral additionality (BA) has been proposed to describe a persistent change in the pattern of collaboration (Antonioli and Marzucchi 2010; Gok and Edler 2011). Building on previous literature aiming at operationalize the idea of BA (Capuano et al. 2013), this study uses social network analysis (SNA), in particular pre-specified blockmodeling (Doreian et al. 2005), to explore structural changes in collaborative behaviors in the TDs. At each year point, the distance between the observed network of collaboration and a theoretical core-periphery configuration is measured using pre-specified blockmodeling. Such distance allows to identify a main trajectory of development and to appreciate the structural changes occurred in the pattern of cooperative behaviours dynamically.

The paper is organized as follows: in Sect. 2 the method proposed for analyzing social networks over time with research hypotheses is briefly discussed; Sect. 3 illustrates the strategy by using data from a TD in Southern Italy. Section 4 presents the main blockmodeling results and the two-mode network validation. Section 5 presents some preliminary concluding remarks.

2 Describing Evolution in Collaboration Patterns: Research Hypotheses and Method

A growing stream of research uses SNA to study actual patterns of collaboration among scientists or firms. The working hypothesis emerging from this stream of literature indicates that spontaneous scientific collaborations fit a core-periphery configuration (Goyal 2011; Choi 2012).

We want to investigate to what extent the collaboration in R&D projects developed between organizations involved in a TD similarly exhibits a core-periphery structure as that found in spontaneous scientific collaborations. Building on the core-periphery formulation proposed by Kronegger et al. (2011) for studying co-authorship among Slovenian researchers, two different types of cores are considered in this study: *simple* cores and *bridging* cores. The former are defined as collaborative relations linking actors within a cluster; the latter are, instead, sets of relations linking actors belonging to different clusters in a systematic fashion. Finally, the periphery is defined by scattered relations linking actors within a cluster to simple cores' members but not with each other (for instance actor A and B might have individual collaborations with actor C involved in a simple core, but no direct collaboration linking them).

2.1 Research Hypotheses

More specifically, the following two research hypotheses to evaluate the evolutionary patterns of TDs collaboration network structures are considered:

- H1. *A core-periphery structure with simple cores (single or multiple) explains the collaboration pattern in R&D projects among members of a TD in the start-up phase*. We expect multiple simple cores in the network, given that at each time there will be numerous active projects within the district. As the number of active projects increases, by definition, the number of simple cores should also increase.
- H2. *Network configuration evolves into a pattern with bridging cores, whose members also collaborate in a systematic fashion with members of other cores*. The multiplication of bridging cores over time indicates that new collaborations are created between disjoint research groups rather than within established simple cores. These cross-cutting collaborations can therefore signal that a process of knowledge diffusion is taking place within the TD.

2.2 Pre-specified Blockmodeling

Blockmodeling is a type of clustering for relational data, aiming at reducing a network into a simpler graph in which clusters are identified on the basis of positional equivalence and ties indicate relations between positions. More formally, as defined in Doreian et al. (2005, p. 221):

> A blockmodel is an ordered quadruple $M = (Z, K, T, \pi)$ where:
>
> - Z is a set of positions obtained by shrinking clusters into nodes;
> - $K \subseteq Z \times Z$ is a set of connections between positions;
> - T is a set of predicates used to describe the types of connections between clusters in a network;
> - a mapping $\pi : K \rightarrow T \setminus null$ assigns predicates to connections.

While traditional blockmodeling is principally an exploratory technique, in this study we use blockmodel in a confirmatory way to evaluate the correspondence between a theoretical configuration and the observed data structure.

A theoretical blockmodel is pre-specified using a criterion function defining the type and the location of the blocks expected. This criterion function is subsequently minimized using the local relocation algorithm embedded on the data[1] (de Nooy et al. 2011). The value of the criterion function is the sum of all inconsistencies between the observed data and the theoretical blockmodel defined. Such value therefore can be used as an indication of the goodness of fit of the model hypothesized.

The solutions with the lowest value of the criterion function need however to be further validated. In this study, blockmodel validation is done using the original two-mode network data and other attribute information.

3 Collaboration Patterns in Italian TDs: A Case Study

In 2003, TDs were instituted in Italy by two national laws (L. 317/91 and L. 140/99) with the aim of stimulating both collaborations between firms and research centers and investments in R&D throughout the country. The Italian Ministry for University and Research (MIUR) identified several geographical areas as potential locations in which to establish TDs. Moreover, collaborations between organizations involved in TDs and global partners were encouraged through funding and public support. A question arises whether public funds have succeeded in creating persistent and structural changes sustainable in the long run.

The paper focuses on an Italian TD located in Campania region, named IMAST. This TD achieved important results in its field of specialization, i.e. composite materials and polymers engineering, obtaining prominence both nationally and globally. The population under analysis is constituted by the associated members and partners involved in the TD's activities. The network data describing the institutional interactions between organizations are taken from administrative archives. In particular, we have defined the collaboration networks focusing on the co-participation in research projects that encompasses several R&D projects funded by both national grants (MIUR) and European Commission (EC) from 2005 to 2012.[2]

It is indeed straightforward to think about inter-organizational collaboration among organizations as a social network, where the actors' ties are represented by the co-participation to a common R&D project.

Let $\mathscr{N}$ be the set of n members (*associated members* and *partners*) and $\mathscr{P}$ be the set of the p R&D projects observed on the n members over time, we obtain an affiliation matrix $\mathbf{A}$ $(n \times p)$ with entry a_{ik} equal to 1 if $i \in \mathscr{N}$ participates in the project $k \in \mathscr{P}$, 0 otherwise. To observe the evolution of research collaborations, we derive from $\mathbf{A}$ separate adjacency matrices of size $(n \times n)$ by the aggregation of $\mathscr{N}$ members and $\mathscr{L}$ links present for each year. The dynamic networks at time t

---

[1]In this study we use a direct approach consisting of analyzing the original data without transforming them into dissimilarities/similarities measures.

[2]We thank the IMAST's administrative staff who helped us updating the data to July 2012. The network at time 1 (2005) has not been included in the further analysis, given that only one project has been started.

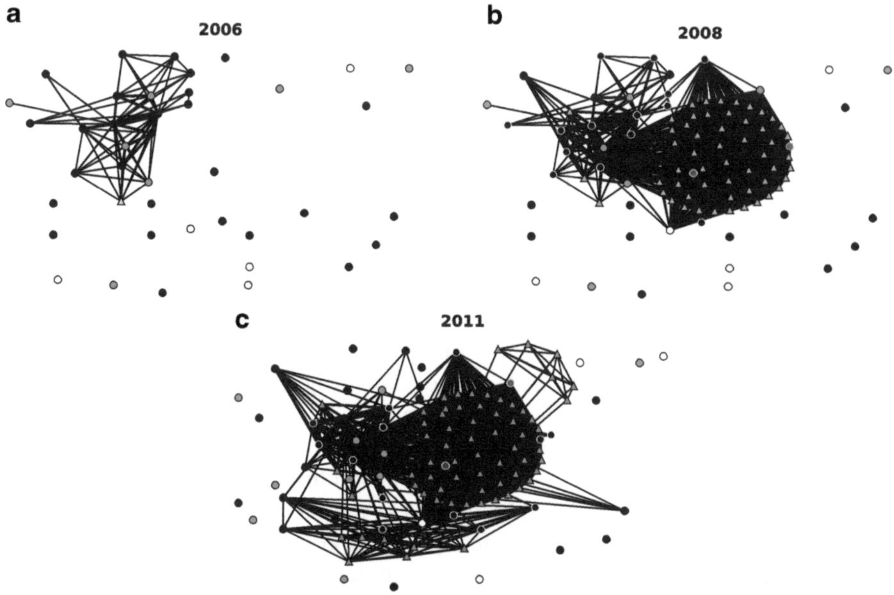

Fig. 1 Three time points of collaboration networks among members involved in IMAST TD. Node's shape: associated member (*circle*); partner (*triangle*). Node's color of associated members: firm (*gray*), research center (*black*) and other organizations (*white*). Software Pajek (de Nooy et al. 2011) is used for the visualization of networks

can be described as a graph $G_T(\mathscr{N}, \mathscr{L}, \mathscr{T})$, with $\mathscr{T}$ the set of ordered time points $t \in \mathscr{T}$ in which the actors $n \in \mathscr{N}$ and the links $l \in \mathscr{L}$ could be not active in all time points.[3]

It is important to note that the derivation of **G** (describing a one-mode network) through **A** (describing a two-mode network) has two notable drawbacks: on the one hand, structural information are lost (as discussed in Everett and Borgatti 2013), in our case regarding the particular kind of R&D projects; on the other hand, cohesiveness in **G** is artificially increased by this transformation. Indeed, all actors involved in a project in the two-mode network become, by definition, linked each other in the one-mode network. Conscious of these problems, we will proceed in the analysis baring in mind these limitations.

Sample network visualizations are provided in Fig. 1. As mentioned, groups of actors involved in projects can be easily distinguished as cohesive groups. At t_2, the network (Fig. 1a) is characterized by a core of denser activity (one or more projects linking the same actors) and a group of less connected or isolated associated members; at t_4 (Fig. 1b) and at t_7 (Fig. 1c) the network opens up to include

---

[3]In our case it is possible to know the exact year when a link is created/terminated and when a member is entered or withdrawn from the network according to the time duration of each research project.

non-members (partners) in an European project in which the TD co-participates. This co-participation can be read as an international acknowledgement of the district support by the members.

4 Main Results: Evolving Network Structure

Following the core-periphery hypotheses discussed in Sect. 2.1, for the blockmodeling analysis we specify the types of blocks we expect to find for the TD's observed temporal networks described in Sect. 3. Specifically, on the main diagonal we expect complete blocks indicating that actors in the corresponding clusters all collaborate with each other (i.e., they participate to the same project). These actors constitute simple cores. The last block on the main diagonal is specified to be null,[4] indicating that the actors in the corresponding clusters have no active collaborations among themselves, but they collaborate with actors in other clusters. These actors represent the periphery. Finally, we left off-diagonal blocks unspecified to understand how the patterns of collaboration between clusters (bridging cores) change over time.

The number of clusters were progressively increased from 3 to 9 to find the best fitting solution using 500 repetitions. The blockmodel has been tested at each year obtaining for each pre-specified number of clusters an array of 7 results.

In Table 1 a summary of results is presented. In particular, for each period we report the number of clusters defined, the number of active projects at each year, the number of inconsistencies found between the observed network and the ideal core-periphery structure, and the typology of projects started by year.

The graphs in Fig. 2 show the main results of the blockmodeling analysis, where nodes represent clusters of structurally similar actors, links represent collaboration patterns between clusters and loops represent collaborations within members in the same clusters.

A clear trajectory can be seen in the evolution of the network as *bridging cores* multiply over time. These results indicate that actors within the TD are motivated to participate in multiple projects and to start fresh collaborations with those actors that were previously involved in different projects (simple cores).

4.1 *Composition of Bridging Cores and Two-Mode Network Validation*

In Fig. 3 the partition obtained using pre-specified blockmodeling analysis is validated on the original two-mode network. Tree most significant time periods are presented: t_2, t_4 and t_7:

[4] A penalty of 100 was imposed on this block to force the model to be as consistent as possible with a theoretically defined core-periphery configuration. Results both with and without this penalty are not too different.

Table 1 Blockmodeling inconsistencies and specifications over time

Time periods	Years	Active projects	Clusters	Inconsistencies	Funding bodies
t2	2006	7	5	10	MIUR
t3	2007	12	7	14	MIUR
t4	2008	13	6	86	MIUR-EC
t5	2009	14	9	58	MIUR-EC
t6	2010	12	9	52	MIUR-EC
t7	2011	11	9	68	MIUR-EC
t8	2012	8	9	42	MIUR-EC

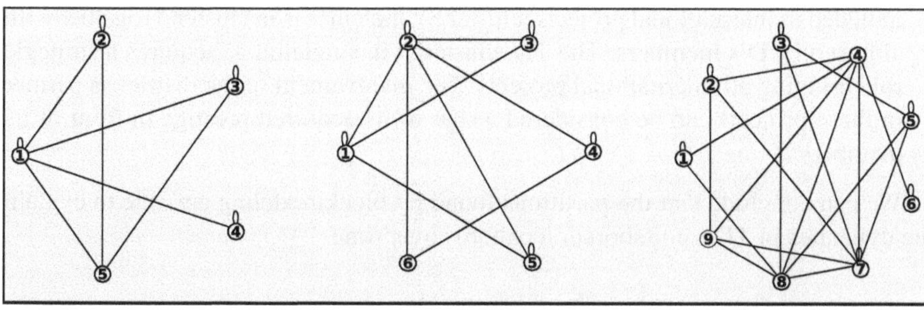

Fig. 2 Reduced graphs obtained using pre-specified blockmodeling in t_2 (2006), t_4 (2008) and t_7 (2011)

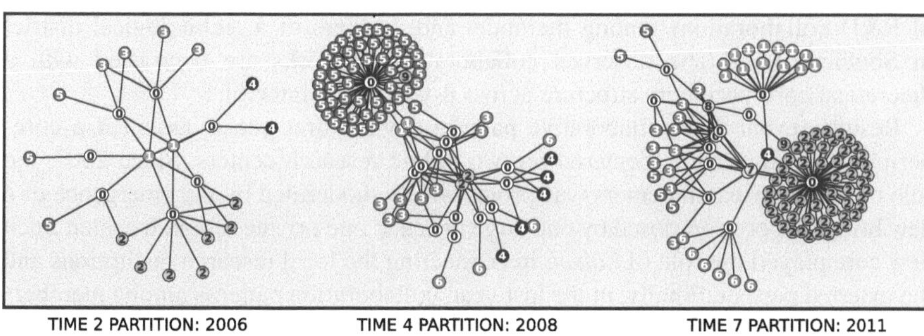

Fig. 3 Visualization of the two-mode networks over time using blockmodeling partitions in t_2 (2006), t_4 (2008) and t_7 (2011)

- at time t_2 (2006), the network has a sort of star configuration with two public research centers bridging all the projects. This two actors will be referred to as local brokers. Similarly, actors classified into clusters 2 and 3 have participated to two distinct pair of projects.
- at time t_4 (2008), the network has dramatically changed its structure. At the top of the graph the cloud of partners in an EC project is presented. The cloud is

connected to the local network by two actors (one firm and one private research center) that will be hereafter referred to as global brokers. These actors were, at t_2, in cluster 3, which has now notably enlarged to include more actors and more projects. A step below the upper line of projects, there are the two actors which represented the main brokers at t_2. They are still very central taking part in almost all the projects active in this period. Actors in the periphery (cluster 4) are last on the right. They collaborate into smaller team, only to a selected number of projects;
- at time t_7 (2011), the periphery has disappeared. Cluster 6 includes only partners of an industrial project. The two global brokers (cluster 7) still maintain at this time their role. Local brokers have conversely differentiated: only one of them is included in international projects (cluster 8); the other is in cluster 3 together with almost all TD's members. The TD, clustered in singleton 4, acquires a strategic role bridging all international projects. The involvement of the district as partner in these projects can be considered a sign of its acquired prestige in front of its members.

We can conclude that the partitions found by blockmodeling are able to explain the dynamics in TD's collaboration patterns over time.

5 Conclusions

A pre-specified blockmodeling analysis has been performed to study the evolution of R&D collaborations among members and partners of a technological district in Southern Italy. The observed collaboration networks are contrasted with a theoretical core-periphery structure across 8-years time interval.

Results reveal that collaborative patterns, in the first phase, assumed a core-periphery configuration, centered on two public research centers. Since 2007, the role of public research centers was progressively moderated by the emergence of a new bridging core composed by one large firm and one private research center. Such new core played the role of *liaison* in connecting the local research institutions and the external partner. Finally, in the last year, collaboration patterns among members intensified to become a *quasi* complete network. This evolution suggests that a process of informal and non-codified information diffusion can taking place among members.

Acknowledgement Work supported by REPOS project "Reti, Politiche pubbliche e Sviluppo". POR Campania FSE 2007–2013 (manager: M.R. D'Esposito). The authors would like to thanks the IMAST technological district for data availability.

References

Antonioli, D., & Marzucchi, A. (2010). *The behavioural additionlity dimension in innovation policies: a review*. Working Papers 201010, University of Ferrara, Department of Economics.

Capuano, C., De Stefano, D., Del Monte, A., D'Esposito, M. R., & Vitale, M. P. (2013). The analysis of network additionality in the context of territorial innovation policy: the case of Italian technological districts. In P. Giudici, S. Ingrassia, & M. Vichi (Eds.), *Statistical models for data analysis* (pp. 81–88). Heidelberg: Springer.

Choi, S. (2012). Core-periphery, new clusters, or rising stars?: International scientific collaboration among 'advanced' countries in the era of globalization. *Scientometrics, 90*, 25–41.

de Nooy, W., Mrvar, A., & Batagelj, V. (2011). *Exploratory network analysis with Pajek*. Cambridge: Cambridge University Press.

Doreian, P., Batagelj, V., & Ferligoj, A. (2005). *Generalized blockmodeling*. Cambridge: Cambridge University Press.

Everett, M. G., & Borgatti, S. P. (2013). The dual-projection approach for two-mode networks. *Social Networks, 35*, 204–210.

Gok, A., & Edler J. (2011). The use of behavioural additionality in innovation policy-making. Working paper University of Manchester.

Goyal, S. (2011). Social networks in economics. In P. J. Carrington & J. Scott (Eds.), *The SAGE handbook of social networks* (pp. 67–79). New York: Sage.

Keynes, J. M. (1936). *The general theory of employment, interest and money*. London: Macmillan.

Kronegger, L., Ferligoj, A., & Doreian, P. (2011). On the dynamics of national scientific systems. *Quality & Quantity, 45*, 989–1015.

The *RCI* as a Measure of Monotonic Dependence

Emanuela Raffinetti and Pier Alda Ferrari

Abstract In this paper a statistical interpretation of a recent measure, called "*Rank-based Concordance Index*" (*RCI*), in terms of monotonic dependence relationship between a non-negative dependent variable and a quantitative independent one is provided. Due to its rank-based construction, the measure presents properties and features that make it suitable also in an ordinal context of analysis. In applied research many data sets contain observations from ordinal variables rather than continuous ones. In such situations, the study of dependence relationship among variables represents an interesting issue, since ordinal variables are not specified according to a metric scale. The proposal discussed here can thus contribute to solve this problem.

Keywords Concordance curve • Dependence relationship • Ordinal covariates

1 Introduction

Dependence relations between variables is one of the most widely studied topics in statistics theory. Non-meaningful statistical models can be constructed, unless that specific assumptions are made about the dependence relationship.

A first investigation about the existence of dependence relationships between two variables is typically achieved by resorting to standard dependence measures, such as the Pearson's r-correlation coefficient (see e.g. Pearson 1907) and the Spearman's r_S-correlation coefficient (see e.g. Spearman 1904). Typically, the Pearson's r-correlation coefficient results as a measure of the linear dependence

E. Raffinetti (✉) • P.A. Ferrari
Department of Economics, Management and Quantitative Methods, Università degli Studi di Milano, Via Conservatorio 7, 20122 Milano, Italy
e-mail: emanuela.raffinetti@unimi.it; pieralda.ferrari@unimi.it

D. Vicari et al. (eds.), *Analysis and Modeling of Complex Data in Behavioral and Social Sciences*, Studies in Classification, Data Analysis, and Knowledge Organization, DOI 10.1007/978-3-319-06692-9_25,

relationship between two quantitative variables, while the rank-based construction of the Spearman's r_S-correlation coefficient makes it appropriate when the two considered variables have ordinal nature. The purpose of this paper is to introduce a novel monotonic dependence measure which accomplishes both the situations. More in detail, by exploiting some relevant contributions in concordance analysis provided by Muliere (1986) and subsequently by Raffinetti and Giudici (2012), here we focus on the approach followed by the latter authors and propose a new interpretation of their *RCI* ("Rank-based Concordance Index") in terms of monotonic dependence measure.

Due to its construction and interpretation, the measure is also proposed for the dependence relationship analysis between a quantitative dependent variable and an ordinal independent one. It is well known that in case of ordinal variables, the distance between categories is not defined. For the analysis, researchers often assign numerical scores to categories. This requires attention and good judgment from the users of the scale, but it provides benefits in the variety of methods available for data analysis (see e.g. Agresti 2010). However, the subjective choice of scale can lead to arbitrary conclusions especially with regard to the dependence study. In this scenario, the proposed measure seems pretty suitable, since it is based only on ordering and not on specific values of the independent variable. Given its properties, this measure can be applied to all contexts where the dependent variable is quantitative and the independent one is either continuous or ordinal.

2 Background on the *RCI* Formalization

RCI was proposed by Raffinetti and Giudici (2012) as a measure of multivariate concordance when a non-negative quantitative response variable and a set of k covariates are considered. In more detail, let Y and $X_1, \ldots, X_k$ be variables linked by a multiple linear regression model of Y on $X_1, \ldots, X_k$, where Y has quantitative nature. Let L_Y be the response variable *Lorenz curve* characterized by the set of pairs $(i/n, (1/(nM_Y)\sum_{j=1}^{i} y_{(j)}))$, where $y_{(i)}$ are the observed values of Y ordered in increasing sense, for $i = 1, \ldots, n$, and M_Y is the Y mean value, and $L_Y^{'}$ the corresponding *dual Lorenz curve* characterized by the set of pairs ordered in reverse sense, i.e. $(i/n, (1/(nM_Y)\sum_{j=1}^{i} y_{(n+1-j)}))$.

The concordance curve is a similar curve obtained using the Y values reordered according to the ranks of the $\hat{Y}$ values given by the least squares regression model $\hat{Y} = \beta_0 + \beta_1 X_1 + \ldots + \beta_k X_k$. Such values are denoted by y_i^*. The set of pairs $(i/n, (1/(nM_Y)\sum_{j=1}^{i} y_j^*))$ defines the C *concordance curve*.

The *RCI* index is then defined as follows:

$$RCI = \frac{\sum_{i=1}^{n-1}\{i/n - (1/(nM_Y))\sum_{j=1}^{i} y_j^*\}}{\sum_{i=1}^{n-1}\{i/n - (1/(nM_Y))\sum_{j=1}^{i} y_{(j)}\}} = \frac{2\sum_{i=1}^{n} i y_i^* - n(n+1)M_Y}{2\sum_{i=1}^{n} i y_{(i)} - n(n+1)M_Y}. \tag{1}$$

Note that (1) can be also expressed as the ratio between two correlation coefficients, $corr(i, y^*)$ and $corr(i, y)$. The proof is immediate. Since $\sum_{i=1}^{n} i = \frac{n(n+1)}{2}$, it results that:

$$\frac{corr(i, y^*)}{corr(i, y)} = \frac{\frac{\left(\frac{\sum_{i=1}^{n} iy_i^*}{n} - \frac{\sum_{i=1}^{n} i}{n}\frac{\sum_{i=1}^{n} y_i^*}{n}\right)}{\sqrt{\sigma_i}\sqrt{\sigma_{y^*}}}}{\frac{\left(\frac{\sum_{i=1}^{n} iy_{(i)}}{n} - \frac{\sum_{i=1}^{n} i}{n}\frac{\sum_{i=1}^{n} y_{(i)}}{n}\right)}{\sqrt{\sigma_i}\sqrt{\sigma_y}}} = \frac{\frac{1}{2n}\left(2\sum_{i=1}^{n} iy_i^* - n(n+1)M_Y\right)}{\frac{1}{2n}\left(2\sum_{i=1}^{n} iy_{(i)} - n(n+1)M_Y\right)}, \tag{2}$$

which corresponds to (1).

Through specific inequalities, one can prove that the C concordance curve always lies within the convex region bounded by L_Y and $L_Y^{'}$ and also that $-1 \leq RCI \leq +1$ (see Muliere 1986).

3 Proposal

This section discusses about our developments concerning the *RCI* interpretation and its application in case of ordinal covariate.

More precisely, Sect. 3.1 is focused on the interpretation of *RCI* in terms of monotonic dependence relationship. Section 3.2 regards its attitude in capturing information about the existing dependence relationship when the covariate has ordinal nature. Finally, in Sect. 3.3 the *RCI* robustness in case of outliers is shown.

3.1 The RCI Interpretation

RCI given in (1) is here interpreted as a monotonic dependence measure of a variable from another one. In order to better clarify that, we first consider the case of a simple linear regression model with both quantitative variables and three extreme scenarios.

Let Y and X be respectively the dependent and the independent variables and C the concordance curve defined in Sect. 2. If we suppose the existence of a perfect direct linear relationship of Y from X, the C curve, represented by the continuous-pointed curve (Fig. 1), overlaps with the response variable Lorenz curve, also defined in Sect. 2, leading through (1) to $RCI = +1$. If, on the contrary, a linear inverse relation between Y and $\hat{Y}$ occurs, a perfect overlapping between the C curve and the dual Lorenz curve is obtained (Fig. 2) and $RCI = -1$.

Suppose now that in the linear regression model $E(Y|X) = E(Y)$. Such situation represents the case of uncorrelation between Y and X, then the $\hat{Y}$ values are all equal and a problem related to the original values reorder arises. Conventionally and consistently with the described construction and some proposals already presented in literature (for example, Spearman), we suggest to substitute

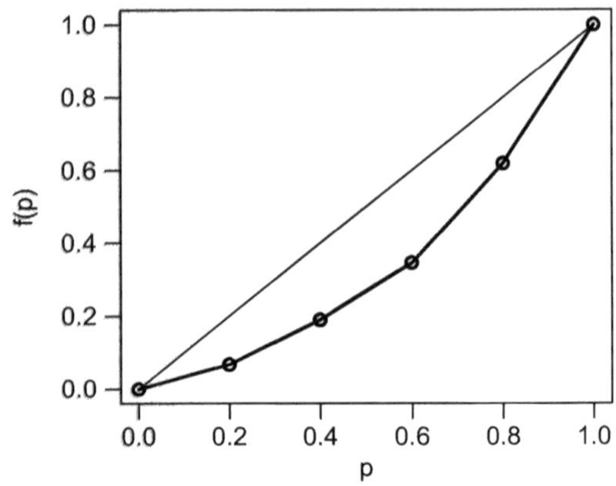

Fig. 1 $C \equiv L_Y$

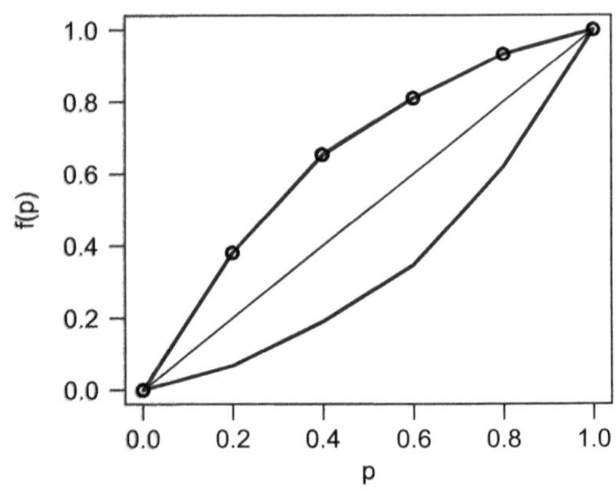

Fig. 2 $C \equiv L_Y^{'}$

to each of these observed values their mean, making the order trivial. Thus, the concordance curve graphically connects the points of coordinates $(i/n, i/n)$ for $i = 1, \ldots, n$ (Fig. 3). We name it *uncorrelation curve*, since it describes the situation where Y and X are uncorrelated. In all other intermediate situations the C curve lies between L_Y and $L_Y^{'}$ leading to an *RCI* value within the range $(-1, +1)$.

Even if our proposal has been illustrated with regard to linear relation, the interpretation of *RCI* holds for any monotonic relation as shown by the following brief example. Let Y and X be linked by the dashed relationship $Y = e^{5.43+0.6X}$, represented in Fig. 4, and let Y be a quantitative variable.

Let us suppose to choose the first ten integer values for the explicative variable X and determine the corresponding Y values. Through the least squares linear regression model, the expected $\hat{Y}$ values are also calculated. The resulting C curve is represented in Fig. 5. Since $\hat{Y}$ values maintain the same order of the corresponding Y values and the C curve coincides with L_Y, *RCI* achieves value $+1$; it is worth noting that the corresponding r-correlation coefficient is 0.830. This result shows the real attitude of our index to perfectly catch any monotonic dependence relation

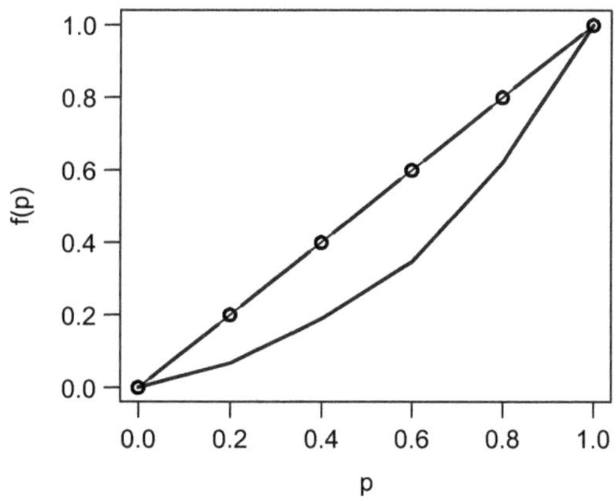

Fig. 3 $C \equiv p$

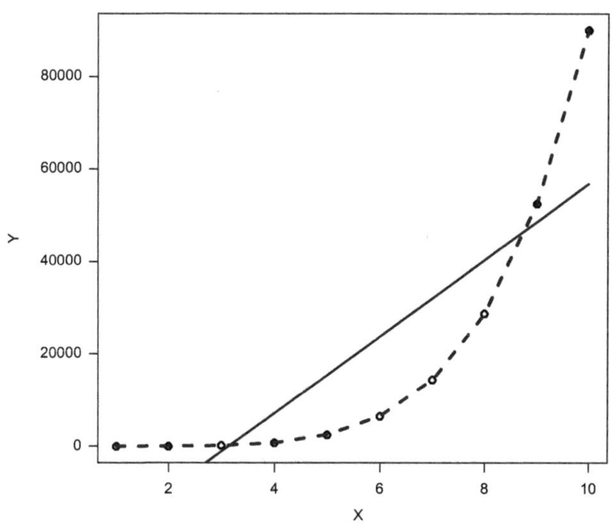

Fig. 4 Data: *dashed line*, lin. regr.: *continuous line*

between Y and X, while the r-correlation coefficient perfectly captures only the linear one.

Due to its meaning, the concordance or C curve can be called *dependence curve* for underlying the fact that it is built on the original response variable values reordered according to the existing dependence relation with the explicative variable.

Similar reasons support this proposal also in case of ordinal covariate, as it will be discussed in Sect. 3.2.

Fig. 5 $C \equiv L_Y$

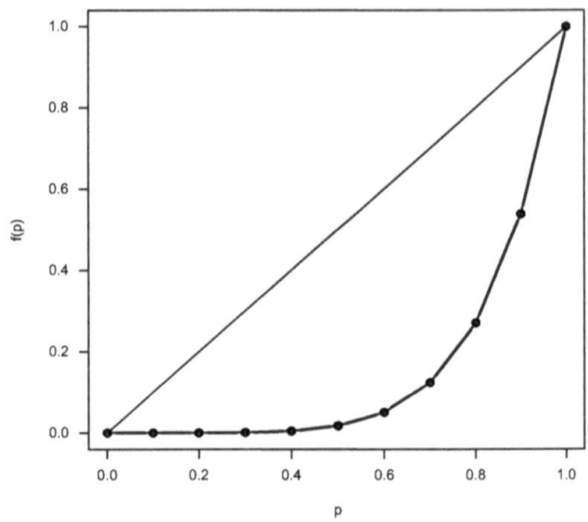

3.2 Ordinal Data

In many fields several relevant phenomena are described by ordinal variables. In such situations, the study of dependence relationship among variables generally represents an open issue, since ordinal variables are not specified according to a metric scale and for this reason the corresponding assigned values can affect the obtained results. In fact, results changing with respect to the different adopted scales highlight the need of defining novel measures leading to the same conclusions in terms of existing dependence relationships, i.e. invariant with regard to quantification of the variable categories.

The proposed *RCI* features allow us to satisfy this condition since *RCI* is based only on the dependent quantitative variable original Y values reordered according to ranks of its corresponding estimated values $\hat{Y}$. We call this property *scale invariance* property. Let us consider the example described in Sect. 3.1 and let us modify the original codes for ordinal covariate values $x = \{1, 2, 3, 4, 5, 6, 7, 8, 9, 10\}$ in $\tilde{x} = \{11, 67, 123, 170, 234, 321, 359, 402, 421, 543\}$. Now the values assigned to X are changed but the ordering is maintained. Thus, the corresponding dependence curve is equivalent to the one represented in Fig. 5, where the covariate X was characterized by the first integer ten values and *RCI* had the same value +1, verifying the invariance property. This does not occur for the r-correlation coefficient; in this case its value is changed moving from 0.830 to 0.568. In fact, as just shown, it is affected by the choice of the adopted values.

When analysis refers to ordinal data, a relevant issue also arises, that is the problem of tied categories. The term *tied data* is used to point out data described by a frequency distribution. It may occur that some or all the X values are characterized by a frequency greater than one, then all the related regression estimates $\hat{Y}$ have the same values. This condition causes some difficulties in establishing the corresponding order of the original Y values. In such a context, the rule proposed

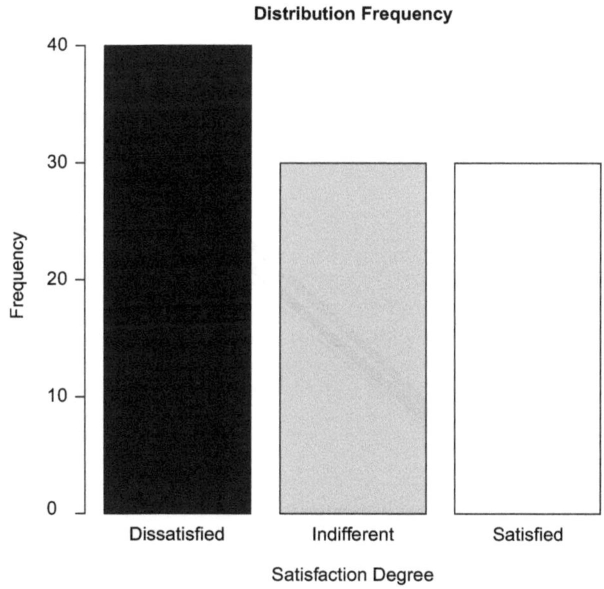

Fig. 6 X frequency distribution

in Sect. 3.1 for the uncorrelation case can still once be adopted and the Y values corresponding to the equal Y^* values are substituted by their mean. The already mentioned *RCI* invariance with respect to different chosen subjective scales holds also in case of ordinal tied data. To better understand this aspect, let us generate, for instance, an independent variable X expressing the customer satisfaction degree towards a service. Let X be a categorical variable taking the following ordered values: *Dissatisfied* (encoded as 1), *Indifferent* (encoded as 2) and *Satisfied* (encoded as 3) and let us run a simulation study. For this purpose, Ferrari and Barbiero (2012) illustrated a novel procedure to simulate samples from ordinal variables with pre-specified correlation matrix and marginal distributions and implemented this method in `R` through the package `GenOrd`. Here, the scenario is a little bit different since in this case one variable has ordinal nature and the other one is quantitative. For this reason, to define data to be used, we resorted to a specific `R` package, named `mvtnorm`, for obtaining samples from a bivariate normal distribution with a pre-specified pairwise ρ-correlation coefficient (here $\rho = 0.2$). A sample $n = 100$, by maintaining the value of variable Y and by discretizing the variable X according to the above categories (whose frequency distribution is represented in Fig. 6), was generated. The obtained data led to the dependence curve shown in Fig. 7 and an $RCI = 0.172$ which results really good if we consider that when one of the two correlated continuous variables is transformed into a categorical one, the corresponding correlation coefficient shrinks.

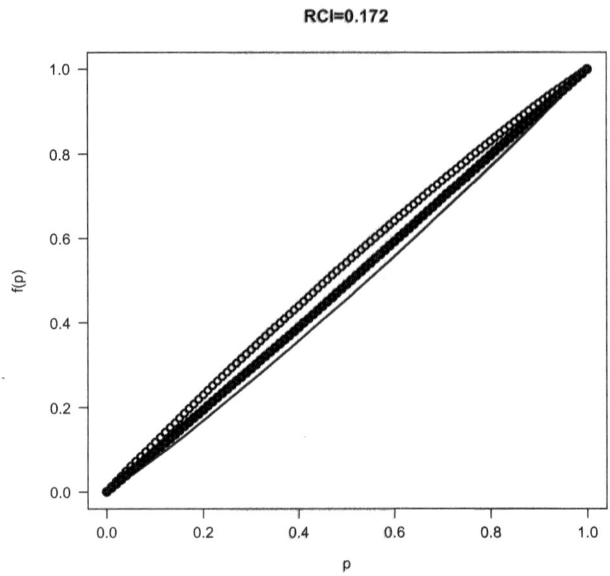

Fig. 7 *C* curve—codes 1, 2, 3

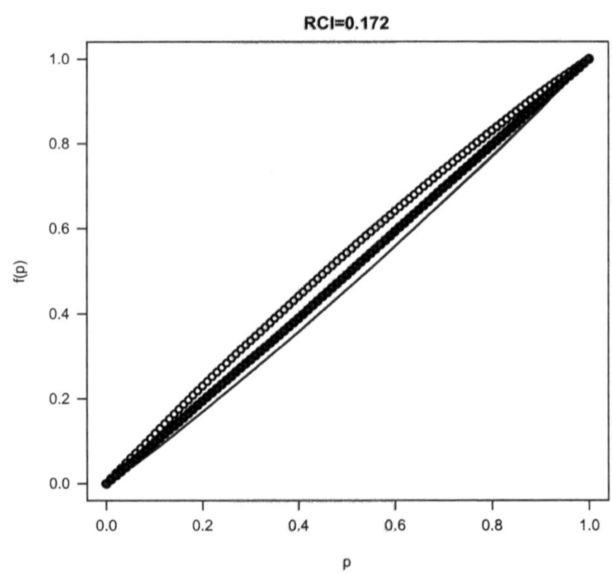

Fig. 8 *C* curve—codes 2, 3, 4

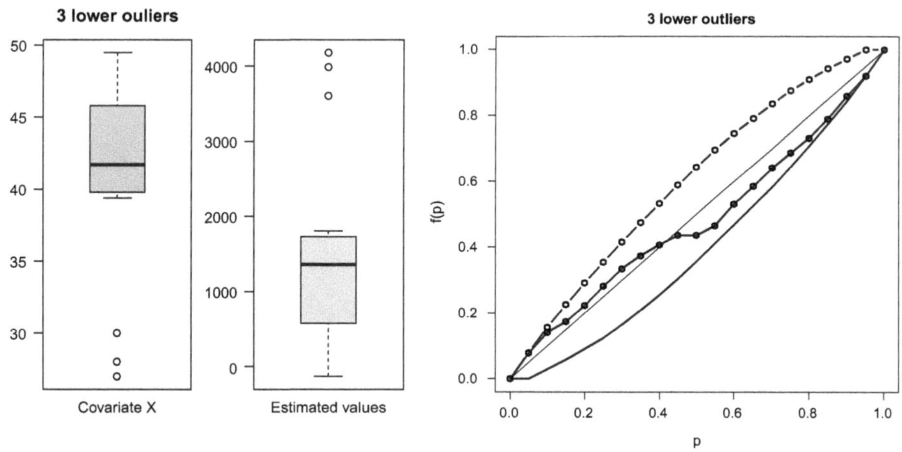

Fig. 9 Box plots of X and $E(Y|X)$ and dependence curve in case of three lower outliers

Let the previous ordered categories now be re-encoded as: *Dissatisfied* (encoded as 2), *Indifferent* (encoded as 3) and *Satisfied* (encoded as 4). As it is evident by Fig. 8, the dependence curve maintains the same position of that represented in Fig. 7 and *RCI* takes the same value (0.172), showing that differences between the scales used to express the X ordered categories do not affect the *RCI* attitude in capturing the existing dependence relationship.

3.3 The RCI Behavior in Case of Outliers

In literature there is much debate regarding what to do with extreme or influential data points, considered as outliers. Outliers can have deleterious effects on statistical analyses, as stated for instance by Osborne and Overbay (2004). Mainly, they can seriously bias or influence estimates that may be of substantive interest. In many cases, outliers are relevant part of data and they can not thus be removed without losing piece of information. For this reason, researchers sometimes use *robust* procedures to protect their results from being distorted by the presence of outliers (see e.g. Barnett and Lewis 1994).

Due to its properties, *RCI* results as a robust dependence measure in case of outliers. In fact, being built on the reordered y_i's it is not affected by anomalous values in X or $\hat{Y}$. Let us discuss about this topic by considering a quantitative variable X whose distribution can be at first characterized by three lower outliers (Fig. 9), then by three upper outliers (Fig. 10) and finally by six outliers (Fig. 11). To show the *RCI* robustness, we compare the three previous considered scenarios with the situation of non-outliers, obtained by removing the outliers (Fig. 12).

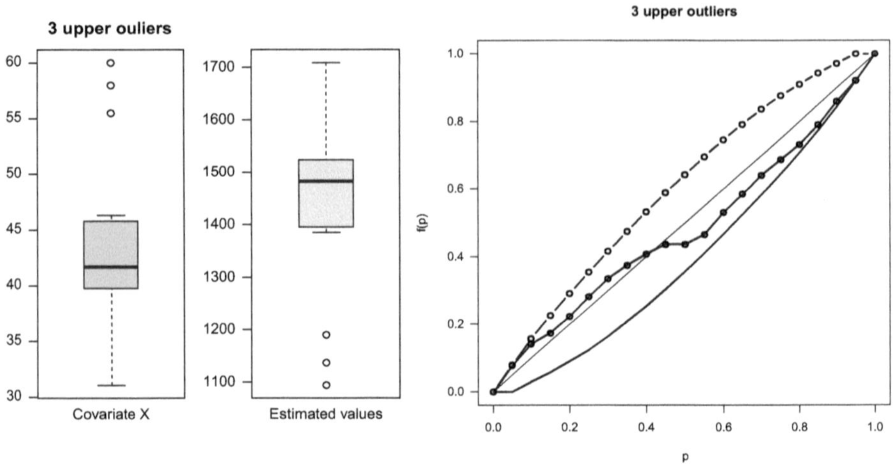

Fig. 10 Box plots of X and $E(Y|X)$ and dependence curve in case of three upper outliers

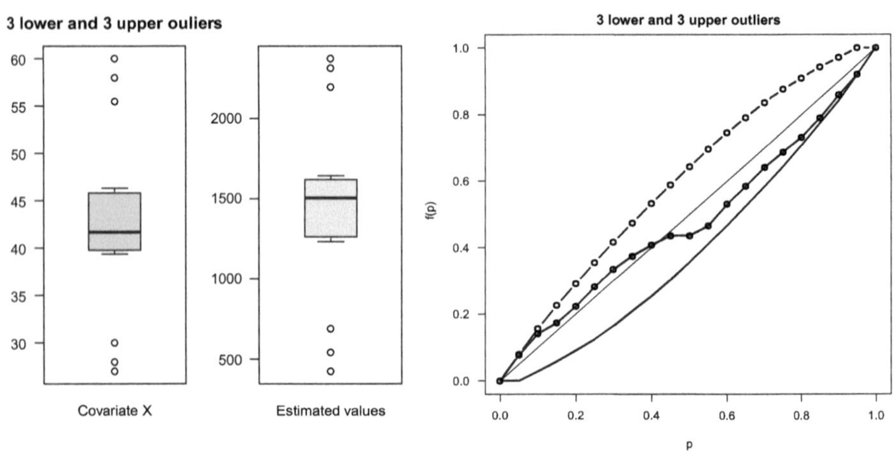

Fig. 11 Box plots of X and $E(Y|X)$ and dependence curve in case of three lower and three upper outliers

Focusing on Figs. 9, 10, and 11, the corresponding dependence curves in presence of outliers have the same behavior of that one in Fig. 12 implying an *RCI* which preserves its original value. These results strongly support the *RCI* dependence measure adequacy in case of outliers.

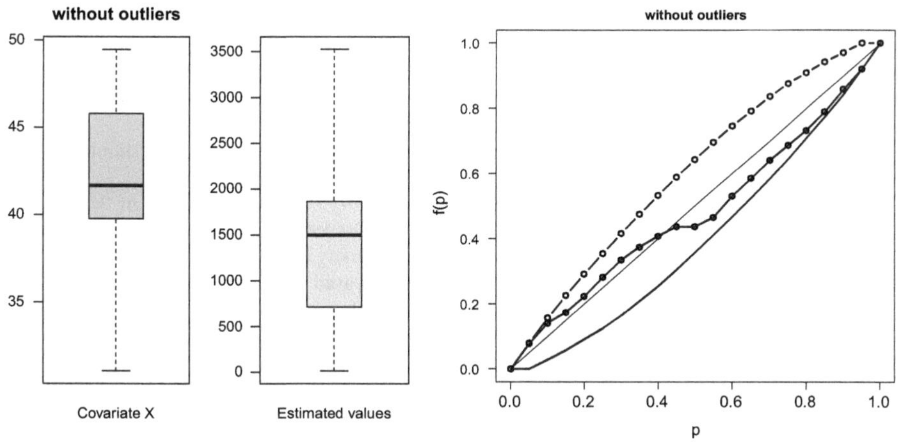

Fig. 12 Box plots of X and $E(Y|X)$ and dependence curve in case of non-outliers

4 Concluding Remarks

This paper proposes a new interpretation of the *Rank-based Concordance Index* (*RCI*), in order to catch the existence of any monotonic dependence relationship between a non-negative dependent variable and an independent one either continuous or ordinal. Even if the *RCI* employment has been discussed here only with regard to a positive quantitative response variable and an unique covariate, our proposal can be expanded to contexts involving more than one covariate and including covariates of different nature. Our research activity will be devoted to the *RCI* application to any real-valued response variable in order to overcome the original non-negativeness condition required by the classical Lorenz curve definition. Furthermore, we are carrying out further investigations about the comparison between the *RCI* index and the Spearman's r_S-correlation coefficient, in order to detect possible similarities or dissimilarities. Our attention will also be focused on developments concerning the *RCI* extension to an inferential perspective, by building an appropriate test for assessing the monotonic dependence relationship significance. Finally we will intend to consider the multivariate case.

Acknowledgement The authors acknowledge financial support from the European Social Fund Grant (FSE), Lombardy Region.

References

Agresti, A. (2010). *Analysis of ordinal categorical data* (2nd ed.). Hoboken: John Wiley & Sons.

Barnett, V., & Lewis, T. (1994). *Outliers in statistical data* (3rd ed.). New York: John Wiley & Sons.

Ferrari, P. A., & Barbiero, A. (2012). Simulating ordinal data. *Multivariate Behavioral Research, 7*(4), 566–589.
Muliere, P. (1986). Alcune osservazioni sull'equità orizzontale di una tassazione (Some notes about the horizontal equity of a taxation), *Estratto da "Scritti in onore di F. Brambilla"* (vol. II, pp. 551–559). Milano: Edizioni di Bocconi Comunicazione.
Osborne, J. W., & Overbay A. (2004). The power of outliers (and why researchers should always check for them). *Practical Assessment, Research & Evaluation, 9*(6), 1–12.
Pearson, K. (1907). *Mathematical contributions to the theory of evolution: XIV*. On the general theory of skew correlation and non-linear regression (Drapers' Company Research Memoirs, Biometric Series, No. II). Cambridge: Cambridge University Press.
Raffinetti, E., & Giudici, P. (2012). Multivariate Ranks-based concordance indexes. In A. Di Ciaccio et al. (Eds.), *Advanced statistical methods for the analysis of large data-sets*. Studies in theoretical and applied statistics (pp. 465–473). Berlin: Springer.
Spearman, C. (1904). The proof and measurement of correlation between two things. *American Journal of Psychology, 15*, 72–101.

A Value Added Approach in Upper Secondary Schools of Lombardy by OECD-PISA 2009 Data

Isabella Romeo and Brunella Fiore

Abstract In the last decade a great deal of interest at the national and international level has been shown in measuring the school impact on student achievement. Standardized tests and the Value Added Methodology have emerged as the appropriate instruments for this purpose. The aim of this paper is to find a value added measure for upper secondary schools of the Lombardy region from the OECD-PISA 2009 data. The initial cognitive level of the student, which is necessary for the analysis, has been obtained by summarizing different teachers' evaluations from a Rasch analysis. A multilevel model has been fitted to control the student and school factors effecting the reading results. In particular, even the reading enjoyment variable has been considered, since it explains a high variability of student performance. The ranking of the upper secondary schools based on the value added measures is compared with the one obtained using raw data, showing significantly different results.

Keywords HLM • Longitudinal studies • OECD-PISA data • Rasch analysis • VAM

I. Romeo (✉)
Department of Statistics and Quantitative Methods, University of Milano-Bicocca, Via Bicocca degli Arcimboldi, 8, Milano, Italy
e-mail: isabella.romeo@unimib.it

B. Fiore
Department of Sociology and Social Research, University of Milano-Bicocca, Via Bicocca degli Arcimboldi, 8, Milano, Italy
e-mail: brunella.fiore@unimib.it

D. Vicari et al. (eds.), *Analysis and Modeling of Complex Data in Behavioral and Social Sciences*, Studies in Classification, Data Analysis, and Knowledge Organization,
DOI 10.1007/978-3-319-06692-9_26,

1 Introduction

During the past several years there has been growing interest in the use of standardized test data to measure the impact of various educational institutions on student progress. In particular, the attention of researchers and policy makers today focuses on schools and their quality: the objective is to assess how schools affect their students' improvement of knowledge. The Value Added Model (VAM) is the methodology recognized in the educational literature on this issue (Goldstein et al. 2000), which exploits student achievement data in a longitudinal perspective. In this paper, beside the usual variables considered in this context, as the initial level of knowledge and the cultural and socioeconomic status, even the reading enjoyment of the student is taken into account given its importance in explaining differences in performance.

This work identifies the upper secondary school contribution in increasing students' performance in the Lombardy region, using the OECD-PISA 2009 data. The paper is structured as follows: Sect. 2 focuses on the Value Added Models (VAMs), Sect. 3 introduces the data, Sect. 4 focuses on reading enjoyment, Sect. 5 addresses the methodology used and finally Sect. 6 shows the results obtained.

2 Value Added Models (VAMs)

During the last decade a growing interest in the development of appropriate indicators of school effectiveness has been shown. The simplest way to compare schools consists in considering the unadjusted mean of student performances for each school. It does not represent a good practice since students enter schools with large mean differences in results (Raudenbush 2004). Since the 1980s (Hill and Rowe 1996; Raudenbush and Willms 1995) researchers have tried to isolate the school's influence on students' progress, leaving out all factors that are beyond school control. The models recognized for this purpose are the Value Added Models (VAMs). In literature various types of models are recognized depending on the variables used to control for differences among schools. The simplest form of VAM takes into account only the prior educational level. Most recent models isolate the school effect, adjusting for both prior achievement and student background characteristics. The resulting value added measure is called "Type A" effect (Raudenbush 2004), which includes both school context and school practice. The former involves the composition of the school and the social environment. The latter is influenced by school leaders and teachers. In order to isolate only the effect of school practice a "Type B" model (Raudenbush and Willms 1995) can be used. This model not only controls student characteristics, but also the effect of compositional variables, such as average prior achievement and socioeconomic status of students, and context variables such as school size, as well as school type (Timmermans et al. 2011). The empirical comparison of various value added models has been already discussed by many researchers (Timmermans et al. 2011; Tekwe et al. 2004).

3 Educational Data

PISA is a comprehensive and rigorous international programme, promoted to assess student performance and to collect data on students, families and institutional factors that could help explain differences in performance. In particular, information from this survey focuses on how well students are prepared to meet the challenges of life, rather than to examine how well they perform a particular curricula specified by the school system. This survey collects information about the reading, mathematics and science results of 15-year-old students. The survey is done every 3 years with a different major subject area. Reading represents the focus in the first survey conducted in 2000 and also in the one conducted in 2009. Mathematics is the focus in the surveys of 2003 and 2012,[1] whereas science is the focus of 2006. In this work the PISA 2009 data and in particular, reading has been considered. This survey has a cross-sectional feature, since the students change at each survey. This implies the lack of measures on the cognitive level for each student over time, which would be useful in constructing a measure of student prior achievement. In order to overcome this limit of the survey, the PISA 2009 data have been linked to another dataset with both the final evaluation of the lower secondary education and the grade reported in the first year of upper secondary school, collected on the same students of the Lombardy region. Therefore, only the Lombardy region has been taken into account. Only liceo, vocational and technical schools of the PISA sample have been studied. The number of students and schools in the sample are 1,132 and 46, respectively. The matching procedure leads to a sample of 930 students and 33 schools. This final sample is representative of the original one, thus the missing values can be considered missing at random (MAR).

In all the analysis and computational process it is necessary to consider the particular structure of the PISA dataset which considers the five plausible values[2] (PVs) for parameter estimation and the replicates[3] for standard error estimation. The analysis has been supported by Winstep and SAS programs (SAS macros from OECD PISA 2009 have been exploited).

[1]PISA 2012 data are not available yet.

[2]The PVs are meant to prevent biased inferences, which can occur as a result of measuring the not directly observable student skill. Instead of directly estimating it, a probability distribution is estimated. Then the PVs are random draws from this distribution. The required statistic and its respective standard error have to be computed for each plausible value and then put together (PISA 2012).

[3]Given the PISA complex sample design, the use of replicates is needed to obtain reliable sampling variances. The Fay's variant of the Balanced Repeated Replication (BRR) is used (PISA 2012).

4 The Reading Enjoyment

Various studies have found that reading enjoyment is one of the student's characteristics that better explains the student's reading performance (Brozo et al. 2001; PISA 2010). In the PISA 2009 data, a measure of student enjoyment of reading[4] is included in the dataset. The PISA studies (PISA 2002) have shown that reading enjoyment accounted for twice as much of the difference in performance as the socioeconomic status. This result represents an important issue, given that the socioeconomic status is recognized in literature as one of the variables that best explains student performance most of all. Therefore, reading enjoyment represents a good performance predictor. Reading enjoyment is strictly correlated with an important variable in the educational literature: the reading motivation (PISA 2002). The latter has had little attention in the Value Added Models literature. It is possible to identify two main reasons; on one hand, it is not usual to find this type of information in educational datasets, and on the other hand, this variable is not easily measurable, since it is not directly observable. Furthermore, it represents a particularly complex variable, given that there is a circular association between motivation and achievement as on one side, reading motivation leads to greater engagement which in turn leads to better results and on the other side, better performance leads to greater motivation (Stanovich 1986). Even if reading enjoyment could be considered an aspect of the more complex reading motivation, this variable does not suffer of this problem of endogeneity. Indeed, reading enjoyment can lead student to read more which in turn can lead to achieve better results, but good results have little effect on reading enjoyment (Wang and Guthrie 2004). Thus, in order to evaluate its impact on both performance and school ranking, reading enjoyment is considered.

5 Methods

In order to obtain school contribution to student improvement, "Type B" value added model is considered. In this way, the effect of school practice is isolated by controlling prior achievement, available student and school characteristics.

Prior achievement level is obtained as a summary of the final evaluation of the lower secondary education and the final mark reported in the first year of upper secondary school. A Partial Credit Model (PCM) (Bond and Fox 2001) has been used to place the two marks on the same common logit scale given their different ordinal scale: the final evaluation of the lower secondary education expressed by "Excellent", "Good", "Discrete" or "Sufficient" and the final mark of the first year of upper secondary education expressed on a numeric scale from 1 to 10. It is important to remark that PCM scores are fallible measures of the true values, given

[4]This variable (JOYREAD index) is derived by OECD, putting together eleven items (PISA 2012) by a scaling procedure (Item Response Theory).

the measurement errors implied by the model (Ferrao and Goldestain 2009). Both measures have been used since only the final grade of lower secondary education does not seem to be discriminant enough for student achievement as it always represents a positive judgment and assumes few categories. A measure of the contribution of the school to students' educational growth is then identified as a residual of a two-level multilevel model that takes into account the hierarchical structure of the data that is students grouped into schools (Snijders and Bosker 1999). Let Y_{ij} be the value of reading score for the i-th student of the j-th school, with $i = 1, \ldots, n_j$ and $j = 1, \ldots, J$. The number of students of the j-th school is denoted by n_j.

We specify the following two level model for student i in class j:

$$Y_{ij} = \alpha + \boldsymbol{\beta x}_{ij} + \boldsymbol{\gamma w}_j + u_j + e_{ij}$$

where $\boldsymbol{x}_{ij}$ is the vector of student's covariates and $\boldsymbol{w}_j$ is the vector of school-level covariates. Student-level errors e_{ij} are assumed independent across students, and school-level errors u_j are assumed independent across schools. The errors e_{ij} are independent from the errors u_j.

At the student level (level 1), gender, immigration status, cultural and socioeconomic status (ESCS[5]), grade repetition, use of the computer at home for school and the initial prior achievement measure have been considered. Furthermore, also the quadratic effect of the variable use of the computer at home (PISA 2011) and the reading enjoyment variable have been introduced in the model. All variables considered at the student level have been aggregated and included at the school level (level 2). Moreover, also the type of secondary school, school size (number of students attending the school) and school location[6] (town or city) at the school level have been considered.

Once the model is established it is possible to determine the residuals for each student, comparing the estimated performance with the observed one. The value added measure for each school is obtained by using the shrinkage school residual. The school ranking is derived ordering school residuals. It represents the strength of the school value added measure, other things being equal. Standard errors and then confidence intervals have been drawn through replicates (Sect. 3). A measure significantly above zero represents schools that give their students a greater contribution than other schools on average. On the contrary, schools measuring significantly below zero give their students a lower contribution in respect to the other schools. In order to understand the value added effect on school ranking, the comparison between school rankings obtained from the raw score and the residuals is made.

---

[5]This variable was created by OECD on the basis of the occupational and educational level of the student's parents, home educational and cultural resources (PISA 2012).

[6]The variable assumes value one if the school is located in a town with fewer than 100,000 people otherwise it assumes value zero.

6 Results

Interesting results emerge from the empty multilevel model and the one fitted with the aforementioned variables. From the empty model, a value of 45 % of the Intraclass Correlation Coefficient (ICC) (Snijders and Bosker 1999) is obtained. It underlines how strongly units in the same group resemble each other, justifying the multilevel approach. This value is coherent with the well-known differences in performances among the different types of Italian secondary schools: technical, vocational and liceo.[7]

Table 1 reports the estimates of the model parameters and their standard error (S.E.) with an indication of their statistical significance obtained by a Wald test. At the school level, the prior achievement and the ESCS index are significant ($\alpha = 0.01$) in the model with a positive effect on performance. Moreover, students attending schools in town obtain better results compared with students attending schools in city. This can be explained by both a minor presence of immigrants and a major homogeneity of ethnic minorities in towns. Finally, schools with more students have usually more resources at disposal. This implies more activities in the school that support students in obtaining better results. At the student level, reading enjoyment and the prior achievement have a positive effect on performance, whereas immigration status and grade repetition have a negative effect. For what concern the use of the computer at home for school, both the linear and the quadratic effects result negative. This suggests a mountain shaped relation, with a threshold after which an increase in computer use is associated with a decrease in learning performances (PISA 2011).

In order to evaluate the adequacy of the model and the importance of the reading enjoyment variable, also the full model without this variable is fitted to screen distinct intercept variances for each level. Furthermore, given the well-known great importance in literature of the ESCS in explaining achievement, also the full model without ESCS variable is build. The empty model is used as a baseline to compare the random effect variance reduction and the decrement of log likelihood (in particular the Likelihood Ratio (LR) chi-square test is considered, that is minus two (i.e., -2) times the difference between the starting and ending log likelihood) to the all models considered in Table 2. As said above, particular attention is given to student reading enjoyment: the variability reduction obtained by shifting from the full model without this variable to the one with it is about 10.9 % at level 1 and 0.2 % at level 2. While this variability reduction is greater than the one associated with the ESCS at the student level, it is lower than the one associated with the ESCS at the school level (Table 2). LR chi-square test confirms the importance of reading enjoyment in explaining differences in performances. The model built without the variable of pleasure for reading shows a significant effect of gender and vocational school. This result suggests as the pleasure for reading attenuates the performance

[7] In Italy there are many types of liceo: classical, scientific, socio-pedagogical.

Table 1 Multilevel model estimates

	Variable	Coefficient	S.E.
Student	Intercept	489.9	13.8
	Female (ref. male)	3.9	3.5
	Immigrant (ref. Italian)	−48.2***	8.8
	ESCS	4.3*	2.3
	Student repeating a year (ref. student not repeating the year)	−26.1**	5.8
	Prior achievement	3.3***	0.8
	Use of the PC at home for school	−10.5***	3.0
	Use of the PC at home for school (quadratic effect)	−4.3***	2.0
	Reading enjoyment	24.0***	1.8
School	Percentage of girls	0.2	0.2
	Percentage of immigrants	−0.4	0.3
	Mean ESCS	52.9***	14.2
	Percentage of students repeating the year	0.2	0.3
	Mean prior achievement	10.4***	3.3
	Percentage of students using the PC at home for school	8.6	7.7
	Mean reading enjoyment	−0.8	11.5
	Technical school	−6.3	8.5
	Vocational school	15.5	14.2
	School size	24.1***	2.0
	Town (ref. city)	14.5***	4.0

Significance levels: (*) $\alpha = 0.1$, (**) $\alpha = 0.05$, (***) $\alpha = 0.01$

Table 2 Comparing the variance parameter estimates of different models

	Empty	Complete	Without Joyread	Without ESCS	Without ESCS and Joyread
Variance between	3,052.5	216.9	223.5	273.6	351.5
Reduction of variance between		92.9 %	92.7 %	91.0 %	88.5 %
Variance within	3,702.8	2,732.8	3,136.8	2,747.1	3,163.1
Reduction of variance within		26.2 %	15.3 %	25.8 %	14.6 %
−2loglikelihood	10,817	10,468	10,600	10,477	10,614
LR test		349	218	340	196

differences due to these variables. When the ESCS variable is not considered in the model the type of school becomes significant. These two variables are strictly correlated: students with higher levels of ESCS usually attend liceo, while students with lower levels of ESCS usually attend vocational schools.

Figure 1 shows the school ranking based on value added measures and the number of gained/lost positions in respect of the raw ranking for each school.

Looking at the gain/loss positions in Fig. 1 is pointed out that the two rankings are significantly different. The ranking obtained from the value added model underlines that only a limited number of schools is significantly different from the mean value. While liceo schools are all above the average line when raw scores are

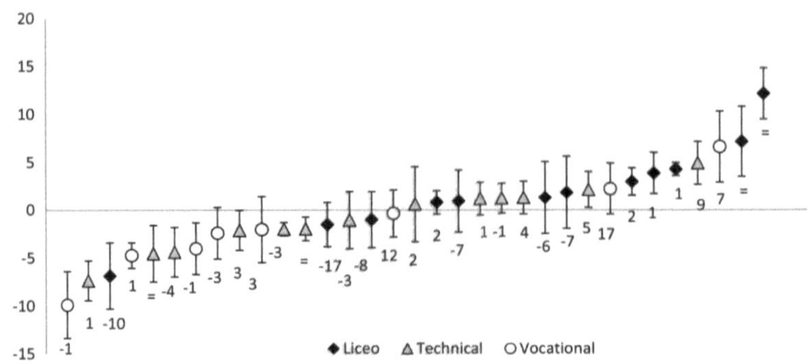

Fig. 1 School ranking based on the value added measure

taken under consideration, some of them are positioned under the average line, after considering value added measures. Similarly, while vocational schools are all in the lower part of the tail when the raw ranking is taken under consideration, they gain many positions in the residual ranking. Definitely, some schools that could appear more disadvantaged, as vocational and technical schools, could show a greater school effect. Finally, a different ranking is obtained when the value added measures are related to the model built without the reading enjoyment. It highlights a more defined clustering of schools in respect of the type of school. These differences underline the importance of the reading enjoyment variable in explaining performances.

In conclusion, it has been found that in order to properly evaluate the school impact on its students, it is necessary to consider a value added measure to avoid invalid ranking. In addition, it has also been highlighted that motivational variable should be considered in VAM, given its impact on school ranking.

References

Bond, T. G., & Fox, C. M. (2001). *Applying the Rasch model: Fundamental measurement in the human sciences*. London: Lawrence Erlbaum Associates Publishers.

Brozo, W. G., Shiel, G., & Topping, K. (2001). Engagement in reading: Lessons learned from three PISA countries. *Journal of Adolescent & Adult Literacy 51*, 304–314.

Ferrao, M. E., & Goldestain, H. (2009). Adjusting for measurement error in the value added model: Evidence from Portual. *Qualitative and Quantitative, 43*, 951–963.

Goldstein, H., Huiqi, P., Rath, T., & Hill, N. (2000). *The use of value added information in judging school performance*. London: Institute of Education.

Hill, P. W., & Rowe, K. J. (1996). Multilevel modelling in school effectiveness research. *School Effectiveness and School Improvement, 7*, 1–34.

OECD, PISA (2002). *Reading for change? Performance and Engagement across countries*. OECD Publishing. Paris: France.

OECD, PISA (2009). *PISA data analysis: SAS* (2nd ed.). OECD Publishing. Paris: France.

OECD, PISA (2010). *PISA 2009 results: Learning to learn - student engagement, strategies and practices* (Vol. III). OECD Publishing. Paris: France.

OECD, PISA (2011). *PISA 2009 results: Students on line: Digital technologies and performance* (Vol. VI). OECD Publishing. Paris: France.

OECD, PISA (2012). PISA 2009 Technical Report. OECD Publishing. Paris: France.

Raudenbush, S. W. (2004). What are value-added models estimating and what does this imply for statistical practice? *Journal of Educational and Behavioral Statistics, 29*, 121–129.

Raudenbush, S. W., & Willms, J. (1995). The estimation of school effects. *Journal of Educational and Behavioral Statistics, 20*, 307–335.

Snijders, T., & Bosker, R. (1999). *Multilevel analysis. An introduction to basic and advanced multilevel modeling*. New York: Sage.

Stanovich, K. (1986). Matthew effects in reading: Some consequences of individual differences in the acquisition of literacy. *Reading Research Quarterly, 21*, 360–407.

Tekwe, C. D., Carter, R. L, Ma, C. X., Algina, J., Lucas, M. E., Roth, J., et al. (2004). An empirical comparison of statistical models for value-added assessment of school performance. *School Effectiveness and School Improvement, 29*, 11–36.

Timmermans, A. C., Doolaard, S., & De Wolf, I. (2011). Conceptual and empirical differences among various value-added models for accountability. *School Effectiveness and School Improvement, 22*, 393–413.

Wang, J. H. Y., & Guthrie, J. T. (2004). Modeling the effects of intrinsic motivation extrinsic motivation, amount of reading, and past reading achievement on text comprehension between U.S. and Chinese student. *Reading Research Quarterly, 39*, 162–186.

Algorithmic-Type Imputation Techniques with Different Data Structures: Alternative Approaches in Comparison

Nadia Solaro, Alessandro Barbiero, Giancarlo Manzi, and Pier Alda Ferrari

Abstract In recent years, with the spread availability of large datasets from multiple sources, increasing attention has been devoted to the treatment of missing information. Recent approaches have paved the way to the development of new powerful algorithmic techniques, in which imputation is performed through computer-intensive procedures. Although most of these approaches are attractive for many reasons, less attention has been paid to the problem of which method should be preferred according to the data structure at hand. This work addresses the problem by comparing the two methods *missForest* and *IPCA* with a new method we developed within the forward imputation approach. We carried out comparisons by considering different data patterns with varying skewness and correlation of variables, in order to ascertain in which situations a given method produces more satisfying results.

Keywords Forward imputation • Iterative PCA • missForest • Missing data

1 Missing Data Treatment

Missing data treatment is frequently invoked when performing data analysis. There exists no field of quantitative research where missing information is not a problem, and an optimal choice of an imputation procedure should be a guarantee of

N. Solaro (✉)
Department of Statistics and Quantitative Methods, Università di Milano-Bicocca, Milan, Italy
e-mail: nadia.solaro@unimib.it

A. Barbiero • G. Manzi • P.A. Ferrari
Department of Economics, Management and Quantitative Methods, Università di Milano, Milan, Italy
e-mail: alessandro.barbiero@unimi.it; giancarlo.manzi@unimi.it; pieralda.ferrari@unimi.it

D. Vicari et al. (eds.), *Analysis and Modeling of Complex Data in Behavioral and Social Sciences*, Studies in Classification, Data Analysis, and Knowledge Organization,
DOI 10.1007/978-3-319-06692-9_27,

reliable statistical analyses. In modern missing data handling, two broad taxonomies dominate recent literature: (1) parametric and nonparametric methods; (2) single and multiple imputation (Little and Rubin 2002). In parametric methods, likelihood-based procedures (e.g. the EM algorithm) are applied starting from a distributional assumption on the missing part of data in order to obtain estimates of missing values according to their generating model. Nonparametric missing data procedures are model-free methods that do not require distributional assumptions on the data. Imputation is thus performed by learning from the data structure at hand. While single imputation is concerned with the problem of assigning a single value to each missing datum, multiple imputation aims at accounting for the uncertainty implicit in the fact that the imputed values are not the actual values. This is achieved by deliberately adding sources of error during the imputation process, thus giving rise to a multitude of estimates for each missing datum from which standard errors and confidence intervals can be computed.

Among nonparametric single imputation techniques, methods based on computer-intensive iterative statistical procedures seem the most promising in producing reliable imputations. In this work, attention is specifically drawn to three different logics of imputing, based on the use of random forest (Stekhoven and Bühlmann 2012), iterative PCA (Nora-Chouteau 1974) and the forward (Ferrari et al. 2011) procedures respectively. In particular, Stekhoven and Bühlmann's method (*missForest*, Stekhoven and Bühlmann 2012) is an iterative technique for the imputation of continuous and/or categorical data based on a random forest, which is a random classifier introduced in the context of machine learning (Breiman 2001). The Iterative Principal Component Analysis (*IPCA*) (Nora-Chouteau 1974; Greenacre 1984) imputes missing values simultaneously by an iterative use of the principal component analysis. It has recently been subject to renewed interest as it is at the core of the multiple imputation technique with PCA, a component of a more general methodology (*missMDA*) introduced by Josse et al. (2011) for imputing missing data with multivariate data analysis techniques. The Forward Imputation (*ForImp*) by Ferrari et al. (2011) is a sequential procedure designed for extracting a latent dimension from ordinal variables in the presence of missing data. The nonlinear PCA (NLPCA) and the nearest-neighbour imputation (NNI) method are alternated in a step-by-step process that recovers the missing ordinal categories and then extracts the latent dimension.

Although grounded on distinct logics, *IPCA* and *ForImp* both depend on factorial methods, which are widely used also in contexts where the incompleteness of information requires a different approach from a purely imputation perspective. This is the case of data fusion and data grafting procedures which, allowing databases from different sources to be combined together by recovering mismatches of variables and/or units, can be regarded as special cases of missing data imputation (Saporta 2002; Aluja-Banet et al. 2007).

This work has two objectives. The first is to re-formulate *ForImp* as an imputation technique for quantitative variables. Indeed, in its original version *ForImp* was not expressly developed as an imputation method, but rather as a method for missing data handling in NLPCA in alternative to commonly used standard options, such as passive treatment (Ferrari et al. 2011). The second is to offer a critical comparison of the thus revised *ForImp* with *missForest* and *IPCA* based on various configurations of quantitative data as given by different patterns of skewness and correlation of variables.

2 The Forward Imputation for Quantitative Variables

Since our goal is to re-design the *ForImp* method as a pure imputation technique, we specifically focused on missing data handling in the case of quantitative variables. Accordingly, we relied on the traditional linear PCA to build up the new version of the method, which will be termed Forward Imputation with the PCA (*ForImpPCA*). Although the logic behind *ForImpPCA* is very similar to the original *ForImp* (Ferrari et al. 2011), it is characterized by several features. Since the dimensionality reduction problem is not the primary concern, the PCA method is merely involved as a tool functional to the imputation exercise. In particular, the same number of principal components are extracted as the number of variables in the starting data matrix, in order to produce convenient synthesis indicators that are more or less related to the original variables.

The *ForImpPCA* method assumes an $n \times p$ quantitative data matrix $\mathbf{X}$ with x_{ij} values ($i = 1, \ldots, n, j = 1, \ldots, p$) with at least p rows free of missing values and the other $n - p$ rows with at most $p - 1$ missing values ($n > p, p \geq 2$). Then, in a preliminary phase, data are prepared by splitting $\mathbf{X}$ into a complete submatrix $\mathbf{X}_0$ and K submatrices $\mathbf{X}_k$, where index k denotes the number of missing values potentially contained in each row ($k = 1, \ldots, K \leq p - 1$). Should k identify a submatrix without elements, we would set: $\mathbf{X}_k = \mathbf{X}_{0 \times p}$, and then jump to the submatrix corresponding to the subsequent k. The core steps of the *ForImpPCA* algorithm are the following:

- Set $k = 1$.

1. *PCA step*: Perform a PCA on the complete $\mathbf{X}_{k-1}$ from either its own variance-covariance matrix or correlation matrix, assumed of full rank, and obtain eigenvalues $\lambda_s^{(k-1)}$ and eigenvectors $\boldsymbol{\omega}_s^{(k-1)}$ with generic element $\omega_{js}^{(k-1)}$ from it, ($j, s = 1, \ldots, p$).
2. *PPC step*: Compute so-called Pseudo Principal Components (PPC) for both the complete $\mathbf{X}_{k-1}$ and the incomplete $\mathbf{X}_k$ by involving only common variables without missing values and eigenvectors obtained at the previous step, in order to obtain artificial variables free of missing values for both complete and incomplete units. We denote by ι the set formed by those among the k-combinations of the p indices of variables containing missing values in the rows of $\mathbf{X}_k$. Then PPCs, denoted by $\tilde{C}$, are given by linear combinations of the original variables outside the ι set with coefficients given by the element in the corresponding eigenvectors:

$\tilde{C}_{s(\iota)}^{(k)} = \sum_{\substack{l=1 \\ l \notin \iota}}^{p} \omega_{ls}^{(k-1)} X_l^{(k)}$ for submatrix $\mathbf{X}_k$, and: $\tilde{C}_{s(\iota)}^{(k-1)} = \sum_{\substack{l=1 \\ l \notin \iota}}^{p} \omega_{ls}^{(k-1)} X_l^{(k-1)}$ for submatrix $\mathbf{X}_{k-1}$, $s = 1, \ldots, p$.

3. *Donors' selection step*: PPCs represent common, complete information for the comparison of complete and incomplete units. PPCs are accordingly used to compute the Minkowski distance d_r of order r, $(r \geq 1)$, between each incomplete unit $u_i^{(k)}$ in $\mathbf{X}_k$ and the complete units $u_c^{(k-1)}$ in $\mathbf{X}_{k-1}$:

$$d_r(u_i^{(k)}, u_c^{(k-1)}) = \left\{ \sum_{s=1}^{p} \left| (\tilde{c}_{s(\iota),i}^{(k)} - \tilde{c}_{s(\iota),c}^{(k-1)}) w_s^{(k-1)} \right|^r \right\}^{1/r}, \quad c = 1, \ldots, n_{k-1}, \quad (1)$$

where the weights: $w_s^{(k-1)} = \sqrt{\lambda_s^{(k-1)} / \sum_{m=1}^{p} \lambda_m^{(k-1)}}$, being the square root of normalized eigenvalues, are used to strengthen (weaken) the role of PPCs derived from principal components with higher (smaller) variances. Thereafter, donors are detected as an opportune percentage of the complete units nearest to a specific incomplete unit. Formally, donors $u_{\delta,i}^{(k)}$ for unit $u_i^{(k)}$ are given by the first $q100\,\%$ complete units $u_c^{(k-1)}$ corresponding to the q-th quantile $d_{q,i}$ of the distances d_r, $(0 < q < 1; i = 1, \ldots, n_k)$.

4. *Imputation step*: Once the donors have been identified, their values in the original data matrix are used for imputation by means of a weighted average. Weights are given by the reciprocals of the distances between donors and each specific incomplete unit in order to put more (less) emphasis on less (more) distant donors. For a missing value on variable X_j and unit $u_i^{(k)}$ the imputed value is therefore given by:

$$\tilde{x}_{ij}^{(k)} = \frac{\sum_{\delta=1}^{n_\delta} x_{\delta j}^{(k-1)} \frac{1}{d_{\delta i}}}{\sum_{\delta=1}^{n_\delta} \frac{1}{d_{\delta i}}}, \qquad \forall j \in \iota,$$

where n_δ is the total number of donors for $u_i^{(k)}$ and $d_{\delta i}$ is the distance between the δ-th donor and unit $u_i^{(k)}$ as computed in step 3.

- Set $k = k + 1$ and jump to the *PCA step* until $\mathbf{X}$ is completely imputed.

3 A Data Structure-Driven Simulation Study for Comparison

A Monte Carlo simulation study was carried out to assess the performance of the *ForImpPCA* method by comparing it with *missForest* and *IPCA* in the presence of different data patterns and Missing Completely At Random (MCAR) generated missing values (Little and Rubin 2002). In this study, attention was specifically addressed to skewed data structures, in order to verify whether and to what extent

Table 1 Experimental conditions in the simulation study (1,000 runs for each scenario)

Common set of experimental conditions:	
– Number of variables in $\mathbf{X}$	$p = 3; 5; 10$
– Number of units in $\mathbf{X}$	$n = 500; 1{,}000$
– Percentage of MCAR missing values	$5\,\%; 10\,\%; 20\,\%$
Data generation from $N_p(\mathbf{0}, \mathbf{R})$:	
– Correlation coefficient	$\rho = 0; 0.3; 0.7$
Data generation from $\mathrm{MSN}_p(\boldsymbol{\Omega}, \boldsymbol{\alpha})$:	
– Skewness parameter	$\alpha = 1; 4; 10; 30$
– Correlation parameter in $\boldsymbol{\Omega}$	$\omega = 0; 0.5; 0.8$

skewness could affect the imputation capability of the three methods. Accordingly, complete data matrices were randomly generated from both the multivariate normal (*MVN*) distribution and the multivariate skew normal (*MSN*) family of distributions, the latter being an extension of the multivariate normal distribution allowing for the presence of skewness (Azzalini and Dalla Valle 1996; Azzalini and Capitanio 1999). To better understand the role of *MSN* parameters involved in the simulation study, it is worth recalling that a p-dimensional random vector $\boldsymbol{X}$ is $\mathrm{MSN}_p(\boldsymbol{\Omega}, \boldsymbol{\alpha})$ distributed if its density function (d.f.) can be expressed as:

$$f(\mathbf{x}; \boldsymbol{\Omega}, \boldsymbol{\alpha}) = 2\phi_p(\mathbf{x}; \boldsymbol{\Omega})\Phi(\boldsymbol{\alpha}^t \mathbf{x}), \tag{2}$$

where: $\phi_p(\mathbf{x}; \boldsymbol{\Omega})$ is the $N_p(\mathbf{0}, \boldsymbol{\Omega})$ d.f., with $\boldsymbol{\Omega}$ a correlation matrix of full rank; $\Phi(\cdot)$ is the $N(0, 1)$ distribution function, and $\boldsymbol{\alpha}$ is a p-dimensional parameter vector regulating the skewness. In particular, if: $\boldsymbol{\alpha} = \mathbf{0}$, then the d.f. (2) reduces to a multivariate normal: $\boldsymbol{X} \sim N_p(\mathbf{0}, \boldsymbol{\Omega})$.

We generated data from both *MVN* and *MSN* distributions according to the simulation settings reported in Table 1, for a total number of, respectively, 54 scenarios in the case of *MVN*, and 216 in the case of *MSN*. Specifically, in each scenario a complete data matrix $\mathbf{X}^*$ was generated from an *MVN* or an *MSN* distribution, and then 1,000 matrices $\mathbf{X}_t$ were formed from it with a given percentage of MCAR missing data, $t = 1, \ldots, 1{,}000$ (Table 1). Then, *missForest*, *IPCA* and *ForImpPCA* were applied with the following options. For *missForest*, the maximum number of iterations was increased from 10 (the default in the R library *missForest*, Stekhoven and Bühlmann 2012) to 50. For *IPCA*, the number of extracted principal components was fixed to the maximum possible, i.e. $p - 2$, with $p \geq 3$ (R library *missMDA*, Josse et al. 2011). For *ForImpPCA*, we considered the Euclidean distance ($r = 2$ in formula (1)), and the first q-th quantile of such distances with $q = 0.05; 0.1; 0.15; 0.2$ in order to detect donors.

Simulation results were synthesized, and comparisons among the three methods performed, through the Relative Mean Square Error (*RMSE*) computed as a function of the difference between the complete data matrix $\mathbf{X}^*$ and the imputed data matrix $\tilde{\mathbf{X}}_t$ at the t-th simulation run: $RMSE_t = \sum_{j=1}^{p} \frac{1}{n\sigma_j^2}(\mathbf{x}_j^* - \tilde{\mathbf{x}}_{j,t})^t(\mathbf{x}_j^* - \tilde{\mathbf{x}}_{j,t})$, where $\mathbf{x}_j^*$ is the j-th column vector of $\mathbf{X}^*$, $\tilde{\mathbf{x}}_{j,t}$ is the j-th column vector of $\tilde{\mathbf{X}}_t$, and σ_j^2 is the variance of the j-th variable in $\mathbf{X}^*$, ($t = 1, \ldots, 1{,}000$). Codes of *ForImpPCA* were implemented and simulations performed in the R environment (R Development Core Team 2012).

3.1 Simulation Results

Figure 1 shows line plots of *RMSE* median values, plotted against the percentages of MCAR missing values (5 %; 10 %; 20 %), obtained for the three methods (*ForImp-PCA* with $q = 0.1$) under a subset of the scenarios considered, with the number of variables varying ($p = 3; 5; 10$), number of units fixed to $n = 1{,}000$, and data generated from *MVN* (with $\rho = 0; 0.3; 0.7$) and *MSN* (with $\omega = 0; 0.5; 0.8$ and $\alpha = 4; 30$). The other omitted results exhibit the same trend. Two remarks are worth making. First, as expected, *RMSE* increases as the complexity of the data increases, that is, the number of variables and the proportion of missing values. Moreover, *ceteris paribus*, *RMSE* tends to decrease as the correlation between variables increases, thus indicating that the imputation process is more effective if variables are closely related. Second, the three methods produce very similar *RMSE* values with a low percentage of missing values, whereas they display a noticeably different performance in the presence of higher proportions of missing data. In particular, *IPCA* turns out to be the best imputation method in the case of normally distributed data (1st row of panels, Fig. 1), and highly correlated variables (2nd and 3rd rows, last column, Fig. 1), while *ForImpPCA* tends to perform best with skew distributions and variables with small/medium correlations (2nd and 3rd rows, first two columns, Fig. 1). Finally, *missForest* tends to produce the highest *RMSE* values in most scenarios considered, although it must be remembered that it is designed especially for imputation in the case of mixed-type data.

Figure 2 displays a more detailed picture of the results achieved in the specific scenarios with $p = 5$ variables, $n = 1{,}000$ units, and 20 % of missing data. In addition to *missForest* and *IPCA*, boxplots of *RMSE* distributions are shown also for *ForImpPCA* with different donors' quantiles ($q = 0.05; 0.1; 0.15; 0.2$), in order to check their effect on the imputation task. The above remarks concerning *IPCA* and *ForImpPCA* can now be understood more clearly. The best performance of *IPCA* can be observed in the first row of panels, while 2nd to 4th rows in the first two columns highlight the best performance of *ForImpPCA*. Moreover, a comparison among boxplots of *ForImpPCA* pertaining to different donors' quantiles suggests that, overall, having a high percentage of donors is not a convenient choice if variables are highly correlated (last column of panels, Fig. 2), while having few donors is not suitable if variables are uncorrelated or little correlated (1st column, Fig. 2). This would seem to indicate that a good choice is to select donors that correspond to the first $q = 0.1$ or $q = 0.15$ quantile of Euclidean distances.

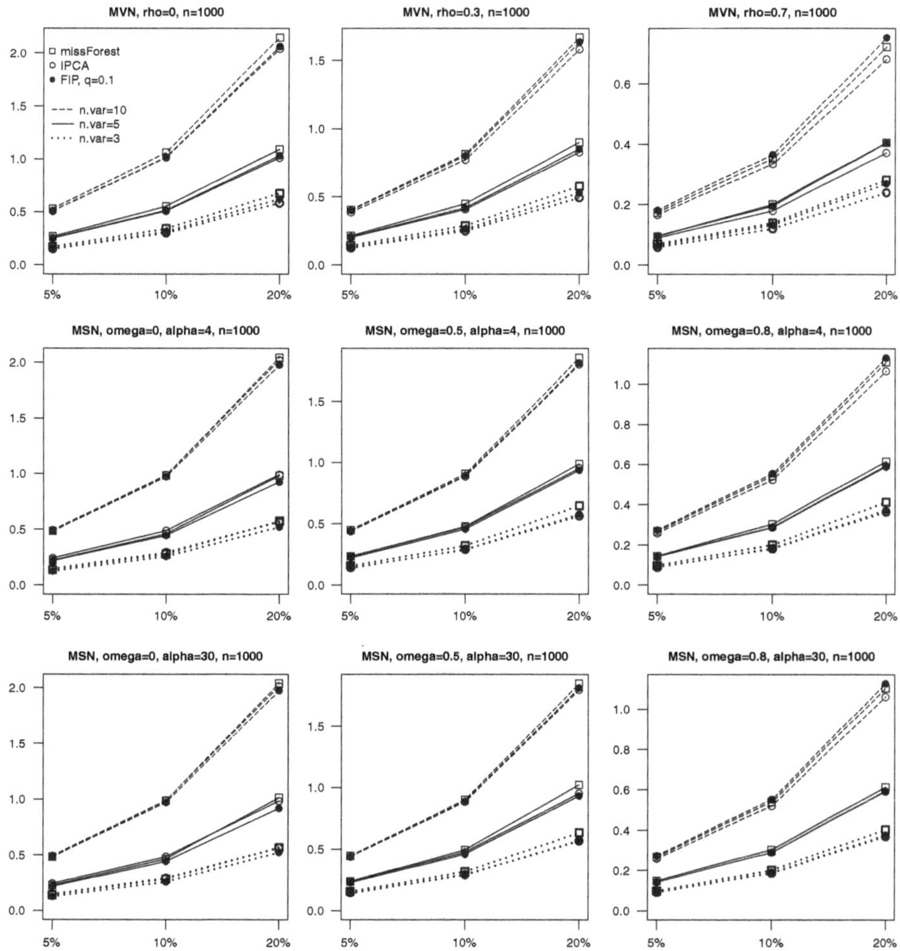

Fig. 1 Line plots of *RMSE* median values of *missForest*, *IPCA*, and *ForImpPCA* (FIP), plotted against percentages of MCAR missing data with $p = 3; 5; 10$ variables and $n = 1{,}000$ units

4 Discussion and Future Work

In the light of our current results, *ForImpPCA* seems to be promising as a single imputation method. It performs best with skew distributions and variables which are not highly correlated, characteristics typically encountered in real data. Nonetheless, further studies would help investigate the performance of *ForImpPCA* more thoroughly. For example, the results obtained indicate that it would be useful to examine *ForImpPCA*, and to then compare it with other methods, in the presence of data contaminations such as multivariate outliers, or a different generating mechanism of missing data, such as MAR (Little and Rubin 2002). From a methodological point of

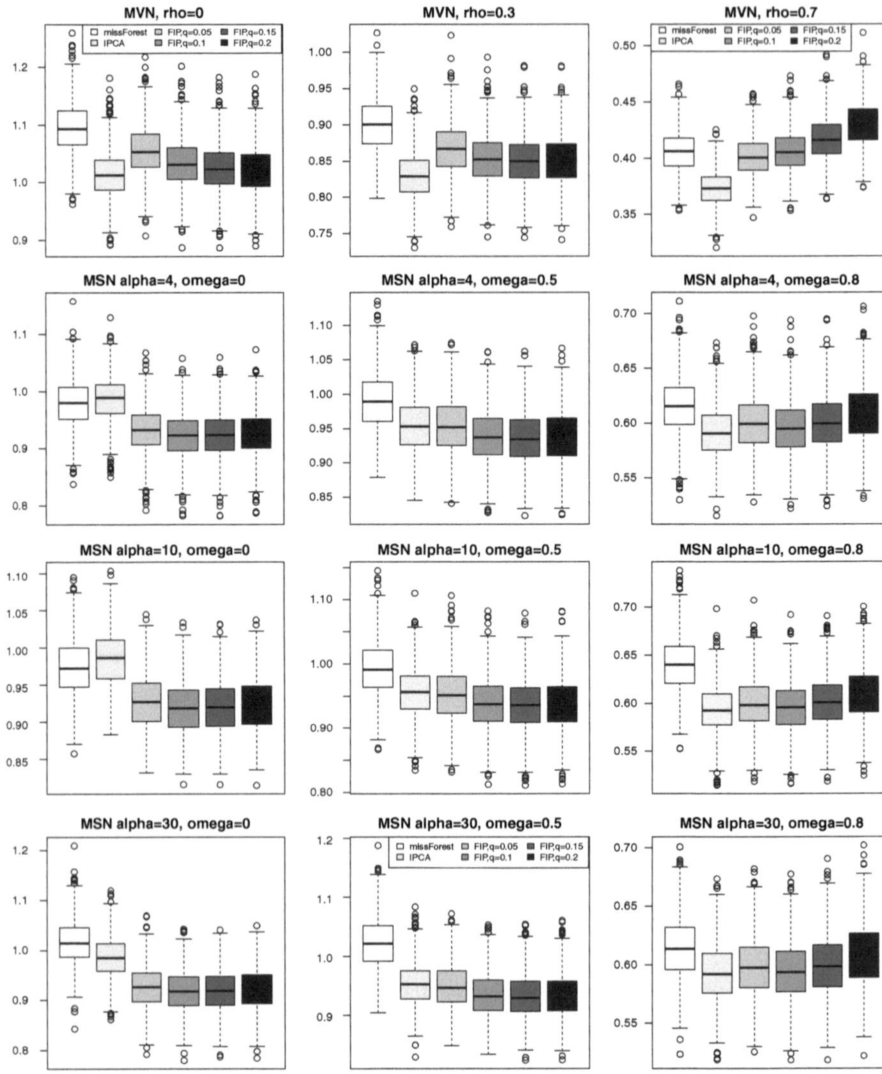

Fig. 2 Boxplots of *RMSE* distributions of *missForest*, *IPCA*, and *ForImpPCA* (FIP) with $q = 0.05, 0.1, 0.15, 0.2$ donors' quantile, under the scenarios with $p = 5$ variables, $n = 1{,}000$ units, and 20 % of MCAR missing data

view, the potentially optimal properties of *ForImpPCA* along with its performance in cases of more complex data structures need to be further investigated in order to highlight the capacity of *ForImpPCA* to manage different skew distributions better.

References

Aluja-Banet, T., Daunis-i-Estadella, J., & Pellicer, D. (2007). GRAFT, a complete system for data fusion. *Computational Statistics & Data Analysis, 52*, 635–649.

Azzalini, A., & Capitanio, A. (1999). Statistical applications of the multivariate skew normal distribution. *Journal of the Royal Statistical Society: Series B, 61*(3), 579–602.

Azzalini, A., & Dalla Valle, A. (1996). The multivariate skew-normal distribution. *Biometrika, 83*(4), 715–726.

Breiman, L. (2001). Random forests. *Machine Learning, 45*(1), 5–32.

Ferrari, P. A., Annoni, P., Barbiero, A., & Manzi, G. (2011). An imputation method for categorical variables with application to nonlinear principal component analysis. *Computational Statistics & Data Analysis, 55*, 2410–2420.

Greenacre, M. (1984). *Theory and applications of correspondance analysis*. London: Academic.

Josse, J., Pagès, J., & Husson, F. (2011). Multiple imputation in principal component analysis. *Advances in Data Analysis and Classification, 5*, 231–246.

Little, R. J. A., & Rubin, D. B. (2002). *Statistical analysis with missing data* (2nd ed.). New York: Wiley.

Nora-Chouteau, C. (1974). Une méthode de reconstitution et d'analyse de données incomplètes. Ph.D. thesis, Université Pierre et Marie Curie.

R Development Core Team (2012). *R: A language and environment for statistical computing*. Vienna: R Foundation for Statistical Computing.

Saporta, G. (2002). Data fusion and data grafting. *Computational Statistics & Data Analysis, 38*, 465–473.

Stekhoven, D. J., & Bühlmann, P. (2012). MissForest - non-parametric missing value imputation for mixed-type data. *Bioinformatics, 28*(1), 112–118.

Changes in Japanese EFL Learners' Proficiency: An Application of Latent Rank Theory

Naoki Sugino, Kojiro Shojima, Hiromasa Ohba, Kenichi Yamakawa, Yuko Shimizu, and Michiko Nakano

Abstract In the present study the authors compared achievements of Japanese learners of English as a foreign language (EFL) at the end of 6 years of formal instruction based on their test performance in the English section of the National Center Tests for University Admissions administered in 1990, 1997, and 2004. Direct comparisons were made possible by equating the scales of these three tests using the common subject design. In addition to 121 Japanese EFL learners who took the tests prepared by the researchers for the equating purpose, 10,000 cases were randomly sampled from each year's actual test-takers. Their test performance was rendered into analysis based on the Latent Rank Theory (Shojima, Neural

---

This study is supported by the Grant-in-Aid for Scientific Research (B) (22320114) from the Japan Society for the Promotion of Science (JSPS).

N. Sugino (✉) • Y. Shimizu
Ritsumeikan University, Kusatsu, Shiga, 525-8577, Japan
e-mail: gwisno@is.ritsumei.ac.jp; yukos@ec.ritsumei.ac.jp

K. Shojima
National Center for University Entrance Examinations, Meguro-ku, Tokyo, 153-8501, Japan
e-mail: shojima@rd.dnc.ac.jp

H. Ohba
Joetsu University of Education, Joetsu, Niigata 943-0815, Japan
e-mail: hohba@juen.ac.jp

K. Yamakawa
Yasuda Women's University, Asaminami-ku, Hiroshima, 731-0153, Japan
e-mail: kyamakaw@yasuda-u.ac.jp

M. Nakano
Waseda University, Shinjuku-ku, Tokyo, 169-8050, Japan
e-mail: nakanom@waseda.jp

D. Vicari et al. (eds.), *Analysis and Modeling of Complex Data in Behavioral and Social Sciences*, Studies in Classification, Data Analysis, and Knowledge Organization, DOI 10.1007/978-3-319-06692-9_28,

test theory: A latent rank theory for analyzing test data (DNC Research Note, 08-01). Retrieved from http://www.rd.dnc.ac.jp/~shojima/ntt/Shojima2008RN08-01.pdf, 2008; Shojima, Neural test theory. In: K. Shigemasu, A. Okada, T. Imaizumi, & T. Hoshino (Eds.), *New trends in psychometrics*, pp. 417–426, 2009. Tokyo: University Academic Press; Shojima, Neural test theory. In: M. Ueno & K. Shojima (Eds.), *Gakushu Hyoka no Shin-choryu [New trends in evaluation of learning]*, pp. 83–111, 2010. Tokyo: Asakura Shoten). The results indicate that the test-takers in 1990 are unique both in their membership to the latent ranks and in the knowledge that characterizes the high-achievers. Implication of the present study will be discussed in the last section.

Keywords EFL (English as a foreign language) proficiency • Language education policy • Latent rank theory • The National Center Test for University Admissions

1 Introduction and Background

The year 1989 witnessed a major shift of focus in the Japanese EFL education towards communicative language use of the target language. This shift was explicitly stated in the *Course of Study*, a set of national standards governing curriculum development and authorized textbook compilation for the primary and secondary education in Japan. The focal point of the 1989 version of *Courses of Study* was the development of communicative competence in EFL subjects. The subsequent revisions in 1998–1999 and in 2003–2004 took the same route, reforming the subjects, listing concrete situations for the target language use and language functions, and incorporating EFL as compulsory subjects into the primary school curriculum.

However, possible impact of these nation-wide curriculum changes on Japanese EFL learners is under-researched. In order to explore such an impact, a group of researchers (Yoshimura et al. 2005) turned to the English section of the National Center Tests for University Admissions (NCTs) for the period 1990–2004. Every year, about half a million prospective 18-year-olds take the NCT, for most of the national and municipal universities as well as some private ones require the NCT score for admission. It can be safely stated, therefore, that an NCT is a large-scale and high-stake test, and as such, is designed rather as an achievement test at the end of the secondary education (see Watanabe 2013 for a more detailed account of the NCT). The researchers' contention was that, by comparing the ability estimates as measured by the NCTs, the impact of the curriculum changes can be explicated as reflected in attained proficiency of the learners. As a result, the study revealed that there was a sharp decline of the ability estimate in 1997, which coincided with the year when the first cohort who had been taught under the 1989 version of *Course of Study* took the university entrance examinations. The study, however, compared the overall ability estimate and detailed analysis of the decline was not within its scope of investigation.

Along with these successive curriculum revisions outlined above, the Japanese Ministry of Education, Culture, Sports, Science, and Technology has advocated to establish "natural learning attainment targets in the form of 'can-do lists'" (Commission on the Development of Foreign Language Proficiency 2011). Consequently, the prevalence is observed throughout Japan of such lists, which are similar to or adapted from Common European Framework of Reference for Languages (henceforth, CEFR) (Council of Europe 2001). Setting goals, giving feedbacks, and/or providing guidelines for material development in terms of ability descriptors can be taken positively as a shift towards more qualitative characterization of a learner's achievement. However, some researchers (e.g., Alderson et al. 2006; Long et al. 2012) have pointed out that such scales tend to lack sufficient consistency and specification necessary for test construction and for providing learners with specific guidance for their further study. Efforts to bridge the gap are urgently called for in order to make the best use of such scales to benefit the learners.

What is needed is to explicate and monitor what learners can and cannot do with their attained proficiency based on empirical data. To attain this goal, the authors will first equate the scales of the three NCTs administered in 1990, 1997, and 2004, employing the Latent Rank Theory (henceforth, LRT), a test theory newly developed by Shojima (2008, 2009, 2010). Based on the yielded output, we will then describe and compare what learners at each of the latent ranks can do by analyzing specifications of those items that they are able to answer correctly. Also compared is the distribution of test-takers across the latent ranks in order to explicate the possible causes of the decline observed in Yoshimura et al. (2005).

2 The Present Study

The objective of the present study is twofold; (a) to see if any decline in achievements among the three NCTs is observed, and (b) to account for the decline, if observed, in terms of the aspects of test-takers' attained proficiency and of their membership to different latent ranks.

2.1 Test Booklets and Participants

An NCT comprises six testlets and the basic overall structure has remained the same since its inception: the first three testlets consist of discrete items, which test language elements (Domains II, III, and IV) and the latter three are to test reading comprehension (Domain I). The items were classified into the four domains for further analysis as displayed in Table 1. As is shown, although the overall

Table 1 Domains of knowledge/skills and the number of items measured by the NCTs

Domain	Aspects	1990	1997	2004	Total
I	Reading skill	19	17	17	53
II	Vocabulary and grammar	19	18	13	50
III	Phonology	7	6	6	19
IV	Function	6	6	11	23
	Total	51	47	47	145

structure was maintained, minor modifications were made in the numbers of items distributed among the domains. Furthermore, the communicative shift discussed earlier is reflected in the relative proportion of the items pertaining to the discourse and language functions.

Each year's test was divided into two parts, Testlets 1-3 and 4-6, and compiled into six versions of the booklets by combining the parts from different years. For instance, one of the booklets, called Form A, was comprised of the first three testlets of the 1990 test and the latter three testlets of the 1997 test, and another one, Form C, consisted of the first half of the 1997 test and the latter half of the 1990 test.

One hundred and twenty-one undergraduates and graduates at three universities in Japan participated in the study. The six versions of test booklets were randomly distributed at all of the universities so that equal number of participants would answer each booklet. In addition to these 121 students, 10,000 cases from the actual test-takers of 1990, 1997, and 2004 tests were randomly sampled. Altogether, 30,121 sets of responses are rendered into an LRT-based analysis.

2.2 *Analysis*

Three types of information are yielded in an LRT-based analysis. One type of information is about the test items called Item Reference Profile (IRP). The IRP of a particular test item is the probability with which learners at a given latent rank can correctly answer the item. In the present study, as will be discussed later, IRPs are used to describe what have been attained by the learners at a certain latent rank.

The second type of information is Test Reference Profile (TRP), which characterizes the test by expressing the number of correct responses by learners at each latent rank. This index is important in confirming the scale obtained is actually ordinal.

The third set of indices pertains to the test takers. Rank Membership Profile (RMP) is the probability with which each test-taker belongs to each of the latent ranks. The sum of the RMPs is the Rank Membership Distribution (RMD), which shows the frequencies of the test-takers at each latent rank.

Table 2 Goodness-of-fit indices when the number of the latent ranks is set at eight

Fit index	Value
Chi-square	60, 360.157
EDF (effective degrees of freedom)	2, 798.638
IFI (incremental fit index)	0.800
CFI (comparative fit index)	0.799
RMSEA (root mean square error of approximation)	0.026
AIC (Akaike information criterion)	54, 762.881
BIC (Bayesian information criterion)	31, 497.864

3 Results

The number of the latent ranks can be determined by examining the goodness-of-fitness indices and by considering the content validity of the obtained sets of items. Generally speaking, fitness improves as the number of latent ranks increases. Table 2 shows the model fit relatively well when the number of the latent rank was set at eight. From the viewpoint of information criteria, predictive efficiency with eight latent ranks stood comparison with those obtained with other numbers of ranks. Furthermore, each rank was uniquely characterized with the obtained sets of items, thus, with the content validity also taken into consideration, the number of latent ranks was fixed at eight. It was confirmed that the obtained latent rank scale was actually ordinal because the TRP increases monotonically and most of the IRPs also increased.

3.1 Distribution of the Test-Takers

Figure 1 shows the percentages of each year's test-takers at each rank. As can be seen, the RMDs of the test-takers of 1997 and 2004 were almost identical except at Rank 8, where slightly less 2004 test-takers were placed. Until Rank 5, frequencies of the 1997 and the 2004 test-takers were higher than that of the test-takers in 1990, but from Rank 6, this tendency was reversed.

3.2 Comparison of the Rank Characteristics

Each of the 145 items was assigned with the IRPs and other statistical information, such as the ratio of correct responses and the item-total Pearson correlation, on the equated scale. As a way of illustration, let us compare the two items in Table 3, both form the 1990 test. Item 47's IRP exceeded 0.700 at Rank 7, while Item 40,

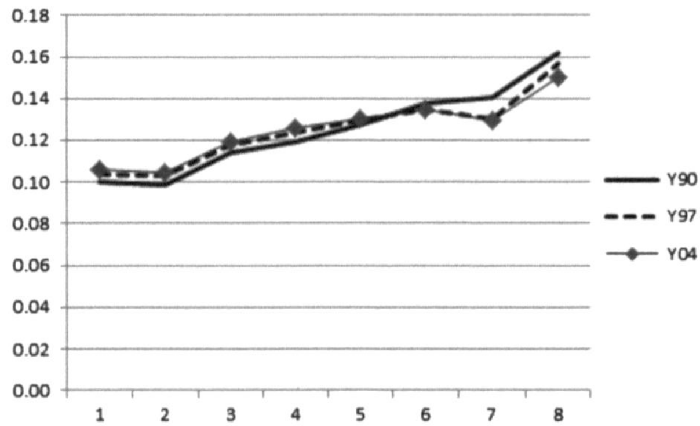

Fig. 1 Rank Membership Distributions (RMDs) of the test-takers to the latent ranks

Table 3 Sample items and their IRPs

Item	R1	R2	R3	R4	R5	R6	R7	R8
y90i47	0.271	0.227	0.201	0.251	0.394	0.566	0.714	0.817
y90i40	0.664	0.809	0.901	0.919	0.933	0.962	0.983	0.995

at Rank 2. Thus, Item 47 was considered to be within the knowledge and ability of learners at Rank 7 and Rank 8, while Item 40 was within the attained level of learners at Rank 2 or higher.

Sorting the items according to the ratios of correct responses produced groups of items with similar IRPs; those with the IRP of 0.700 or higher at a given rank were grouped together as reflecting the knowledge and skills the learners at that rank and below had attained. By specifying the knowledge and skills required for the group of items, the attained level of proficiency at that rank was described.

Due to the limitation of space, however, it is impossible to present the detailed list of can-dos for each of the latent ranks. For the purpose of the present study, it would suffice to see what items differentiate high-achievers and low-achievers in each year's test. Figure 2 shows the distribution of test items with IRPs of 0.700 or higher by the domains.

It can be observed form Fig. 2 that 32 items out of 53 in Domain I concentrated at Ranks 3–5, while 30 items out of 50 in Domain II at Ranks 6–8. This tendency was most apparent in 1990: In the 1990 test, 16 out of 19 Domain I items were within the achieved level if the learners are at Rank 5, although the Rank 5 learners could successfully handle only 2 of 17 Domain II items. Although this tendency became less clear in 1997 and 2004, the majority of Domain II items (13 of 18 in 1997 and 8 of 13 in 2004) were beyond Rank 5. It should also be noted that in the 2004 test, 8 items out of 11 in Domain IV, which requires discourse competence, characterized learners at Ranks 6–8.

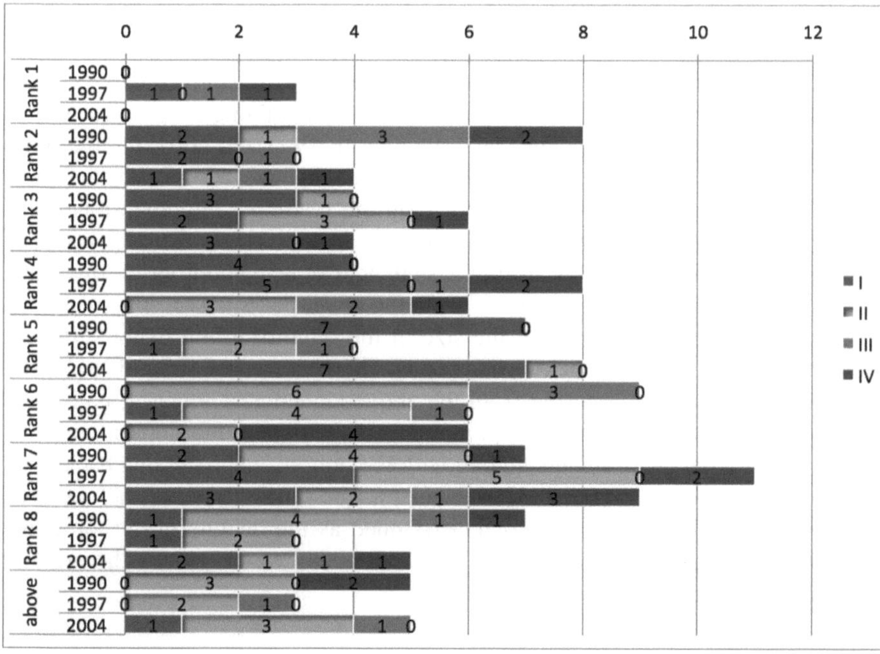

Fig. 2 Distribution of test items with IRPs of 0.700 or higher by the domains

4 Discussion

From the obtained results, the decline in overall proficiency observed in 1997, as reported in Yoshimura et al. (2005), can be accounted for in two terms; in terms of the decreasing proportion of the high-achievers and of the differences in what learners can do at a certain latent rank. From the viewpoint of the test-taker distribution, the decline can be explained by the smaller proportion of the high-achievers (those at Rank 6 and above) in 1997 and 2004, as indicated by the differences among the RMDs of the three test administrations. This comes in accordance with the changes in the test-takers' populations across the administrations. Despite the shrinking 18-year-old population, the NCTs have been taken by the constant numbers of test-takers. This is accounted for by the increase of the participating universities National Center for University Entrance Examinations (2013): In 1990, 95 national, 35 municipal, and 16 private universities participated in the NCT. The number of participating municipal universities increased to 53, and that of private universities to 152 in 1997. The number of participating private universities more than doubled in 2004, when 387 private universities participated. Furthermore, nine municipal and 88 private 2-year colleges newly joined the NCT in the same year. Thus, the growing range of learners resulted in the decrease in the proportion of high-achievers.

In terms of differences in achieved proficiency, the overall decline can be accounted for by less secure bottom-up processing, which is consequent on less accurate understanding of texts. The lower rank learners are characterized by their improving proficiency in reading comprehension skills, although this characteristic of the lower rank learners becomes less apparent in 1997 and 2004. On the other hand, the common characteristics of those high-achievers across the three test administrations are that they display, not only the reading comprehension skills, but also their mastery in vocabulary and grammar. In addition, in 2004, their ability to develop coherent passages by arranging sentences or using appropriate discourse markers differentiate the high-achievers from the lower peers.

The authors are well aware that the size of the monitor group is not sufficient in equating the scales, and acknowledge that there are some limitations in the following extrapolations. Inferred from the above tendencies, however, is that the learners first rely on the top-down processing in getting at the correct answers, and then substantiate their understanding with the bottom-up processing backed up by their knowledge of grammar and vocabulary. The proficiency decline observed in Yoshimura et al. (2005) can thus be explained as the decline in the top-down processing among the test-takers in 1997 and 2004, which led to the smaller population in the higher ranks in these 2 years.

5 Conclusion and Further Research Directions

As an attempt to explicate the possible causes of the decline in overall proficiency observed and reported in Yoshimura et al. (2005), the present study equated the scales of three NCTs. The results show that the decline can be explained in terms of test-takers' distribution across eight latent ranks and in terms of their attained level of knowledge and subskills needed for accurate understanding of texts. In this respect, one of the implications of the present study is that this itself is a demonstration of how LRT can be applied in clarifying what cannot be explicated otherwise.

For further research, we are planning first to increase the number of the monitor test-takers to overcome the shortcomings of the present equating. Also, we are to add other types of scales to be equated, that is, CEFR descriptors and listening comprehension tests, so that a learner's achievement can be assessed from a wider range of angles. In addition, underway is relating the NCT-based descriptors to acquisition of linguistic features, such as *wh*-constructions, unaccusatives and unergatives. By so doing, what is envisioned is creating a common ground where each of the three assessment tools supplements the others in order to provide practically useful information for language teachers and students.

References

Alderson, J. C., Figueras, N., Kuijper, H., Nold, G., Takala, S., & Tardieu, C. (2006). Analysing test of reading and listening in relation to the common European framework of reference: The experience of the Dutch CEFR construct project. *Language Assessment Quarterly, 3*, 3–30.

Commission on the Development of Foreign Language Proficiency (2011). Five proposals and specific measures for developing proficiency in English for international communication.

Council of Europe (2001). *Common European framework of reference for languages: Learning, teaching, assessment*. Cambridge: Cambridge University Press.

Long, M. H., Gor, K., & Jackson, S. (2012). Linguistic correlates of second language proficiency: Proof of concept with ILR 2-3 in Russian. *Studies in Second Language Acquisition, 34*, 99–126. doi:10.1017/S0272263111100519.

National Center for University Entrance Examinations (2013). Daigaku Nyushi Sentaa Youran 2013 [A bulletin of the National Center for University Entrance Examinations 2013]. Retrieved from http://www.dnc.ac.jp/modules/dnc/content0007.html.

Shojima, K. (2008). Neural test theory: A latent rank theory for analyzing test data (DNC Research Note, 08-01). Retrieved from http://www.rd.dnc.ac.jp/~shojima/ntt/Shojima2008RN08-01.pdf.

Shojima, K. (2009). Neural test theory. In K. Shigemasu, A. Okada, T. Imaizumi, & T. Hoshino (Eds.), *New trends in psychometrics* (pp. 417–426). Tokyo: University Academic Press.

Shojima, K. (2010). Neural test theory. In M. Ueno & K. Shojima (Eds.), *Gakushu Hyoka no Shin-choryu [New trends in evaluation of learning]* (pp. 83–111). Tokyo: Asakura Shoten.

Watanabe, Y. (2013). The National Center Test for University Admissions. *Language Testing, 30*, 565–573. doi:10.1177/0265532213483095.

Yoshimura, O., Shojima, K., Sugino, N., Nozawa, T., Sihmizu, Y., Saito, E., et al. (2005). Daigakunyushisentashiken kishutsu mondai wo riyoshita kyotsu-hikenshakeikaku niyoru eigogakuryoku no keinenhenka no chosa [Examination of changes over years in English proficiency based on the NCT results in the common subject design]. *Journal of Japan Association for Researches in Testing, 1*, 51–58.

Robustness and Stability Analysis of Factor PD-Clustering on Large Social Data Sets

Cristina Tortora and Marina Marino

Abstract Factor clustering methods were proposed to cluster large data sets. Among them factor probabilistic distance clustering (FPDC) shows interesting performance. The method is based on two main steps: a Tucker3 decomposition of the distance array and probabilistic distance (PD) clustering on the resulting factors. The aim of this paper is to apply FPDC on behavioral and social data sets of large dimensions, to obtain homogeneous and well-separated clusters of individuals. The scope is to evaluate the stability and the robustness of the method dealing with these data sets. Stability of results is referred to the invariance of results in each iteration of the method. Robustness is referred to the sensitivity of the method to errors in data. These characteristics of the method are evaluated using bootstrap resampling.

Keywords Bootstrapping • Factor clustering • Robustness • Stability

1 Introduction

Clustering methods easily deal with a large number of units, however, they can became unstable, or can fail in finding the true clustering structure, when the number of variables is large. Factor clustering methods were proposed to cluster large data sets; these methods are based on two main steps: linear transformation of original variables and clustering on transformed variables. The two steps are iterated until the convergence is reached. In literature this approach firstly appeared

C. Tortora (✉)
University of Guelph, 50 Stone Road East, Guelph, ON, Canada
e-mail: ctortora@uoguelph.ca

M. Marino
Università degli Studi di Napoli Federico II, via Cinthia 40, Napoli, Italy
e-mail: mari@unina.it

D. Vicari et al. (eds.), *Analysis and Modeling of Complex Data in Behavioral and Social Sciences*, Studies in Classification, Data Analysis, and Knowledge Organization,
DOI 10.1007/978-3-319-06692-9_29,

in 1984 as simple two-step clustering (Lebart et al. 1984), the two-steps were not re-iterated. The method was called *tandem analysis* (Arabie and Hubert 1994). However, the two-steps optimize different criteria and the first factor step can mask the real clustering structure (De Soete and Carroll 1994). Factor clustering methods are an improvement of two-step methods; the two steps optimize a common criterion iteratively until the convergence is reached. A wide range of factor clustering methods have been proposed (Vichi and Kiers 2001; Timmerman et al. 2010; Rocci et al. 2011; Ghahramani and Hinton 1997; McLachlan and Peel 2003). Among them factor probabilistic distance clustering (FPDC) (Tortora et al. 2013) presents interesting characteristics. The two main steps are: a Tucker3 decomposition (Kroonenberg 2008) of the distance array and probabilistic distance (PD) clustering on the Tucker3 factors (Ben-Israel and Iyigun 2008). A simulation study has shown that the algorithm has good performance dealing with: outliers, different number of elements in each cluster and variance changing among clusters Tortora et al. (2013). In this paper FPDC is applied on high dimensional continuous data sets, related to behavioral and social fields, to obtain homogeneous and well-separated clusters of individuals. The main object is to verify the stability and the robustness of FPDC dealing with these data sets. Different executions of the same method on the same data set can bring to different solutions; stability of results is referred to the invariance of results in different executions of the method. Robustness is referred to the sensitivity of the method to errors in data. The performance of an algorithm should not be affected by small deviation from the assumed model and it should not deteriorate due to noise and outliers. Bootstrap resampling technique is used to evaluate these aspects.

The paper has the following structure. In Sect. 2, PD-clustering method is briefly introduced. In Sect. 3, FPDC is presented. Section 4 contains a discussion about the stability and the robustness of clustering methods, and presents some way to test the issues. Section 5 is devoted to the application of FPDC on real and simulated data sets, and Sect. 6 offers some conclusions.

2 Probabilistic Distance Clustering

Given a set of n statistical units described by J continuous variables, PD-clustering is a non-hierarchical clustering method that assigns the n units to K clusters according to their belonging probability to the cluster.

We introduce PD-clustering (Ben-Israel and Iyigun 2008) according to Ben-Israel and Iyigun notation. Given X a generic data matrix, a center matrix C of generic element c_{kj}, and given K clusters that are assumed not empty, PD-Clustering is based on two quantities: the distance of each data vector x_i from the K vectors of cluster centers c_k, $d(x_i, c_k)$, and the probabilities for each point to belong to a cluster, $p(x_i, c_k)$ with $k = 1, \ldots, K$ and $i = 1, \ldots, n$. The relation between them is the basic hypothesis of the method, the probability of any point belonging to any cluster is assumed to be inversely proportional to the distance from the cluster centers. Let us consider the general term x_{ij} of X with $j = 1, \ldots, J$, the distances between each point and all centers can be computed according to different criteria. The probability $p(x_i, c_k)$ of each point belonging to a cluster is computed according to the following assumption: for any k, given i, the product between the distance $d(x_i, c_k)$ and the probability $p(x_i, c_k)$ is a constant $F(x_i)$ depending on x_i. For short we use $p_{ik} = p(x_i, c_k)$ and $d_k(x_i) = d(x_i, c_k)$. PD-clustering basic hypothesis is: $p_{ik} d_k(x_i) = F(x_i)$. We notice that as the distance of the point from the cluster center decreases, the probability of the point belonging to the cluster increases. The constant depends only on the point and does not depend on the cluster k. The quantity $F(x_i)$ is a measure of the closeness of x_i from all cluster centers. It measures the classifiability of the point x_i with respect to the centers c_k. If the point coincides with one of the cluster centers $F(x_i)$ is equal to zero; in that case the point belongs to the class with probability 1. If all the distances between the point x_i and the centers of the classes are equal to d_i, $F(x_i) = d_i/k$ and the probabilities belonging to each cluster are equal: $p_{ik} = 1/K$. The smaller the $F(x_i)$ value, the higher the probability for the point to belong to one cluster. Defining with JDF, *Joint Distance Function*, the sum of $F(x_i)$ over i, the whole clustering problem consists in the identification of the centers that minimize the JDF. In Iyigun (2007) it is demonstrated that the centers that minimizes the JDF can be computed using the following formula: $c_k = \sum_{i=1}^{n} \left(\frac{u_k(x_i)}{\sum_{j=1,\ldots,N} u_k(x_j)} \right) x_i$, where $u_k(x_i) = \frac{p_{ik}^2}{d_k(x_i)}$.

For the sake of brevity we don't go into distance choice details; in this paper we consider: $d_k(x_i) = \sum_{j=1}^{J} |x_{ij} - c_{kj}|$. Starting from this formula the matrix of distances D of order $n \times K$ is defined, where the general element is $d_k(x_i)$. Indicating with c_k the generic center, the final solution $\widehat{JDF}$ is obtained minimizing the quantity:

$$\text{JDF} = \sum_{i=1}^{n} \sum_{k=1}^{K} d_k(x_i) p_{ik}^2 = \sum_{i=1}^{n} \sum_{j=1}^{J} \sum_{k=1}^{K} |x_{ij} - c_{kj}| p_{ik}^2, \tag{1}$$

where $p_{ik} \in [0, 1]$ and $\sum_{k=1}^{K} p_{ik} = 1$. It is worth to note that the optimized function corresponds to the function optimized by the *fuzzy c-means* method (Bezdek 1974), however, the method differs in the computation of centers (Ben-Israel and Iyigun 2008).

An iterative algorithm allows one to compute the solution of PD-clustering problem. The algorithm properties are illustrated in Iyigun (2007), where the author demonstrates the convergence, too. Each unit is then assigned to the kth cluster according to the highest probability that is computed a posteriori.

3 Factor PD-Clustering

PD-clustering is stable dealing with a large number of units but it becomes unstable as the number of variables increases. A widely used strategy to cope with this problem is variable transformation; a linear transformation of original variables into a reduced number of orthogonal ones can significantly improve the algorithm performance. The linear transformation of variables and the PD-Clustering need to optimize a common criterion (De Soete and Carroll 1994). The FPDC consists in an integrated procedure based on the Tucker3 decomposition (Kroonenberg 2008) and PD-Clustering. It has been demonstrated (Tortora et al. 2013; Gettler Summa et al. 2011) that the liner transformation of variables that minimize the problem in (1) corresponds to the Tucker3 decomposition of the 3-way array of distances G of general element $g_{ijk} = |x_{ij} - c_{kj}|$, where $i = 1, \ldots, n$ indicates the units, $j = 1, \ldots, J$ the variables and $k = 1, \ldots, K$ the occasions. For any c_k a $n \times J$ G_k matrix of distances is defined. The Tucker3 method decomposes the array G in three components, one for each mode, in a full core array Λ, and in an error term E:

$$g_{ijk} = \sum_{r=1}^{R} \sum_{q=1}^{Q} \sum_{s=1}^{S} \lambda_{rqs}(u_{ir} b_{jq} v_{ks}) + e_{ijk}, \tag{2}$$

where λ_{rqs} and e_{ijk} are respectively the general term of the three way array Λ of order $R \times S \times Q$ and E of order $n \times J \times K$; u_{ir}, b_{jq} and v_{ks} are respectively the general term of the $n \times R$ matrix U, $J \times Q$ matrix B and $K \times S$ matrix V. Note that R, Q and S are the number of components respectively of U, B, and V. The number of components must be user defined. Factor axes in the Tucker3 model are sorted according to their explained variability. The first factor axes explain the greatest part of the variability; the latest factors represent the ground noise. According to Kiers and Kinderen (2003), the choice of the parameters R, Q and S is a ticklish problem as they define the overall explained variability. Interested readers are referred to Kroonenberg (2008) for the theoretical aspects concerning that choice. We use an heuristic approach based on the eigenvalues scree plot to cope with this crucial issue. The coordinates x^*_{iq} of the generic unit x_i into the space of variables obtained through a Tucker3 decomposition are computed according to the following expression: $x^*_{iq} = \sum_{j=1}^{J} x_{ij} b_{jq}$. Finally, on the x^*_{iq} coordinates PD-Clustering is applied to solve the clustering problem (1). Let us start considering the expression (1); it is worth noting that minimising the quantity:

$$\text{JDF} = \sum_{i=1}^{n} \sum_{j=1}^{J} \sum_{k=1}^{K} |x_{ij} - c_{kj}| p^2_{ik} \qquad \text{s.t.} \quad \sum_{i=1}^{n} \sum_{k=1}^{K} p^2_{ik} \leq n, \tag{3}$$

is equivalent to computing the maximum of $-\sum_{i=1}^{n} \sum_{j=1}^{J} \sum_{k=1}^{K} |x_{ij} - c_{kj}| p^2_{ik}$, under the same constrain. In Tortora et al. (2013) it is demonstrated that, given p_{ik} and c_{ik}, b_{jq} is obtained according to a Tucker3 transformation of the array of distances G

minimising the JDF. An iterative algorithm alternatively calculates c_{kj} and p_{ik} on one hand, and b_{jq} on the other hand, until the convergence is reached. Defining with t the number of iterations, the algorithm can be summarized in the following steps: (1) random initialization of matrix C of elements c_{kj}, and initialization of JDF^t; (2) computation of the array of distances G: $g_{ijk} = |x_{ij} - c_{kj}|$; (3) Tucker3 decomposition of $G = U\Lambda(V' \otimes B')$; (4) projection of data point in the new space $X^* = XB$; (5) PD-clustering of X^* uploading C and JDF^{t+1}; (6) if $JDF^{t+1} < JDF^t$ go to step 2, else stop. The minimisation of the quantity in the formula (3) converges at least to local minima, it can be empirically demonstrated.

4 Robustness and Stability Analysis

The social data sets used in the present paper are characterized by a large number of variables. Many clustering algorithms have stability problem dealing with these kind of data; different iterations of the same method, on the same data set, can bring to different solutions. The issue can be due to different causes; among them the effect of the choice of the initial solutions and/or the presence of local minima (Bubeck et al. 2012). More generally the stability of clustering solutions requires that solutions are similar for two different data sets that have been generated by the same source (Lange et al. 2004). To test the stability of a clustering method, a similarity measure between clustering partitions is needed. A wide range of index exists in the literature, we choose the Calinski-Harabasz index because of is simplicity and large diffusion (Vendramin et al. 2009). The index is equal to the ratio between the trace of the between clusters variance matrix and the trace of the within cluster variance matrix, multiplied for $(N - K)/(K - 1)$.

Dealing with large data sets two main situations are available: a large number or a small number of units. When the number of unit is large the stability of the method can be tested reiterating the method on the same data sets several times. However, when the number of units is small, even smaller than the number of variables, the stability can be tested using a resampling method. In the literature it exists a wide range of examples of use of resampling methods to test cluster stability. Some examples can be found in: Monti et al. (2001), Ben-Hur et al. (2002), Dudoit and Fridlyand (2002), Bryan (2004), Grün and Leisch (2004), Lange et al. (2004), Tibshirani and Walther (2005), and Hennig (2007). We choose to use bootstrap resampling method, it allows us to avoid distortion obtaining data sets of the same size. The strategy consists in resample new data sets from the original one using bootstrap method and than apply FPDC to them. The results are compared using the Calinski-Harabasz index (CHI).

To be useful in practice, clustering methods must be even robust: the performance of an algorithm should not be affected by small deviation from the assumed model and it should not deteriorate due to noise and outliers (Devé and Krishnapuram 1997). As Huber said: large deviation from model assumption should not cause a catastrophe (Huber 1981). Maronna and Zamar (2002) proposed a technique to contaminate a data set introducing outliers. The contamination is obtained by generating $x_i \sim N_k(ra_0\sqrt{J}, G_k)$ where a_0 is a unitary vector generated orthogonal to $(1, 1, \ldots, 1)^T$ and r measures the distance between the outliers and the cluster center.

5 Application on Real and Simulated Data Sets

The method has been applied on a real data set, it is not so big but it can help for the interpretation of results. The data are the education data of European Countries (source Eurostat). The country selected are 43, other countries presented too much missing values, the variables are 22 and they are numeric. The number of factors has been chosen using an empirical procedure based on the explained variability. The explained variability was computed varying the number of factors and represented on a scree plot, the plot allows us to find a cutoff point. For this data set, we have chosen the minimum number of factors that explains the 85 % of the variability, 4 factors for variables, 3 factors for units, and 2 factors for occasions. The algorithm convergence is shown in Fig. 1a, it represents the value of the JDF at each iteration of the FPDC algorithm, the method has been applied 100 times.

The clustering partition is described in the following. Cluster 1: Turkey and Russian Federation; cluster 2: Eastern Europe, Luxembourg and Portugal; cluster 3: Western Europe, Hungary and Ukraine. Countries belonging to the first cluster are characterized by a short duration of primary and pre-primary school. Countries belonging to the second cluster represent the average country for duration of education, however the starting age is higher than the countries of others groups. The last group contains countries characterized by a short duration of secondary school, but they have the higher gross domestic product.

To test the stability of the algorithm on this data set, bootstrap resampling techniques has been applied in order to extract 100 samples, it has been shown that the number of bootstrap replications doesn't have to be very large (Hennig 2007). On each sample the FPDC algorithm has been applied and the CHI measured. The frequency distribution of the CHI on the 100 samples is represented in Fig. 1b. Only in 18 % of cases the value of the index is very higher than the mean, the mode value is obtained in 51 % of cases. We can conclude that the algorithm is stable.

To evaluate the robustness of the algorithm the data set has been contaminated at two fixed levels: 10 % and 20 %. The distance of outliers has been set at two levels, results are shown in Table 1. The error rate is the number of misclassified, respect to the presented solution, divided for the total number of units. The maximum error rate is obtained by adding 20 % of outlier at a minimum distance. The maximum error rate is 11.76 %.

A big data set, with $n \ll J$, has been simulated to show the performance of the method dealing with big data sets. It is characterized by 4 normal distributed clusters with centers randomly generated, $n = 800$, and $J = 10{,}000$. It is worth to

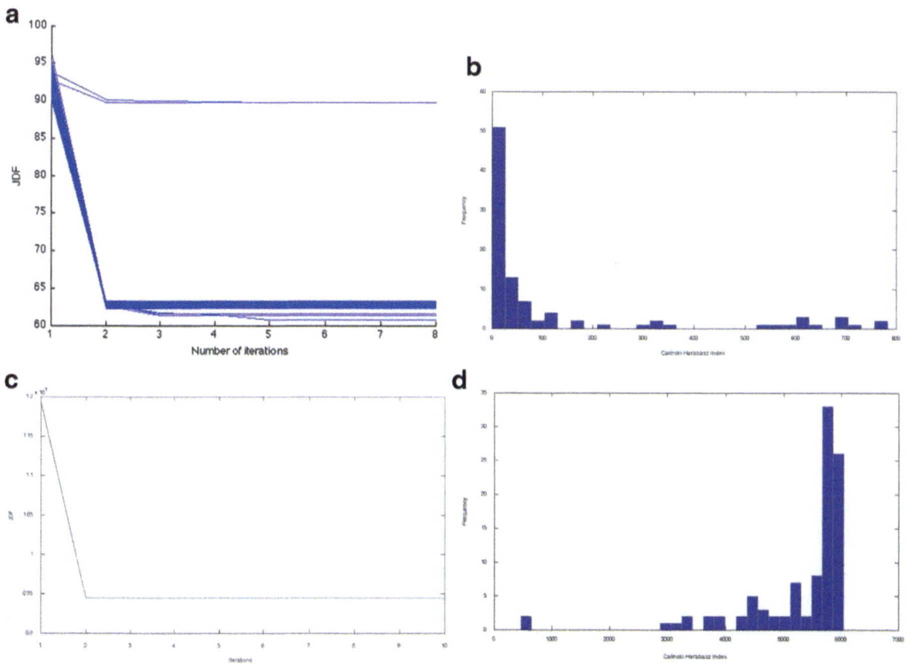

Fig. 1 (**a**) Factor PD-clustering convergence on education data set. (**b**) Frequency distribution of CHI on 100 bootstrap resampling on education data. (**c**) Factor PD-clustering convergence on big data set. (**d**) Frequency distribution of CHI on 100 bootstrap resampling on big data set

Table 1 Robustness of first data set

r	% of outliers	Error rate (%)
Min	10	2.13
Min	20	11.76
Max	10	10.60
Max	20	7.80

note that PD-clustering fail in finding the true clustering structure with an error rate of 50 %.

FPDC found the true clustering structure with error rate 0 % on 100 iterations. The algorithm converges always to the same solution as shown in Fig. 1c. To test the stability, bootstrap resampling technique has been applied in order to extract 100 samples. On each sample the FPDC algorithm has been applied and the CHI measured. The frequency distribution of the CHI on the 100 samples is represented in Fig. 1d. The algorithm is stable, the modal value is obtained in 33 % of cases and in 78 % of cases the index is in the interval [5000; 6000]. The analysis of robustness is shown in Table 2, the maximum error rate is 10.9 %.

Table 2 Robustness of the simulated data set

r	% of outliers	Error rate (%)
Min	10	3.75
Min	20	3.68
Max	10	5.80
Max	20	10.90

6 Conclusion and Perspective

The paper aims at applying FPDC on large data sets to obtain homogeneous and well-separated clusters of individuals. The scope is to evaluate the stability and the robustness of the method. Stability of results is referred to the invariance of results in each iteration of the method. Robustness is referred to the sensitivity of the method to errors in data, as outliers. Bootstrap resampling has been used to test stability: the algorithm is stable and the stability increase for bigger data sets. To test the robustness data was contaminated at some fixed level and the error rate is maximum 12 %. Using FPDC, unit weights are inversely proportional to the distance from the cluster center, thanks to this characteristic, outliers have a low weight in the determination of the centers. The method was applied on continuous data set related to social or behavioural fields, future work will focus on the extension of FPDC for qualitative or frequency data that are quite common in these fields.

References

Arabie, P., & Hubert, L. (1994). Cluster analysis in marketing research. In R. P. Bagozzi (Ed.), *Advanced methods in marketing research* (pp 160–189). Oxford: Blackwell.

Ben-Hur, A., Elisseeff, A., & Guyon, I. (2002). A stability based method for discovering structure in clustered data. *Pacific Symposium on Biocomputing, 7*(6), 6–17.

Ben-Israel, A., & Iyigun, C. (2008). Probabilistic d-clustering. *Journal of Classification, 25*(1), 5–26.

Bezdek, J. C. (1974). Numerical taxonomy with fuzzy sets. *Journal of Mathematical Biology, 1*(1), 57–71.

Bryan, J. (2004). Problems in gene clustering based on gene expression data. *Journal of Multivariate Analyis, 90*, 67–89.

Bubeck, S., Meilă, M., & Von Luxburg, U. (2012). How the initialization affects the stability of the k-means algorithm. *Probability and statistics: PS, 16*, 436–452.

De Soete, G., & Carroll, J. D. (1994). k-means clustering in a low-dimensional euclidean space. In E. Diday, Y. Lechevallier, M. Schader, et al. (Eds.), *New approaches in classification and data analysis*. (pp. 212–219). Heidelberg: Springer.

Devé, R. N., & Krishnapuram, R. (1997). Robust clustering methods: A unified view. *IEEE Transiction on Fuzzy Systems, 5*(2), 270–293.

Dudoit, S., & Fridlyand, J. (2002). A prediction-based resampling method to estimate the number of clusters in a dataset. *Genome Biology, 3*, 0036.1–0036.21.

Gettler Summa, M., Palumbo, F., & Tortora, C. (2011). Factor pd-clustering. Working paper [arXiv:1106.3830v3]

Ghahramani, Z., & Hinton, G. E. (1997). The em algorithm for mixtures of factor analyzers. Crg-tr-96-1, University of Toronto, Toronto.
Grün, B., & Leisch, F. (2004). Bootstrapping finite mixture models. Compstat 2004, *proceedings in Computational Statistics*, 1115–1122.
Hennig, C. (2007). Cluster-wise assessment of cluster stability. *Computational Statistics & Data Analysis, 52*(1), 258–271.
Huber, P. J. (1981). *Robust Statistics*. New York: Wiley.
Iyigun, C. (2007). *Probabilistic Distance Clustering*. Ph.D. thesis, New Brunswick Rutgers, The State University of New Jersey.
Kiers, H. A. L., & Kinderen, A. (2003). A fast method for choosing the numbers of components in tucker3 analysis. *British Journal of Mathematical and Statistical Psychology, 56*(1), 119–125.
Kroonenberg, P. M. (2008). *Applied multiway data analysis*. Hoboken: Ebooks Corporation.
Lange, T., Roth, V., Braun, M. L., & Buhmann, J. M. (2004). Stability-based validation of clustering solutions. *Neural Computation, 16*(6), 1299–1323.
Lebart, A., Morineau, A., & Warwick, K. (1984). *Multivariate statistical descriptive analysis*. New York: Wiley.
Maronna, R. A., & Zamar, R. H. (2002). Robust estimates of location and dispersion for high-dimensional datasets. *Technometrics, 44*(4), 307–317.
McLachlan, G. J., & Peel, D. (2003). Modelling high-dimensional data by mixtures of factor analyzers, *Computational Statistics and Data Analysis, 41*(3), 379–388.
Monti, S., Tamayo, P., Mesirov, J., & Golub, T. (2001). Consensus clustering: A resampling-based method for class discovery and visualization of gene. *Expression Microarray Data, Machine Learning, 52*, 91–118.
Rocci, R., Gattone, S. A., & Vichi, M. (2011). A new dimension reduction method: Factor discriminant k-means. *Journal of Classification, 28*(2), 210–226.
Tibshirani, R., & Walther, G. (2005). Cluster validation by prediction strength. *Journal of Computational and Graphical Statistics, 14*, 511–528.
Timmerman, M. E., Ceulemans, E., Kiers, H. A. L., & Vichi, M. (2010). Factorial and reduced k-means reconsidered. *Computational Statistics & Data Analysis, 54*(7), 1858–1871.
Tortora, C., Gettler Summa, M., & Palumbo, F. (2013). Factor pd-clustering. In U. Alfred, L. Berthold, & V. Dirk (Eds.), *Algorithms from and for nature and life* (volume, in press).
Vendramin, L., Campello, R., & Hruschka, E. (2009). In SDM. *On the comparison of relative clustering validity criteria* (pp. 733–744).
Vichi, M., & Kiers, H. A. L. (2001). Factorial k-means analysis for two way data. *Computational Statistics and Data Analysis, 37*, 29–64.

A Box-Plot and Outliers Detection Proposal for Histogram Data: New Tools for Data Stream Analysis

Rosanna Verde, Antonio Irpino, and Lidia Rivoli

Abstract In this paper, we propose a method for monitoring the evolution of data described by histograms of values. Our proposal consists to define new order statistics on the quantile functions associated with the empirical distributions, represented by the histogram-data. We introduce the Median, the First and the Third Quartile quantile functions, as well as a generalized representation of the box and whiskers plot. For example, the proposed representations and indices are useful for identifying and classifying outliers, arriving along the time in a data stream environment.

Keywords Data stream • Histogram data • Outliers • Quantile functions

1 Introduction

In recent years, there has been an increasing interest for data streams analysis. A data stream is a massive sequence of ordered observations arriving at a very high rate. Typically data streams arise from sensor networks, web traffic logs, financial transactions, security systems. The size of data does not allow one to work on stored data but only to process them *on the fly* (i.e., as they arrive). For this reasons, traditional data mining techniques are not appropriate in the data stream context and new analysis methodologies are needed. A frequent approach consists in detecting suitable summary structures (or *synopses*) which are able to preserve

R. Verde (✉) • A. Irpino
Department of Political Sciences "J. Monnet", Second University of Naples, Caserta, Italy
e-mail: rosanna.verde@unina2.it; antonio.irpino@unina2.it

L. Rivoli
University of Naples "Federico II", Naples, Italy
e-mail: lidia.rivoli@unina.it

D. Vicari et al. (eds.), *Analysis and Modeling of Complex Data in Behavioral and Social Sciences*, Studies in Classification, Data Analysis, and Knowledge Organization,
DOI 10.1007/978-3-319-06692-9_30,

the information contained in observations after raw data have been observed and then lost. For instance, typical syntheses of sequences of data may are obtained by means of moments, sketches, wavelets and histograms (Gama and Pinto 2006).

In this work, we consider to split a data stream in non-overlapping time-windows with equal size and, for each of them, the window-data density distribution is summarized by a histogram. In this paper, we propose to work on the quantile functions, associated with histograms, rather than on the density functions. According to Gilchris (2000), quantile functions (i.e., the inverse of cumulative distribution function) is a useful tool for the analysis of distributions. After computing the quantile functions (or *qf*s) associated with each histogram, an extension of the Quartiles and of the box and whisker plot (Tukey 1977) for the set of *qf*s is provided. This innovative representation for *qf*s is used for monitoring the evolution of histogram data as well as for identifying potential outliers. Furthermore, we also evaluate the dissimilarity of an outlier histogram from the five-histogram summary with respect to location and shape.

The paper is organized in the following way. In Sect. 2, we provide a short introduction about histogram data. Section 3 describes the procedure for obtaining the box-plot of a of quantile functions. In Sect. 4, we present a strategy for detecting the outlier quantile functions. Finally, Sect. 5 shows an application on real data, while Sect. 6 ends the paper with some conclusions and perspectives.

2 Summarization by Histogram Data

Let $Y = \{y_1, \ldots, y_l, \ldots\}$ be a data stream whose observations y_l arrive at a fixed time-stamp t_l. Suppose that Y is split in non-overlapping time-windows with equal size and for each of them, the corresponding histogram is obtained. Thus, the data stream Y is represented by an infinite number of histograms denoted with H_i, $i = 1, 2, \ldots$. Each histogram can be represented as a set of ordered couples $H_i = \{(I_{i1}, f_{i1}), \ldots, (I_{ik}, f_{ik}), \ldots, (I_{iK_i}, f_{iK_i})\}$, $i = 1, 2, \ldots,$ where I_{ik}, $k = 1, \ldots, K_i$ are the bins of the histogram and f_{ik} are the relative frequencies associated with I_{ik}. Within each bin $I_{ik} = [\underline{y}_{ik}, \bar{y}_{ik})$ it is assumed that the values are uniformly distributed, so the (cdf) F_i associated to each H_i can be defined as follows:

$$F_i(x) = \begin{cases} 0 & \textit{if } x < \underline{y}_{i1}, \\ \sum\limits_{l=1}^{k-1} f_{il} + \frac{x - \underline{y}_{ik}}{\bar{y}_{ik} - \underline{y}_{ik}} f_{ik}, \quad k \geq 2 & \textit{if } \underline{y}_{ik} \leq x < \bar{y}_{ik} \\ 1 & \textit{if } x \geq \bar{y}_{iK_i} \end{cases} \tag{1}$$

where, $F_i(x)$ is clearly a piece-wise linear function. Then, the associated quantile function (the inverse of the cumulative distribution function) (*qf*) of H_i is expressed by:

$$F_i^{-1}(t) = \begin{cases} y_{i1} & \text{if } t = 0 \\ \underline{y}_{ik} + \frac{t - w_{ik-1}}{w_{ik} - w_{ik-1}}(\bar{y}_{ik} - \underline{y}_{ik}) & \text{if } w_{ik-1} \leq t < w_{ik} \\ \bar{y}_{iK_i} & \text{if } t = 1 \end{cases} \quad (2)$$

For our scope, we identify each histogram H_i, $i = 1, \ldots N$ by a set of couples $\{(I_{ik}, f_{ik}); k = 1, \ldots, K_i\}$ where $w_{ik} = \sum_{l=1}^{k} f_{il}$, $k = 1, \ldots, K_i$ are the cumulative relative frequencies or *levels* (of cumulated frequencies). For computing the Median *qf*, we perform two-steps algorithm as described in Rivoli et al. (2012).

3 Box and Whiskers Plot for *qfs*

Consider the first N windows of the stream Y and the corresponding histograms H_i, $i = 1, \ldots, N$. We introduce a box-plot for a set of *qf*s extending the classical the box plot. To this end, we define the *Median*, the *First* and the *Third Quartile* and the *whiskers* quantile functions for a set of the *qf*s. Taking into account the classic proprieties of the median, and as shown in Arroyo et al. (2011), the *Median histogram* is defined as the histogram H_M whose quantile function is the solution of the following minimization problem based on ℓ_1 Wasserstein distance:

$$\min_{H_M} \sum_{i=1}^{N} d_1(H_i, H_{ME}) = \min_{F^{-1}(t)} \sum_{i=1}^{N} \int_0^1 \left| F_i^{-1}(t) - F_{ME}^{-1}(t) \right| dt, \quad (3)$$

where F_i^{-1} and F_{ME}^{-1} are the *qfs* associated with H_i and H_{ME} respectively.

It is noteworthy that, just as H_{ME} is the barycenter histogram of the set of histogram data H_i according to the ℓ_1 Wasserstein distance, so the Average histogram is the barycenter according to the ℓ_2 distance (as shown in Verde and Irpino 2008; Irpino and Verde 2006). As in the classic case, in Arroyo et al. (2011) it is highlighted that the definition of Median histogram of an even number of histograms is non unique. However, as in the classic case, we use the mean between the two most central *qf*s.

Considering the nature of the data and the minimization of the function in Eq. (3), in Rivoli et al. (2012) a strategy is presented for obtaining a *level-wise median qf*, that is, a quantile function that, for each $t \in [0; 1]$, identify a value that separates the higher $N/2$ quantiles from the lower ones at a given level t.

Theorem 1. *$F_{ME}^{-1}(t)$ median function is a quantile function and it is unique.*

Proof. Firstly, we prove that *$F_{ME}^{-1}(t)$ is a quantile function.*

Given a set of N histograms H_i, we denote with $F_i^{-1}(\mathrm{t})$ (for $i = 1, \ldots, N$) the corresponding quantile functions (i.e., non-decreasing functions in $[0, 1]$). Let $0 \leq t^* \leq 1$ be a level of the quantile functions and $0 \leq \epsilon \leq 1$ be a number such that

$0 \leq t^* \leq (t^* + \epsilon) \leq 1$. We denote with $\mathbf{y}(t^*) = \{y_1(t^*) = F_1^{-1}(t^*), \ldots, y_N(t^*) = F_N^{-1}(t^*)\}$ the vector of the N quantile values $y_i(t^*)$ at a fixed level t^*. We denote with $ME(t^*) = median(\mathbf{y}(t^*))$ the median of the vector $\mathbf{y}(t^*)$, that is the solution of the problem in Eq. (3) at a level t^*.

Let us define with $\mathbf{y}(t^* + \epsilon) = \{y_1(t^* + \epsilon) = F_1^{-1}(t^* + \epsilon), \ldots, y_N(t^* + \epsilon) = F_N^{-1}(t^* + \epsilon)\}$ the vector of the quantiles of the distributions at level $t^* + \epsilon$, such that $y_i(t^* + \epsilon) \geq y_i(t^*)$ for each $i = 1, \ldots, N$.

Denoting with $ME(t^* + \epsilon)$ the median of the vector $\mathbf{y}(t^* + \epsilon)$, we obtain that $y_i(t^* + \epsilon) \geq y_i(t^*)$ and it follows that: $ME(t^* + \epsilon) \geq ME(t^*)$.

Thus, being $\lim\limits_{\epsilon \to 0} ME(t^* + \epsilon) = ME(t^*)$, we have proved that $ME(t)$ is a non-decreasing function in [0, 1]. Finally being all the $F_i^{-1}(t)$ defined in [0, 1], we can conclude that $ME(t)$ is a quantile function.

For proving that $F_{ME}^{-1}(t)$ is unique, we must distinguish when N is odd or even.

- N is odd. $ME(t^*) = y_{[\frac{N+1}{2}]}(t^*)$, where $[\frac{N+1}{2}]$ denotes the position of the $y_{[i]}(t^*)$ in $\mathbf{y}(t^*)$ after being ordered (from the lower to the higher value), as usually known.
- N is even. According to Eq. (3), being solved for each value between $y_{[\frac{N}{2}]}(t^*)$ and $y_{[\frac{N}{2}]+1}(t^*)$, the analytic solution of $ME(t^*)$ is not unique. In this case, we adopt the empirical solution used in data analysis by computing the median as the arithmetic mean of the two values as follows:

$$ME(t^*) = \frac{y_{[\frac{N}{2}]}(t^*) + y_{[\frac{N}{2}]+1}(t^*)}{2},$$

  and it is unique.

  It is easy to note that being $y_{[\frac{N}{2}]}(t^* + \epsilon) \geq y_{[\frac{N}{2}]}(t^*)$, also in the even case: $ME(t^* + \epsilon) \geq ME(t^*)$. □

Being the qfs functions, we have a level-wise order (for a given t we may order the *qf*s) but not a complete order relation among them. In particular, if for each $t \in [0; 1]$ the order of the *qf*s is always the same, we can extend the level-wise order to a complete order relation, but in general the observed *qf*s tend to intersect each other.

However, using the same arguments in the proof of Theorem 1, we can define the level-wise first and third quartile qf (denoted here with $Q_{(0.25 \cdot N)}(t)$ and $Q_{(0.75 \cdot N)}(t)$, respectively). Indeed, considering that the quartile qfs are computed according to the order of the $\mathbf{y}(t^*)$ vector, it is easy to observe that $Q_{(0.25 \cdot N)}(t^*) \leq Q_{(0.5 \cdot N)}(t^*) = ME(t^*)$ or that $ME(t^*) \leq Q_{(0.75 \cdot N)}(t^*)$. Further, this guarantees that the quartiles and the median qfs does not intersect each other.

The algorithm proposed in Rivoli et al. (2012) introduces some numerical solutions for transforming the continuous problem into a finite number of easy subproblems to solve. Using the algorithm in Rivoli et al. (2012) it is possible to compute the generic $p \cdot N$ ($p \in [0; 1]$) order *qf* statistic and we denote them as $Q_{(pN)}(t)$. The proposed algorithm for the search of order-quantile functions

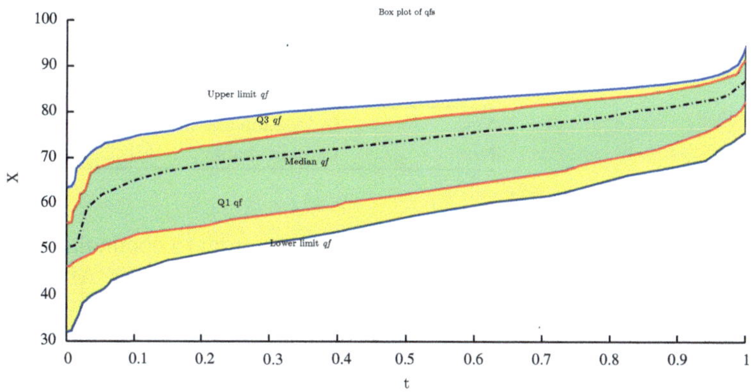

Fig. 1 The quantile function box-plot representation by five *qf*s: the median; the first and the third quartile *qf*s, delimiting the box; an Upper and a Lower bound *qf*, as extremes of the whiskers

guarantees a unique correspondence between histograms and *qf*s, so the First Quartile-histogram H_{Q_1} is associated with the *qf* $Q_{(0.25 \cdot N)}(t)$, the Third Quartile-histogram H_{Q_3} with the qf $Q_{(0.75 \cdot N)}(t)$ and the Median histogram H_M with the *qf* $ME(t)$.

The box-plot of *qf*s is obtained as follows. The *box* can be defined as the region bounded by $Q_{(0.25 \cdot N)}(t)$ and $Q_{(0.75 \cdot N)}(t)$ and the *Inter Quartiles Range* (or *IQR*) is the value of this area computed by Wasserstein distance (Irpino and Verde 2006; Verde and Irpino 2007). The choice of the ending *qf*s of the two whiskers are the $Q_{(0.05 \cdot N)}(t)$ and $Q_{(0.95 \cdot N)}(t)$ functions, i.e., those two functions that, given a level $t_0 \in [0, 1]$, contains the most 90 % central values of all the *qf*s observed in t_0. In fact, this solution is less sensitive to outlying *qf*s related to the choice of the maximum $Q_{(N)}(t)$ and the minimum $Q_{(0)}(t)$ *qf*s.

Finally, in order to obtain the box plot for the set of *qf*s through the five-*qf*s defined above, we consider a graph having the domain of the five-*qf* statistics on the horizontal axis and the set of values assumed by them on the vertical axis. We have named this graph as the ***qf-box-plot*** and an example is shown in Fig. 1.

4 Outlier Detection

The qf-box-plot allows for detecting possible changes in the data distribution over the time and identifying potential outliers. When a new histogram H_{N+j}, $j = 1, 2, \ldots$ is available, its *qf* $F^{-1}_{N+j}(t)$ can be compared with the previous N *qf*s in terms of location and shape. To this end, two further different box-plots are considered: the first one identifies the outlyingness in location and the second one in shape. The first box-plot is a classical box-plot depicting the five mean values μ_{Q_1}, μ_{Q_3}, μ_{ME} and μ_{Low} μ_{Upper} associated with the First and the Third quartile,

Table 1 Types of outliers

Classification of outliers with respect to box-plot of the:		
	Centered Qfs	
Mean values	Yes	No
Yes	In location and in shape	Only in location
No	Only in shape	No outlier

the Median and the Lower and Upper histograms. In order to obtain the second box plot, all qfs are centered by their means: $F_i^{-1c}(t) = F_i^{-1}(t) - \mu_i$. Afterwards, we detect qf-box-plot as described above.

For each H_{N+j}, $j = 1, 2, \ldots$, its mean value and the shape of the centered qf $F_{N+j}^{-1c}(t)$, $j = 1, 2, \ldots$ are evaluated. In particular for qf-box plot representation, we can distinguish four different situations. The $F_{N+j}^{-1c}(t)$ associated to the histogram H_{N+j}, $j = 1, 2, \ldots$ may:

1. be included in the region between the First Quartile and the Third Quartile qf;
2. be included between the lower or the upper whisker qfs;
3. intersect the lower or the upper whisker qf or both of them;
4. be below the lower whisker or above the upper whisker qf.

Obviously, in the fourth case, the histogram has to be considered as an *effective outlier*. In this case, the mean value of $H_{N+j}^{-1}(t)$, $j = 1, 2, \ldots$ is also an outlier for the mean values-box plot. The inverse is not always true. The other cases have to be considered with major attention because a histogram does not necessarily differ in location if its qf is not included in the qf-box plot.

In general, any histogram which is not included between the whiskers of the distribution of the mean values and of the distribution of the centered qfs, can be classified as an outlier. Table 1 illustrates the different cases that may arise:

The three outlying cases described in Table 1 can be defined as *potential outliers*. In particular, for the qfs which differ in shape, we also propose to evaluate the degree of outlyingness by means of a dissimilarity measure.

If t_0 is the x-axis value of the intersection point between the potential outlier qf and the upper (or lower) whisker qf and for $t > t_0$ the $F_{N+j}(t)$ is outside the box-plot, then we define the following measure:

$$RD := \int_{t_0}^{1} \left| F_{N+j}^{-1c}(t) - Q_p^c(t) \right| dt \Big/ \int_{0}^{1} \left| F_{N+j}^{-1c}(t) - Q_p^c(t) \right| dt \tag{4}$$

where the numerator represents the area beyond the box plot (or the distance between $F_{N+j}^{-1c}(t)$ and $Q_p^c(t)$, $p = 0.95$ or 0.05) and the denominator is the total area enclosed by the two qfs. Similarly, if t_0 is the x-axis value of the intersection point between the potential outlier qf and the upper (or lower) whisker qf and for $t < t_0$ the $F_{N+j}(t)$ is inside the box-plot, then

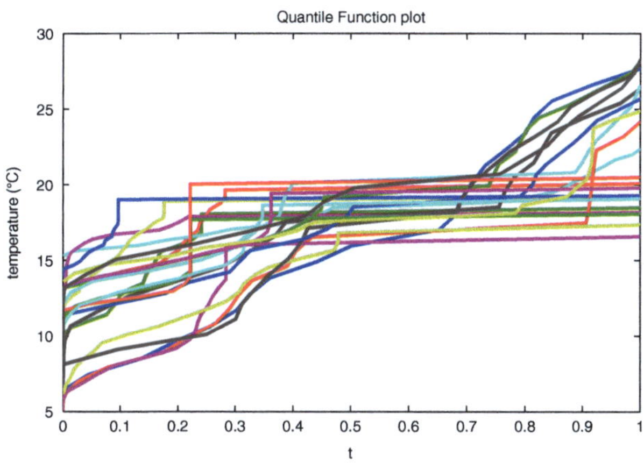

Fig. 2 Plot of the 21 *qfs* associated to histograms of the temperatures in the Italian stations

$$RD' := \int_0^{t_0} \left| F_{N+j}^{-1\,c}(t) - Q_p^c(t) \right| dt \Big/ \int_0^1 \left| F_{N+j}^{-1\,c}(t) - Q_p^c(t) \right| dt,\ j = 1, 2, \ldots \quad (5)$$

is the ratio between the area below the box plot (or the distance between $F_{N+j}^{-1\,c}(t)$ and $Q^c_, p = 0.95$ or 0.05) and the total area enclosed by the two *qf*s.

It is worth noting that, the more this ratio is close to 1 the more $F_{N+j}^{-1\,c}(t)$ presents a different shape with respect to the box-plot. Furthermore, if the $F_{N+j}^{-1\,c}(t)$ intersects both the whiskers, the numerator of *RD* (or *RD'*) is equal to the sum of the areas delimited by $F_{N+j}^{-1\,c}(t)$, with $Q_0^c.05(t)$, and $F_{N+j}^{-1\,c}(t)$, with $Q_0^c.95(t)$.

5 An Application on Real Data

To demonstrate the effectiveness of the proposed approach, an analysis on real data has been performed. The data represent the hourly values of water maximum temperatures recorded by 21 meteorological Italian stations from 1, January, 2009 to 31, December 2009. For each station, data are summarized by a histogram and then, the quantile function associated with each histogram is computed.

The recorded observations range from 5 to 21°C, but, comparing the distribution of the 21 *qf*s, the water temperature presents a higher variability. Figure 2 shows that there are two different groups of *qf*s. A first one consists in a set of *qf*s presenting a graph that is stable from the 50th percentile onwards; the second one shows an increasing trend.

Using the proposed strategy, we have computed the following *qf*s: $Q_{(0.25\cdot N)}(t)$, and $Q_{(0.75\cdot N)}(t)$, $Me(t)$, $Q_{(0.05\cdot N)}(t)$, and $Q_{(0.95\cdot N)}(t)$ for the set of $N = 21$ *qf*s

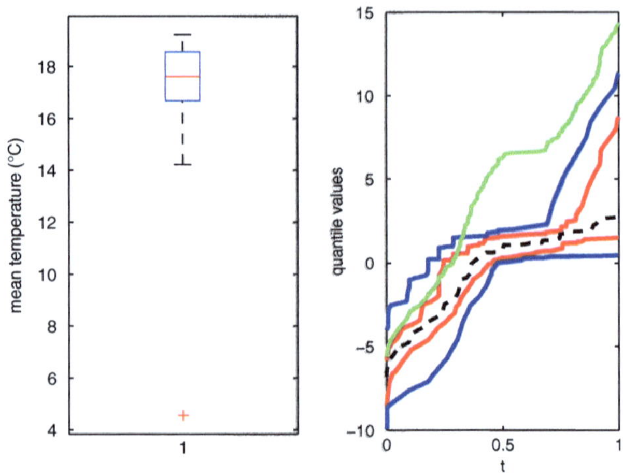

Fig. 3 (**a**) Box-plot of the mean values of the histograms and an outlier observation. (**b**) qf-box-plot of the centered *qf*s with an outlier *qf*

represented in Fig. 3b with a red, dashed and blue curve respectively. Figure 3a,b also illustrates how our approach is able in detecting potential outliers with respect to the location and the shape respectively. In this case, the value of $\mathrm{RD} = 0.83$, for the outlier, is consistent with the fact that the area outside the box-plot is greater than the inside one.

6 Conclusions and Perspectives

Based on an extension of order statistics for quantile functions, we proposed a qf-box-plot that can be a suitable visual tool for identifying potential outlier quantile functions. The classification of their degree of outlyingness, with respect to their location or their shape, may be accomplished by using a classical box plot and a *qf*-box plot representation, respectively. A concrete field of application of the proposed strategy is, for instance, the representation and the monitoring of numeric data streams. In a data stream analysis approach, the frequency of the observed (potential) outliers is a hint of the evolution of the data stream. Furthermore, the different kind of outlyingness (due to the location or to the shape, or both) can suggest a different kind of evolution of the stream. It allows the user different interpretations and different strategies for updating the stream summaries. Further investigations could regard appropriate strategies for updating the qf-box-plot representation when changes occur.

References

Arroyo, J., González-Riviera, G., Maté, C., & Muñoz San Roque, A. (2011). Smoothing methods for histogram-valued time series. An application to value-at-risk. *Statistical Analysis and Data Mining, 4*(2), 216–228.

Gama, J., & Pinto, C. (2006). Discretization from data streams: Applications to histograms and data mining. In *Proceedings of the ACM Symposium on Applied Computing* (pp. 662–667), New York.

Gilchris, W. (2000). *Statistical modelling with quantile functions*. London/Boca Raton: Chapman & Hall/CRC.

Irpino, A., & Verde, R. (2006). Dynamic clustering of histograms using Wasserstein metric. In A. Rizzi & M. Vichi (Eds.), *Advances in computational statistics* (pp. 869–876). Heidelberg: Physica-Verlag.

Rivoli, L., Irpino, A., & Verde, R. (2012). The median of a set of histogram data. In *XLVI Riunione Scientifica della Società Italiana di Statistica*, CLEUP [ISBN 978-88-6129-882-8].

Verde, R., & Irpino, A. (2007). Dynamic clustering of histogram data: Using the right metric. In *Studies in classification, data analysis, and knowledge organization* (vol. I, pp. 123–134).

Verde, R., & Irpino, A. (2008). *Comparing histogram data using a Mahalanobis-Wasserstein distance (COMPSTAT 2008)* (pp. 77–89). Heidelberg: Physica-Verlag.

Tukey, J. W. (1977). *Exploratory data analysis*. Reading: Addison-Wesley.

Assessing Cooperation in Open Systems: An Empirical Test in Healthcare

Paola Zappa

Abstract This paper aims to detect the social mechanisms underlying cooperation in organizational communities. To this purpose, it proposes to apply a longitudinal Social Network Analysis approach based on Stochastic Actor-Oriented Models for network dynamics to Web 2.0 data on interpersonal interaction. The paper claims and demonstrates that such an approach allows alleviating some limitations of current studies. It overcomes the issue of relational missing data. Also, it models directly the network structure as the outcome of actors' counterparts selection in their neighbourhood. Application is on a virtual community of Italian oncologists who collaborate in resolving diagnoses. Using repository and field data, we reconstruct a network, with clinicians as nodes and emails exchanged as ties. Then, we model cooperation longitudinally. Evidence is provided that emergent behaviors are effectively captured and advantages of this approach are discussed.

Keywords Healthcare • Social network analysis • Stochastic actor-oriented models • Virtual communities of practice

1 Introduction

Within the domain of organizational and behavioral sciences, a large stream of literature has addressed team-working and cooperation in several contexts and with various methods. Particularly well established is the use of Social Network Analysis, SNA henceforth (Wasserman and Faust 1994). Mapping a group as a network, whose nodes are the individuals and ties the relations between pairs of them,

P. Zappa (✉)
Dipartimento di Economia, Metodi Quantitativi e Strategie d'Impresa, Università Degli Studi Di Milano-Bicocca, Milano, Italy
e-mail: paola.zappa1@unimib.it

D. Vicari et al. (eds.), *Analysis and Modeling of Complex Data in Behavioral and Social Sciences*, Studies in Classification, Data Analysis, and Knowledge Organization, DOI 10.1007/978-3-319-06692-9_31,

SNA appears as the natural approach for addressing various issues (Krackhardt 1994): describing how interaction takes place, identifying the underlying social processes and detecting individual behaviours. Recent research has mainly evoked the need of examining the informal or emergent network structure, explaining the theoretical social mechanisms that simultaneously drive tie formation (Contractor et al. 2006). In this respect, the cross-sectional or network snapshot-comparison approaches adopted by most papers are not completely satisfactory. They cannot effectively uncover the mechanisms influencing partner selection nor explain the emergent network structure as a consequence of tie creation and dissolution at local (dyadic or triadic) level. This gap between theoretical assumptions and applications could be filled by the adoption of a longitudinal perspective (Monge and Contractor 2003). Longitudinal modelling, however, has been limited by two issues, which have found solution only recently: unavailability of methods and unavailability of data. In terms of data, the diffusion of electronic communication, especially of Web 2.0 applications, has offered new opportunities for social interaction and cooperation. Also, this technology has made available large datasets for tracing back social dynamics, i.e., lists of emails exchanged, messages posted, wikies contributed, etc. Social networks, online communities and virtual communities of practice represent interesting phenomena in themselves. Indeed, they have been proven useful settings for examining dynamics of social interaction in general (Quintane and Kleinbaum 2011). Compared to survey data, electronic datasets offer several advantages. They are fully observed networks, not affected by the presence of missing data, which are a crucial issue in social network studies (van Duijn et al. 2009). They are objectively measured. Finally, they are fine-grained and continuously observed networks, suitable for longitudinal modelling (Opsahl and Hogan 2011). From a methodological viewpoint, various approaches have been proposed in order to deal with panel data. Among them, especially Stochastic Actor-Oriented Models for network dynamics (Snijders et al. 2010), SAOM hereafter, have received attention. SAOM capture the long term evolution, allow a complex modelling of local network configurations and have been proven appropriate for various empirical settings as well as robust to different specifications of tie values, time spans and network effects. This paper applies SAOM to online panel data, demonstrating the capability of such an approach to successfully identify the local mechanisms behind network emergence. Also, we deal with the data specifications that SAOM require and discuss their implications. The paper is organized as follows. Section 2 provides a brief technical introduction to SAOM and specifies the research hypotheses. Section 3 describes empirical setting, data and their treatment. Section 4 presents the results. Finally, Sect. 5 concludes.

2 Method: Stochastic Actor-Oriented Models

SAOM are a recently developed method proposed for examining networks evolution. They allow detecting the emergent network structure as a consequence of actors' choices and reactions to others' choices in actor neighbourhood. Rationale and estimation procedure have been extensively documented in a few papers (Snijders et al. 2010). Here, we report the salient characteristics of SAOM and frame them into our case study.

Formally, SAOM are continuous-time Markov chain models. Let us suppose we have a finite set of actors $I = \{1, \ldots, n\}$ and a relation R observed on them. R is represented by a binary adjacency matrix X and is collected at discrete points in time $t_1, \ldots, t_M$, with $M \geq 2$. The change in network ties is generated by an unobserved process taking place continuously between observation moments. Therefore, there is a continuous underlying time parameter $t \in T$. The changing network $X(t)$ is regarded as the outcome of a Markov process. From the general theory of continuous-time Markov processes (Norris 1997), we know there is a transition rate or intensity matrix Q, whose generic element $q(x, y)$ describes the rate at which $X(t) = x$ tends to transition into $X(t + dt) = y$ as $dt \to 0$. Since each transition consists exactly in actor i updating an outgoing tie, $q(x, y)$ is equal to $q_{ij}(x)$ and:

$$q_{ij}(x) = \lambda_i(x)\, p_{ij}(x) \tag{1}$$

with $(i \neq j)$. At random instants, one probabilistically selected actor i may get the opportunity to change one and only one outgoing tie. The network change process is then decomposed in a sequence of small changes, named micro steps.

1. $\lambda_i(x)$ measures the frequency of the micro steps, i.e., how fast actor i has the opportunity for a change between t_m and t_{m+1}, and depends on the rate function. The rate function for the time period $t_m \leq t < t_{m+1}$ is:

$$\lambda_i(\rho, x, a, m) \tag{2}$$

   $i \in I, x \in X$, ρ is a factor depending on the period m, a is a parameter depending on individual or network characteristics v. $\lambda_i(x)$ can be a constant function within periods, i.e., $\lambda_i(x) = \rho_m$, or not constant, i.e., $\lambda_i(x) = \rho_m exp\left(\sum a_h v_{hi}\right)$.
2. $p_{ij}(x)$ is the probability that i selects x_{ij} as the tie variable to change. The value of $p_{ij}(x)$ depends on the objective function, which measures i's preferred direction of change. In formula:

$$f_i(\beta, x) = \sum\nolimits_{k=1}^{L} \beta_k s_{ik}(x) \tag{3}$$

   $f_i(\beta, x)$ is the value of the objective function for actor i depending on the state x of the network. The terms s_{ik}, i.e., the effects, are statistics dependent on the network x in the neighbourhood of actor i (structural effects) and on other actor's characteristics in this neighbourhood (monadic and dyadic effects). The parameters β_k are the weights that indicate the strength of the effects s_{ik}. The change process can be interpreted as actors' decisions optimizing the objective function $f_i(\beta, x)$ plus a stochastic error term $U_i(t, x, j)$ having a Gumbel distribution (Maddala 1983). This specification implies that the objective function can be alternatively formulated recurring to multinomial logistic regression. Conditional on actor i being allowed to make a change between a set C of possible new states of the network, the probability of going to a specific state x is given by:

$$\frac{exp\left(f_i\left(\beta,x\right)\right)}{\sum_{x'\in C} exp\left(f_i\left(\beta,x'\right)\right)} \tag{4}$$

The estimation procedure of the vector of K parameters $\vartheta = (\rho_1, \dots, \rho_{M-1}, a_1, \dots, a_H, \beta_1, \dots, \beta_L)$ is based on the Method of Moments (Bowman and Shenton 1985) and implemented by computer simulation of the network change process. The MCMC implementation of the MoM uses a stochastic approximation algorithm that is a descendant of the Robbins-Monro algorithm (Robbins and Monro 1951).

2.1 *Research Hypotheses and Specification of Network Effects*

The estimation of SAOM requires to specify the research hypotheses in terms of model effects s_{ik}. Comprehensive overviews of theoretical mechanisms driving cooperation have been offered for organizational settings (Contractor et al. 2006) and also for open systems (Ahuja and Carley 1999). These papers, however, use cross-sectional data. Here, we take an exploratory approach. We build on these studies and test two crucial mechanisms that are expected to shape the network formation. First, balance theory (*Hypothesis 1*), which looks at the tendency of individuals to develop consistent behaviors and knowledge. The tendency toward consistency is generally operationalized as transitivity, which implies a path-shortening process, i.e., the propensity to transform indirect ties into direct ties (positive *transitive triplets* effect). Balance theory also suggests the emergence of a propensity toward generalized exchange of resources and knowledge (positive *three-cycles*). Second, collective action theory (*Hypothesis 2*) sustains the emergence of roles, driven by a preferential attachment mechanism (Barabasi and Reka 1999) which leads central actors in the network to become more and more central (positive *in-degree related popularity*). In this respect, we also assess the correlation between central position in sending and receiving ties, which leads to a core-periphery structure (positive *out-in degree assortativity*). Finally, we include some individual covariates as control effects. Figure 1 reports theoretical mechanisms, corresponding effects and their formal definition and description.

3 Data

We examined a virtual community of practice (VCoP) of clinicians. VCoP are ideal for observing the dynamics of cooperation: they are self-organizating open systems focused on a shared practice and are based on spontaneous participation

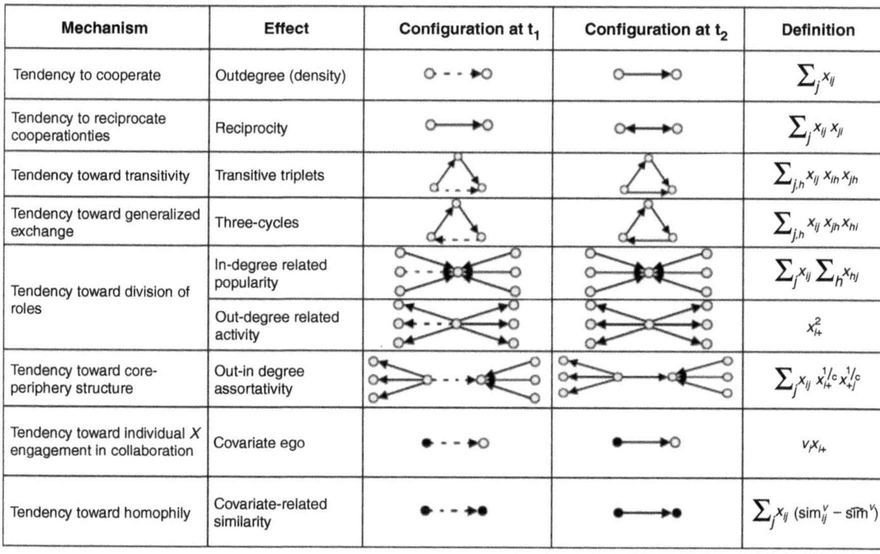

Fig. 1 Model mechanisms and corresponding effects

and on a distributed organization of labour, with various levels of interaction and integration of activities (Wasko and Faraj 2005). Also, VCoP exist primarily through computer mediated communication, which can be easily traced back. In healthcare cooperation has been a crucial topic for long (Valente 1996), but Web 2.0 applications represent a novelty and have rarely been studied from a relational perspective. The setting examined is the Italian Rare Cancer Network, a VCoP of around 600 Italian clinicians, who voluntary cooperate on a Web platform in solving rare cancers diagnoses. Our data are the list of posts exchanged online among the community members in the 2005–2009 period. Each post consists of sender and recipient names, object, patient code and timestamp and the relation investigated is information sharing. In order to be framed into SAOM requirements, data were preprocessed as follows (Ahuja and Carley 1999):

1. Posts with an undefined object and/or many recipients (looking at the distribution shape, threshold for inclusion is set at $n_j \leq 10$) were excluded. As a general rule, we assumed a cooperation tie from i to j exists when i replies to the first specific message m_{ji} of j.
2. Carbon copies (Cc) were included.
3. A minimum number of messages m_{min} posted by the generic actor i was defined for i being considered an active member. If $m_i < m_{min}$, i was excluded. Here, $m_{min} = 2$.

We obtained a list of 106 clinicians and 8,438 posts. For each ordered pair of actors (i, j), x_{ij}, the tie representing the number of messages from i to j, was computed summing up m_{ji} over a convenient timespan. Given the relatively low frequency of

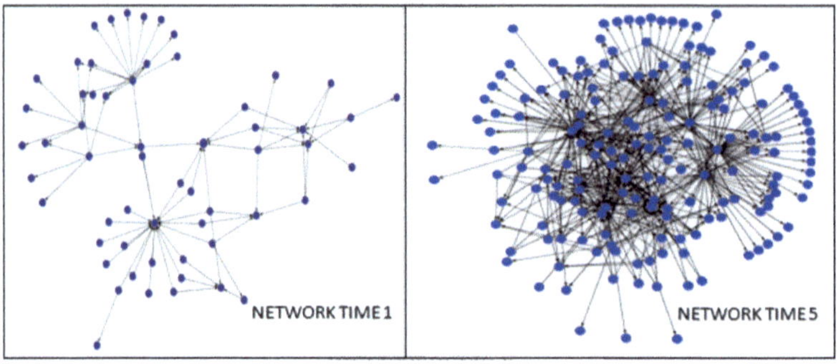

Fig. 2 Community structure over time

interaction, due to the setting examined, we defined $(t_m - t_{m-1}) = 1$ year.[1] Finally, a one-mode network $x(t)$ for each observation moment was reconstructed. In $x(t)$, nodes are the clinicians and ties the messages posted to each other over 1 year. Figure 2 displays the community at t_1 and t_5, underlining a significant increase and structuring in cooperation activity and, thus, justifying the longitudinal modelling.

Being SAOM specified for binary relations,[2] the ties had to be dichotomized. We considered various alternative rules to construct the dichotomous matrix, selecting as cut-off value: (1) the overall median; (2) the overall mean; (3) the value 0. Then, we modeled these dichotomous matrices. No significant differences in the estimates were observed. The results seemed comparable and, therefore, not sensitive to the dichotomization criterion chosen. Since we are interested in detecting the process leading to counterparts selection, not in the intensity of collaboration, we recoded each network at minimum, i.e., all the ties values $x_{ij} > 0$ were set to 1.

The effect of individual attributes on tie formation was accounted for by including survey data. A questionnaire was administered by email to the community members in four waves during the April 2010–January 2011 period (response rate 69.6 %) and missing data were, then, complemented with secondary data sources. Included variables were work-related information (tenure, specified as young and senior clinicians, and role, which distinguishes between clinicians in charge of a department and not) and expertise (Lave and Wenger 1991). Following previous studies (Wasko and Faraj 2005), the latter was captured as perceived expertise.[3] For direct comparability, also an objective measure of expertise was collected.[4] Both measures were set as dichotomous.

[1]The rarety of the cancer forms examined implies also a low frequency of information exchange. Network emergence is therefore assumed to be a slow process.

[2]Extensions to ordinal data have been proposed, but are still to be documented in the literature.

[3]Four degree ordinal scaling, with 1 = not competent at all in the field, ..., 4 = very competent. The scale was dichotomised so that 1 and 2 were recoded as not competent and 3 and 4 as competent.

[4]We scanned the hospital websites and the 2009 Italian White Book on Cancer Treatments. For each hospital, it reports a list of clinicians expert in any form of cancer.

Table 1 SAOM estimation[a]

	Model 1 (null)	Model 2 (restricted)	Model 3 (complete)
Rate function			
Constant rate net (period 1)	4.075 (0.523)*	4.378 (0.561)*	41.016 (24.576)
Constant rate net (period 2)	4.808 (0.473)*	5.138 (0.499)*	19.164 (5.428)*
Constant rate net (period 3)	3.552 (0.327)*	3.736 (0.355)*	6.710 (0.785)*
Constant rate net (period 4)	3.895 (0.343)*	4.061 (0.359)*	6.112 (0.598)*
Rate outdegree	0.129 (0.119)	0.110 (0.016)*	0.095 (0.025)*
Rate indegree	0.006 (0.073)	0.011 (0.020)	−0.026 (0.028)
Objective function			
Outdegree	−1.594 (0.293)*	−1.779 (0.057)*	−2.892 (0.088)*
Reciprocity	2.782 (0.322)*	2.712 (0.087)*	2.402 (0.087)*
Three-cycles			−0.071 (0.072)
Transitive triplets			−0.050 (0.036)
In-degree related popularity			0.016 (0.004)*
Out-degree related activity			−0.006 (0.002)*
Out-in degree assortativity			0.103 (0.011)*
Self-rate expertise ego		0.173 (0.192)	0.171 (0.171)
Self-rated expertise-related similarity		0.177 (0.107)	0.084 (0.100)
Objective expertise ego		0.359 (0.131)*	0.557 (0.130)*
Objective expertise-related similarity		−0.378 (0.096)*	−0.458 (0.091)*
Role ego		−0.401 (0.088)*	−0.217 (0.074)*
Tenure ego		0.403 (0.078)*	0.233 (0.067)*

[a]Two-sided test. *$p < 0.05$ (Standard errors in parentheses)

4 Results

The network was modelled with a forward selection approach. Model 1 includes only the rate function, specified as depending on actors' in- and outdegree, and basic effects of the objective function. Model 2 accounts for the control effects, i.e., individual covariates. Finally, Model 3 assesses the contribution of the target effects, i.e., self-organizing mechanisms which verify our research hypotheses. In respect to the rate function, results (Table 1) indicate that the probability for clinicians to make a change in their outgoing ties is very high in period 1 and then decreases. As the positive *rate outdegree* highlights, very active clinicians have more opportunities to change their partners. Also, the emergent network structure results to be generated by various exogenous (covariates) and endogenous effects. First, we do not find any evidence for balance theory, since either *transitive triplets* or *three-cycles* effects are insignificant. Collective theory, by contrast, is verified. Heterogeneity among clinicians increases over time in respect to being chosen as cooperation partners (positive *popularity* effect), but not with regard to engaging in cooperation (negative *activity* effect). Also, the positive *out-in degree assortativity* coefficient points to a core-periphery structure, with more active doctors strongly cooperating among themselves. The negative effect of *outdegree* indicates clinicians are not likely to cooperate with many colleagues, while the positive *reciprocity* coefficient outlines a tendency toward mutual aid. Individual covariates contribute to clarify the picture. A selection mechanism based on self-rated expertise (positive coefficient for *self-rated expertise ego*) is observed, but information flows between pairs of clinicians with a different level of expertise (negative effect of *objective expertise-related similarity*). Finally, tenured clinicians and those not in charge of any administrative tasks are more active.

5 Discussion and Conclusions

We verify the capability of a SNA longitudinal perspective to identify drivers of cooperation, which is a particular case of communication network. We propose SAOM as a suitable methodology for modelling the selection processes which drive network formation. Also, we suggest the use of electronic datasources for tracing back interaction and dealing with the scant availability of panel network datasets. The good match between methods and data is demonstrated fitting nested SAOM on an original setting, i.e., a VCoP of clinicians. The significant coefficients of some effects demonstrate that the emergent network structure can be effectively represented by a small number of local rules, i.e., actors' behaviors of counterparts selection in their neighbourhood. This aspect is crucial because, compared to cross-sectional methods like ERGM, SAOM allow identifying the causal mechanisms driving network evolution as the result of actors' choices and of their reactions to others' choices. Also, the application points to the usefulness and versatility of Web 2.0 datasets.

References

Ahuja, M., & Carley, K. (1999). Network structure in virtual organizations. *Organization Science, 10*, 741–747.

Barabasi, A. L., & Reka, A. (1999). Emergence of scaling of random networks. *Science, 286*(5439), 509–512.

Bowman, K. O., & Shenton, L. R. (1985). *Encyclopedia of statistical sciences*. New York: Wiley.

Contractor, N. S., Wasserman, S., & Faust, K. (2006). Testing multitheoretical, multilevel hypotheses about organizational networks: an analytic framework and empirical example. *Academy of Management Review, 31*(3), 681–703.

Krackhardt, D. (1994). Graph theoretical dimensions of informal organizations. In K. Carley & M. Prietula (Eds.), *Computational organization theory* (pp. 89–112). Hillsdale: Lawrence Erlbaum Associates.

Lave, J., & Wenger, E. (1991). *Situated learning: Legitimate peripheral participation*. Cambridge: Cambridge University Press.

Maddala, G. S. (1983). *Limited-dependent and qualitative variables in econometrics*. Cambridge: Cambridge University Press.

Monge, P. R., & Contractor, S. N. (2003). *Theories of communication networks*. Oxford: Oxford University Press.

Norris, J. R. (1997). *Markov chains. Cambridge series in statistical and probabilistic mathematics*. Cambridge: Cambridge University Press.

Opsahl, T., & Hogan, B. (2011). *Growth mechanisms in continuously-observed networks: Communication in a Facebook-like community* [arXiv:1010.2141].

Quintane, E., & Kleinbaum, A. M. (2011). Matter over mind? E-mail Data and the Treasurent of Social Networks. *Connect, 31*(1), 22–46.

Robbins, H., & Monro, S. (1951). A stochastic approximation method. *Annals of Mathematical Statistics, 22*(3), 400–407.

Snijders, T. A. B., van de Bunt, G. G., & Steglich, C. E. G. (2010). Introduction to stochastic actor-based models for network dynamics. *Social Networks, 32*, 44–60.

Valente, T. W. (1996). Social network thresholds in the diffusion of innovations. *Social Networks, 18*(1), 69–89.

van Duijn, M. A. J., Gile, K. J., & Handcock, M. S. (2009). A framework for the comparison of maximum pseudo-likelihood and maximum likelihood estimation of exponential family random graph models. *Social Networks, 31*(1), 52–62.

Wasko, M., & Faraj, S. (2005). Why should I share? Examining social capital and knowledge contribution in electronic networks of practice. *MIS Quarterly, 29*(1), 35–57.

Wasserman, S., & Faust, K. (1994). *Social network analysis: Methods and applications*. Cambridge: Cambridge University Press.

MIX
Papier aus verantwortungsvollen Quellen
Paper from responsible sources
FSC® C105338

If you have any concerns about our products,
you can contact us on
ProductSafety@springernature.com

In case Publisher is established outside the EU,
the EU authorized representative is:
Springer Nature Customer Service Center GmbH
Europaplatz 3, 69115 Heidelberg, Germany

Printed by Libri Plureos GmbH
in Hamburg, Germany